THE VIDEO PRIMER

PRODUCTION, AND CONCEPTS

BY RICHARD ROBINSON

quick fox

NEW YORK   LONDON   TOKYO

Acknowledgments

A book of these proportions is always a group effort, written by one person but requiring the talents of others to reach production. I'd like to thank Danny Moses, Herb Wise of Quick Fox Books for their encouragement and interest, Barbara Kelman-Burgower for her remarkable editing and general sense of organization, Milton A. Landau of Panasonic for supplying technical assistance, John Lissner of Grey and Davis, Bill Campeau of AKAI, Edward J. Dudley of RCA who provided a number of photographs of early RCA video equipment which are reprinted here with that company's permission, Pat Masulli for permission to reprint "Operating Instructions" (copyright Four Seasons Publications, Inc.), Bill Narum and John Brumage for their advice, Bob Gruen for his media images, and Lisa Robinson who thinks it's nice I write books like this for a living.

International Standard Book Number: 0-8256-3131-9
Library of Congress Catalog Card Number: 79-63284

Book design by Leslie Bauman
Cover design by Iris Weinstein
Cover photo by Marcia Dover Hoffman

CONTENTS

Since the first edition of *The Video Primer* subtle changes have altered the direction of what was once called "The Video Revolution." Home video machines, inexpensive cameras, and large-screen viewers are finally available, which broadens the potential of do-it-yourself TV. This revised edition encompasses the new technology of Betamax, VHS, and other consumer video formats.

On the other hand, in the intervening years the half-inch portapak era has ended. Various cassette formats have replaced all but the most sophisticated open-reel recorders. The inspired TV techniques developed by independent video makers using Sony portapaks have been co-opted by network television. The Japanese electronics manufacturers—never responsive to the creative potentials of personal video—have all but abandoned the half-inch equipment with which these video makers worked. Instead, they have gone after the financial rewards of more expensive cassette formats favored by local and network TV broadcasters as replacements for 16mm news cameras to provide portable, on-the-spot, live coverage.

There remain great rewards to be gained from exploring yourself and your surroundings with instant TV, but as the second decade of personal video dawns it has become all the more obvious that broadcast television and the men and women who people it don't want to hear that anyone can make his own television, unless, of course, they're looking to "borrow" the hardware potentials and software inspirations of the independent video maker.

The basic concept of TV remains an amazing achievement, and it is hoped that at least some of those who use this book will gain an understanding of basic TV techniques and will seek alternatives that may, someday, make TV communication something more than violence and soap suds.

Richard Robinson
November 1978
New York City

PREFACE
TO
REVISED
EDITION

Vague and insignificant forms of speech,
and abuse of language, have so long passed
for mysteries of science. . . .

John Locke, *An Essay Concerning
Human Understanding*, 1690

The Electronic Awakening

Most people get nothing but a headache trying to make sense out of electronic terminology. The assumptions of faith, the mathematical complexities, and the integration of simple connections to form sophisticated components have resulted in a vast catalog of terms, descriptions, and phrases that are dealt with casually only by TV repairmen, hi-fi salesmen, and Nobel laureates. The confusion the rest of us feel on peeking inside our TV sets is equaled by our sense of mental claustrophobia when we're offered an explanation of what those wires and circuits do.

Electronics is a language like any other. Each word has a meaning which must be understood. Just as you can get around Paris with a couple of years of high school French and the ability to wave your hands, so you can find your way through the world of frequency responses and integrated circuits without too much trouble—if you proceed cautiously and try to make sense of what's happening, one step at a time.

As a culture we've been defeatist about electronics and the language that surrounds it. The joke that the first thing a TV repairman will do on a service call is to make sure the TV set is plugged in is an apt description of the times. We're addicted to electro-mechanical devices which serve us in a variety of ways, from opening cans of soup to filling our leisure hours with entertainment, yet the fewer buttons we have to push to get the job done, and the less we have to understand about the equipment we depend on, the better. This is especially true in video and television. Tv sets that tune themselves and video tape systems in which the user never sees the tape, let alone touches it, are what manufacturers are producing. The fewer buttons and knobs in evidence, the greater the potential sales to the mass market.

Only in the area of moderately to expensively priced hi-fi equipment has this trend been reversed. The hi-fi buff loves the sight of a panel filled with dials that glow red and green, controls that he can slide to shape the sound, and switches that can be thrown. But this is a limited market compared to the billions of dollars spent each year on TV sets. And often all those controls function primarily to satisfy the user's desire to live in a world somewhere between Flash Gordon and HAL.

Our tendency to avoid complexity has led to great sophistication in electronics. Aware that the market favors one button that performs two functions over two buttons, each performing one function, manufacturers have concentrated on this factor in designing and producing equipment. Each level of sophistication reached in eliminating consumer complexities usually signifies a refinement in internal electronics. These refinements result in easier use and im-

INTRODUCTION

proved capabilities of the equipment, but they also place the user one more step away from being able to relate to the device in any terms other than it works or it doesn't. In fifty years we've progressed from crystal sets to color TVs, but this dramatic change in communication devices has been paralleled by the growth of a fanatic aversion to knowing what's inside the box and why it does what it does.

This book is about a new communications technology: inexpensive video equipment used to record and reproduce information. The language of electronics has become the language of communications, and must be used to explain the potentials of the equipment and the methods of utilizing it. Aware that the novice video maker doesn't remember much algebra and hasn't taken a home-study course in TV repair, I've tried to express much of the terminology and information contained here in the language most of us possess through our everyday use of hi-fi equipment and TV sets. An assumption of some familiarity with movie cameras and projectors has also been made. The book is divided into three modular sections, which are more or less independent of each other. *Professional Basics* (Chapters 1–11) provides an insight into the potential of the equipment as you begin to use it and apply your own aesthetics to it. *TV Eye* (Chapters 12–18) covers working with the equipment on a practical level. A *Glossary* is included to further explain the terminology of video equipment and so provide an expanded, ready reference to video tools. If you get lost when reading the text, just stop, refer to the *Glossary,* then reread the material in question. As with any other language, the more you use electronics, the easier it will be for you to relate to it.

Video Form

A number of extremely divergent elements went into creating video; and for that reason the word has developed a variety of meanings, most of which are dependent on where the video person came from and where he or she is going. My definition of video, based on my own experiences with it, has to do with video as a media event. The traditional notions of television as communication have been changed radically by video. Even if you never make a video tape, the fact that you can, that you are able to disturb the flow of network information on your screen with your own visions, is an astounding event at this time. Video as media and all of us as media makers are the motives for this book. On the other hand, video is used throughout the book as a technical term. Both usages are correct, and it's one of the joys of video that the connotations of one definition carry over to the other. The cultural implications of video make the technical practicalities worth mastering; it is crucial to understand what video can do on a technical level since its potentials are what you will apply in turning video into your own media event.

Video began as practical media in April 1956, when the Ampex Corporation—until then a developer and manufacturer of audio tape recorders—started selling the first tape recorder capable of recording and reproducing a television show.

The already booming television industry was caught in a dilemma between an audience living in three different time zones and the fact that network programming originated live from New York and had to be broadcast at the same hour of the day to all of that audience. The original solution was a method devised to store a live television program for later rebroadcast. What was called a **kinescope** was made by pointing a movie camera at a TV screen and actually making a film of the show. Since the television picture tube isn't perfectly flat or square, the resulting film was distorted and looked, at best, makeshift. Add to the lack of quality the fact that it took time to develop the film—there were no instant replays.

Ampex's machine for tape recording video was an instant success, especially at the networks' production centers in New York City. In November 1956, CBS was the first to use the new machine when they video taped *Douglas Edwards and the News* for rebroadcast in the western time zones. The golden era of live television was ending. Within two years, shows were originated on tape regularly, starting with a program called *Confession* on ABC.

The advantages of video tape recordings over film kinescopes were tremendous. There was no developing process, no distortion of the picture signal; in addition, segments could be edited together in various combinations, and it was possible to check back immediately on what was being recorded to make sure it was correct. Also, video signals could be sent electronically over cables to local television stations where they could be recorded for later broadcast to the local areas served by those stations.

Television had originally been developed as a real time event. Something would happen in one place and be seen simultaneously in a number of other places. The first equipment used to accomplish this was heavily mechanical in design, but it was eventually modified into the electric TV camera and electric TV set. The introduction of magnetic recording tape and the video tape recording machine relied on those previous developments. When the video tape recorder was put into use in combination with all-electric cameras and receivers, it changed the nature of television—the video signals became an electronic representation of the original reality, capable of storage, modification, and selective display.

Another aspect of video has to do with the media experience the first generation of video people were undergoing during this time. Born at the end of the second world war, these kids were the first to grow up with television in their homes; they were the first to feel the profound impact of a medium that literally destroyed time and space. Their sensitivity to this box of realistic fantasies was extreme and their experiences mark the beginning of the first video era.

During the 1960s, under the banner of a supposed youth culture, these same young people began to create their own media to suit themselves. For the first time in our society, an inexpensive electro-mechanical technology was available to anyone interested in establishing his own lines of communication. Underground newspapers, rock records, FM radio, and related consciousness styles flourished. The media were capable of propagating and then reinforcing ways of living and thinking. Particular forms took on significance that made their contents secondary. A rock record became a symbol of a way of looking at the world, as did so many other things from American flag bumper stickers to Amerikan flag patches sewn on blue jeans.

Meanwhile, the cheap Japanese transistor radio of the fifties gave way to the cheap Japanese transistor TV of the sixties. As the technology got less and less expensive, the media became more and more available. And in Japan, work was going on to produce inexpensive video tape recorders. Toshiba laid the groundwork with the development of what is called **helical scan video tape recording**. By 1963, Ampex, Sony, and other companies had introduced their own versions of helical scan machines. These video recorders were originally intended for the broadcast industry to replace the more expensive machines that Ampex and RCA were making at the time, but the Japanese had a vision of a much larger marketplace for their video equipment. In 1964 video tape recorders for the home became available in Japan. Sony introduced the first inexpensive video equipment for sale here in the USA in 1966. And, in August 1969, the Electronic Industries Association of Japan (EIAJ) got all the major electronics manufacturers in that country to agree on a standard for their video equipment so that video tapes made on one machine could be played back on another machine.

This equipment is often categorized by the width of the video tape being used on the video tape recorders. The great majority of inexpensive video units use video tape one-half inch wide and carried on a reel, as opposed to a cassette or cartridge. They have come to be known as "half-inch machines" or as complying to the EIAJ "half-inch standard." Most personal video is "half-inch" video.

With portable video tape recorders available for less than two thousand dollars, it became possible for anyone to make his own television. And, for the first post-television generation, it was instantly obvious that they should all make their own television. Why not? After all, they'd found that they could make their own media on other levels. So video became an event as well as a technical definition. It became the opportunity to put yourself on the same television screen on which you'd been watching others. Television could become a looking glass with which you could play Alice anytime you wished.

Two other technical factors helped to solidify video as a new medium: the introduction of video production equipment in the same price range as the Japanese helical scan recorders, and the fact that the steadily growing cable television industry was willing to broadcast recordings made with this equipment to their subscribers' homes. It became possible to have a TV studio in your living room, complete with cameras, switchers, lighting, and recorders, and to produce a program which had all the rudimentary elements of professional television if not the excellent quality the networks had attained with their multimillion-dollar, highly complex equipment. Once an alternative TV program had been made, there was an avenue through which it could be broadcast to a sizable audience.

The immediate future of video is more or less the fulfillment of the promise envisioned by those who have become part of the video movement—being able to make your own television shows. The mass production of miniaturized and sophisticated electronic components combined with the commitment of the Japanese manufacturers in the video marketplace indicates that eventually network-quality programs will be produced with a low-cost line of equipment. In addition, corporate giants such as Sony, Panasonic, MCA, RCA, and Philips are laying the groundwork for consumer video playback equipment which will turn the TV screen into the center of a home entertainment unit with prerecorded programming—movies, specially originated shows, and educational presentations—available to the average consumer as a viewing alternative to commercial television and cable broadcasts. Video, in this sense, must be defined as the potential availability of an infinite amount of information through the TV set.

VIDEO
EQUIPMENT

1

The electric brush and canvas of the video maker are the camera, video tape recorder, and TV screen which are used to observe and display impressions of reality. These are the essential components, although there is a wide range of ancillary equipment available—some of which is desirable if you plan to package your productions along traditional visual lines with titles, technically acceptable edits, a variety of audio information, the format sense of *program* which we associate with movies and TV.

Before examining what particular pieces of video equipment can do for you (and, at times, to you), consider the initial equipment experience: You studied the brochures for months, flipping through them like baseball cards, dreaming and deciding on video gear for your first investment. You want to equip yourself with the basics—at least the basics as presented in their surreal majesty in the brightly colored flyers that Sony, Panasonic, and the rest have spread like some new-wave propaganda. A portable camera and video recorder, an editing deck, a monitor . . . these are, no doubt, first on your list, with a special effects generator (SEG), extra studio camera, and the like second.

The great day comes, you troop home with those neatly labeled boxes, lots of "this side up" and "fragile," and with trembling fingers unpack the camera and recorder. It's a great smell, all that styrofoam and polyethylene packaging, all that vinyl casing.

Once you've got it all out and have found the instruction book, you thread up for your first taste of personal video. You tape whatever's available, rewind, and sit down to watch your first production. The roof caves in. It doesn't look like you'd imagined it would; certainly not like a major network show, and not even as good as the brochures implied. You go back and stare at the brochures, thinking maybe it was more fun just to dream than to actually have the stuff in your hands. Next you begin to think, "Well, if I had a special effects generator and two more cameras it would all be different. Then it'd look like prime time, then the brochures would come through with their promises." If you're prone to anxiety attacks you may just go out and spend the rest of your life savings to get the extra equipment, but the result will be the same: Playback shows something less than you imagined.

The problem isn't the equipment. It can live up to brochure promises and approach prime time, but not by itself. There are any number of extras that aren't really extras. Like lights, microphones, different lenses, audio mixers . . . the list goes on. A $300 portable lighting unit is just as important a piece of video equipment as is the camera; the same goes for at least one good microphone. You can make video without it, but the video you make will not be satisfactory and it won't approach the full potential of the personal video equipment now available.

This book runs down video equipment on a how-to-use-it, how-it-works, how-it-can-be-exploited level. You *can* have "an entire television studio" run on batteries and carried by one person, so long as there's somebody behind you carrying all the extra equipment you need to make high quality video.

1

STRAIGHT
SHOOTING

THE
VIDEO
CAMERA

The camera, more than any other piece of equipment, signifies what the video process is about. It is the eye of the video system; the initial opening through which reality is converted to electric energy. Across a few inches of space it translates what it sees into an original record of a permanently present series of people and events. Of course, it's operative only in conjunction with the TV screen, which displays what it sees, and our belief that what we see on the screen is a factor in our lives.

The first thing you'll realize as you see the results of using a video camera is that it's not a camera in the traditional sense. Perhaps because there is no film inside, because you aren't really limited to how much you can record, because there's no whirring noise telling you and your subject that it's working . . . these factors and the ability to instantly see and hear what you've recorded (sound is an inherent part of the video process) make the video camera more than another way to go out and "shoot" a scene. The whole concept of using a camera to shoot a picture is gone. The act of using the video camera to record information is usually expressed in terms of making a tape or recording an event. The implication is that the event is external to the camera rather than created by it as in still photography and film.

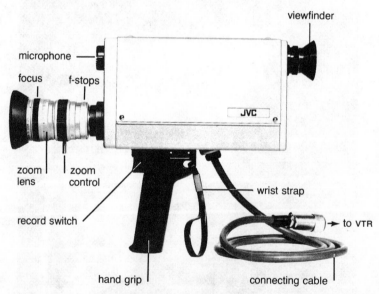

PORTABLE VIDEO CAMERA

viewfinder
microphone
focus   f-stops
zoom lens   zoom control
record switch
wrist strap
to VTR
hand grip   connecting cable

I think of the video camera as an extension of myself. My eye often strays from the viewfinder, and I'll continue recording with only the tactile sense of the camera giving me an indication that it's physically there and in operation. The fact that it's an electrical device rather than a mechanical transcriber is what makes me relate to it differently than I do to pieces of photographic equipment, notebooks, typewriters, and the other paraphernalia of making records of things. The only other units that I use in a like manner are my audio cassette machine and pocket calculator, which have become, through constant use, electric aids.

The video camera is a combination of Tom Corbett's ray gun, a movie camera, a tape recorder, and an external memory unit patched into the cortex. It's totally convenient in its operation; you don't really feel as if you're doing anything—just seeing and absorbing what you normally see and absorb. The trick is that your point of view, when displayed on a TV screen to others, suddenly becomes a much more peculiar view of the universe than you might have supposed.

Function

The purpose of the video camera and its immediate accessories is to be an electric eye. Exposed to a scene and subject through the optics of a lens, the video camera electrically **scans** the light impulses of that scene, which can then be transmitted over great distances as radio waves and displayed on countless television screens, or can be stored on magnetic tape for later transmission and display.

The video camera doesn't contain film. Instead, the light values of the subject on which it is focused are transmitted through the lens where they strike the photosensitive surface of one end of a **vidicon** tube—the heart of the camera system. The function and design of the vidicon tube are explained in detail later in this book (pp. 234–36). For now you should just understand that it is the component of the video camera that sits behind the lens, a vacuum tube that electronically encodes the change in light values from point to point in the scene, changing it into an electronic signal that can be broadcast (or stored on magnetic tape for later broadcast). Eventually these signals are displayed on the screen of a TV set. The screen is a larger version of the camera's vidicon, a vacuum tube performing the opposite function of the vidicon. Whereas one end of the vidicon tube is sensitive to incoming light images and capable of transforming them into an electronic signal, the TV picture tube is sensitive to incoming electronic signals and is capable of transforming them into light images on the screen end of the tube. Don't confuse the video camera with a movie or still camera. The video camera contains no film; it converts light to electronic impulses and these impulses, to be recorded, must be sent to a video tape recorder. The video camera is much more like a sound microphone in its design. To record

THE VIDEO CAMERA

19

sound you need both a microphone and a tape recorder; to record video you need both a "camera" and a tape recorder.

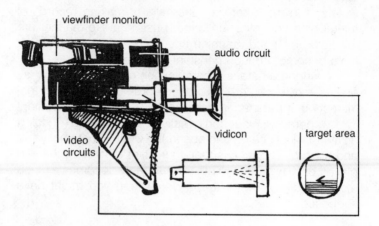

The Portable Camera

Most of us will have video experiences centered around the portable units from Sony, Panasonic, and other manufacturers. They've become the essential video gear and are responsible, due to size and ease of operation, for video being a personal media event. The initial components involved aren't terribly complex—a camera connected to a battery-powered recording deck. Adding an external battery pack, external mike, and a pair of headphones doesn't make it any more complicated, although you will eventually have wires hanging from you every which way. The two basic items you have to learn to manipulate properly are the camera and deck.

The design of all video cameras is essentially the same; however, their size differs dramatically along with their cost and purpose. The portable camera has all the basic features of the larger studio cameras. At the front end is the lens assembly where any **C-mount** movie or television lens can be mounted. At the back end is a viewfinder, which is actually a small television screen that gives the operator a TV-eye view of what the lens is seeing. In between are the internals of the camera: the vidicon tube mounted directly behind the lens to create the video picture signal; amplifiers, limiters, and control circuits, which shape the video picture signal. Most portapak cameras also have a microphone mounted just above the lens assembly to capture the sound of the scene. There is a cable coming from the camera. In the case of the portapak camera, the cable carries video and sound signals from the camera to the recording deck, and control signals and dc power from the deck to the camera.

Cameras have long been associated with a point-and-shoot action. This is not precisely what the video camera does. Rather it's a point-and-collect mechanism. Nonetheless the point-and-shoot philosophy is implied in the camera's design: lens sticking out of one end, viewfinder at the other end, and a pistol-grip handle underneath. Take the handle off if you can—even if you have to go to your video dealer and have him saw it off! The camera is easier to use without the handle, which is bulky, doesn't provide as sure a hold as grasping the camera housing, and gives you less of a feeling of what you're doing. There are some cameras whose handles are not removable. Take this into consideration when you buy a unit.

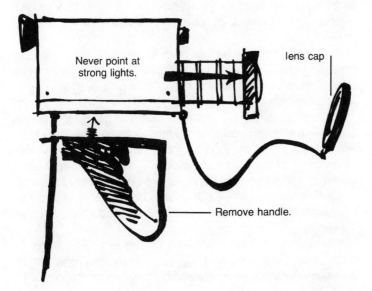

Never point at strong lights.

lens cap

Remove handle.

Camera Tactics

Once you have a general idea of what a video camera is and what it does, you have to start to use it. Connect the cable coming out of the camera to the video recording deck, making sure that the cable plug is securely screwed onto the deck's receptacle post. The umbilical power cord between camera and deck is a constant source of frustration—one moment it's too long, another too short. The camera-to-deck cable can be lengthened, and if you're going to tape in one set area you should get extra cable. This will allow you to set the deck down and just roam around with the camera. If you're on the move with camera *and* deck, you'll probably find the cable as supplied too long. Shorten it to the body-distance between your arm and side. This cable isn't the easiest in the world to rewire yourself so you may want to have a pro do it.

A few words of caution about working with a vidicon video camera:

1. Don't let it get any rough handling.

2. Don't drop it, bounce it, leave it in direct sunlight, the back seats of cars parked in the sun, and so on.

3. Never leave the camera pointing down at the floor for long periods of time. This will damage the vidicon.

4. Always store the camera with the vidicon on a horizontal plane with the floor or tilted slightly upward.

5. One of the best ways to store the portapak camera is on a tripod with a plastic bag covering it.

6. Always keep the lens cap on the lens whenever the camera isn't in use.

7. Never, even with the power off, point the vidicon directly at the sun, a fire, a lighting unit, or any other source of bright, intense light or you'll burn the target area of the vidicon and a black spot (**burn**) will be visible on the display screen.

8. If the vidicon does get burn spots on it, they may be only temporary. Try removing them by pointing the camera at a bright white wall or card for an hour or so with the unit in record.

9. Occasionally you'll find that dust will get on the target area—especially if you change lenses often. This may look like burn, but's it not. Take off the lens and blow across the target area, or very carefully wipe off the vidicon target area with a cotton swab soaked in lens cleaner. Be very careful.

10. If you wear glasses and have trouble seeing through the viewfinder, get a prescription eyepiece made for the viewfinder magnifier.

11. Keep fingerprints and dirt off the lens; when it gets dirty, clean it with lens-cleaning fluid, a lens-cleaning brush, and some lens paper by spreading the fluid on the lens and very lightly wiping the lens with the brush and lens paper. Don't apply any real pressure on the element or you'll mess up the coating put on by the manufacturer.

12. A can of compressed "oil-free" air is great for cleaning the lens, vidicon, and other equipment. Get it at a photo supply store.

Pick up the camera and start to play with it. Carry it around the house for a few days, get used to having it with you and not feeling uncomfortable with it. The camera should be as natural to you as a glove is to a baseball player—extra epidermis. Try to think of it as an absorber of light, a method of collecting what you see.

Mounting the Camera

Most video makers using portable equipment start out trying to hold the camera in their hands. They soon discover that the weight of a small camera, even five or six pounds, is too much to support steadily for more than a few minutes.

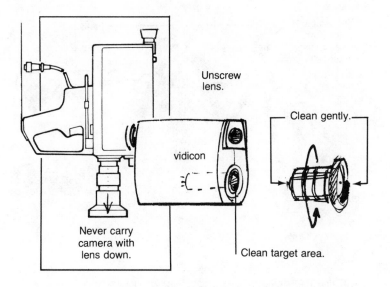

Unscrew lens.

Clean gently.

vidicon

Never carry camera with lens down.

Clean target area.

Also, any small movement of the camera, as might occur when you switch hands to give one a rest, although not noticeable in the viewfinder monitor during taping will show up on playback of the tape. Cameras designed to be hand-held can be mounted on a tripod, body brace, or your hands. The bigger studio cameras must be attached to a tripod—they are too heavy for any other mounting.

Tripods

The main purpose of a tripod is to hold the camera steady at a height that gives the camera a good **angle of view** of the scene. For this purpose alone almost any tripod will do, from the cheapest 35mm still camera tripods sold in photo shops, and the rather overpriced tripods sold by some of the large video equipment manufacturers like Sony, to the expensive tripods you can buy from movie and TV supply houses.

As a general rule, the more sturdily constructed and expensive the tripod, the easier it is to operate. There are two important parts of any tripod: the legs, and the head, or top portion, on which the camera is mounted. The weight of a vidicon video camera is minimal compared to that of the 16mm movie cameras most tripods are designed for, so don't worry about how substantial the tripod legs are. That they can support several hundred pounds of camera is unimportant, and you may be more interested in how much weight they will add if you're planning to lug the tripod around. As for the head, there are two basic types: **friction heads** and **fluid heads**. Friction heads are less expensive, whereas fluid heads provide smoother operation.

The tripod should let you move the camera to follow or concentrate on the action in a scene. This is where the quality of the tripod head comes in. If you want to tilt the

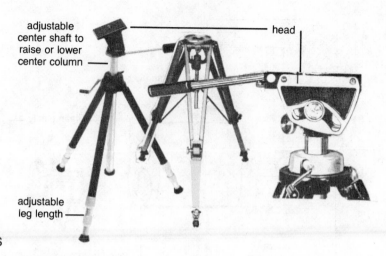

adjustable
center shaft to
raise or lower
center column

head

adjustable
leg length

TRIPODS

camera mounted on the tripod up or down or to the left or right—while the camera is in use—you need to make the transition smoothly. A friction-head tripod has two metal plates, one on top of the other; the upper plate is rotated while the lower stays stationary to effect the side-to-side movement of the camera mounted to the top plate. Cheap tripods have no cushion between these plates to allow smooth rotation. More expensive friction-head tripods have ball bearings or a similar element between the two plates, allowing smoother movement. Fluid-head tripods are the best, since the upper plate to which the camera is attached rests in a bed of fluid—grease or some other viscous matter—making the movement smooth since the camera-top plate "floats" on the fluid as it turns.

Tilting the camera up and down from a horizontal position is another motion of the tripod head. A handle extending from the back of the tripod head is used for this action. In cheap tripods the control is again operated by friction, making it possible to "dump" the camera while tilting it. In other words, the weight of the camera is such that any position off the horizontal will create a center of balance near the top of the tripod-camera assembly, which may cause the whole thing to tip over. In addition, smooth tilting is impossible. Tripods made for 16mm cameras have the advantage over inexpensive video tripods or converted 35mm tripods since they are designed to take the weight of the 16mm camera. The weight of a video camera is considerably less and the controls operate even better. An expensive friction-head tripod will tilt more smoothly because of the gearing used. A fluid head will have the proper internal damping, which means the movement will be smooth and steady, so that when you tilt, it will appear that the scene, not the camera, is tilting.

The third action a tripod performs is to move the camera up or down to various heights. The more expensive tripods can be heightened by adjusting the length of the tripod legs. Some tripods have center shafts on which the heads are mounted and which can be cranked up or down, but these are usually inexpensive tripods that were originally designed for 35mm still cameras. The stability of the whole tripod with a fully extended shaft is much less than that of the more professional tripods without a center shaft. To get a unit that will raise and lower the camera in a smooth, unnoticeable motion (which center shaft tripods won't do), you have to use either a crane or pedestal unit. The crane is a sort of seesaw with weights at one end and the camera and operator at the other. The pedestal works by a hydraulic principle with one shaft within another telescoping up and down. Pedestals start at about $1500; cranes run $14,000 and up.

Attachments are available for most tripods so that a set of wheels, called **dollies,** can be put on the feet of the tripod. They work well, but you must have a solid, heavy-duty tripod and very smooth flooring (such as linoleum) before you'll get rolling action that is imperceptible on your tape.

Hand-held video cameras can be mounted on tripods for studio use. An even better solution to the problem of keeping the camera portable yet stable is to take a tip from 16mm cameramen and get a **body brace.** This is a metal unit *Body Brace* that rests on your shoulder with the camera mounted on it so that the camera is just in front of your face with the viewfinder monitor in line with your eye.

Body braces cost from $50 to $200, depending on their design. Braces made especially for video cameras have been introduced, but you may find that a particular make of 16mm camera brace is more comfortable. They look like iron corsets, but when you put one on you'll find the brace conforms to your body contours and is fairly comfortable. The portable video camera is lightweight to start with—the entire unit will not weigh over ten pounds. You'll be amazed at how soon you get accustomed to a brace, and the results will be fantastic. There will be no camera shake on your tapes; since you have to move your body to move the camera, you'll find that your transitions will be naturally smoother than when you're holding the camera in your hands and trying to move it smoothly. The only real drawback to a brace is that it's difficult to shoot up or down at an angle, since the brace locks your upper torso into a fairly rigid position. The best way to get around this is by practicing, but this disadvantage can never be totally eliminated.

It's helpful to understand camera mounting equipment before getting into the details of camera techniques. If you

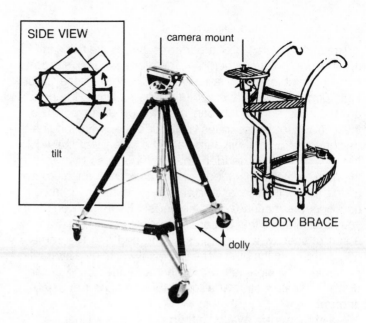

SIDE VIEW

camera mount

tilt

dolly

BODY BRACE

understand what a tripod with a dolly attachment does, you already know all the basic camera moves.

Camera Movements

Angle of View

A camera placed at a given distance from the subject has a number of options for movement. The camera can **tilt up** or **tilt down**, both motions in which the camera remains in one place but its angle of view is changed. The camera is tilted from an imaginary 90° angle parallel to the floor and to other angles either down toward the floor or up toward the ceiling. Tilting changes the perspective, or the angle at which objects in the scene are seen by your eye. Lying down on your back and tilting the camera straight up at the Empire State Building will make the building seem mammoth. The same goes for that angle when used to tape people. A powerful figure looms into view when shot from below with the camera tilted up. On the other hand, the Empire State Building taped from above with the camera tilting down does not have the immediate proportions suggestive of its true size. The same for a human figure taped from above at a distance. These changes in perspective by tilting the camera up or down are exaggerated, dramatic examples, but they apply also in a more normal situation. You will find that human features are more flattering when they're taped from a slight downward camera angle. With the camera set below head level and tilted up, you'll find an element of the grotesque in the sagging chins, flaring nostrils, and bagging eyes that loom into view. Your subjects will riot at the sight of themselves during playback.

Another traditional movement with a stationary camera is

swinging the camera to the right or left. This is known as a **pan.** The camera can either pan left or pan right* to follow the action of the scene or make some point of emphasis.

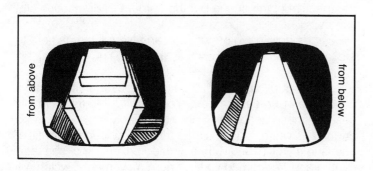

ANGLE OF VIEW

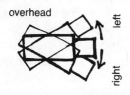

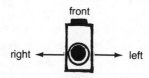

PANNING

One must be careful in panning, as in all other camera actions, to move with a certain "human" pace. Don't swing the camera violently left or right and expect your audience to follow the movement. It will look on the screen like the camera operator was suddenly hit by a truck—an interesting effect, but not a proper pan. Breathe as you move the camera. There must be pacing and rhythm to the pan, or any other camera movement, to allow for the small size of the screen on which that movement will eventually be displayed. Consider the rhythm of the entire sequence of which the pan action is a part; consider the ability of the audience to absorb the events taking place. As you pan the camera you are shifting the perspective of the scene. This must be done with a timing that allows the audience to understand this shift and refocus their perceptions. You may also be creating distance—say a subject walking away from the camera and the camera turning to follow the exit. Since TV is not 3-D but a two-dimensional medium, increasing or decreasing distance from subject to camera while the subject is moving to the left or right requires a sense of timing, not incomprehensibly fast and not exasperatingly slow.

Another tripod action, if the tripod is mounted on a dolly, is to **dolly in** or **dolly out.** This means to roll the camera toward or away from the subject. This in or out movement should not be confused with using a zoom lens to zoom in or

Dollying and Trucking

*All directions are given from behind the camera.

zoom out. Dollying is the actual movement of the camera, so that the perspective of the scene is maintained (although the angle of view changes). The zoom lens alters the perspective of the scene but the camera itself remains stationary.

Besides dollying in or out you can also roll the camera to the left or right. This is known as **trucking**. You can **truck left** or **truck right**.

You can duplicate all of these tripod-based actions with a hand-held or braced camera. Your body is the tripod in this case so these actions will not necessarily be as rigid and angular as when you're using a tripod. You can pan or tilt at any angle you want with a liquid motion. You can add rhythm and freedom to your camera movements, which will create an air of naturalness. One such pattern is the **surround shot**: The camera enters into the middle of an event and wanders through it, giving the viewer a sense of being encompassed by the event. It is much easier to manipulate a hand-held camera so that it seems like just another eye involved in the event it is perceiving than it is to move a tripod-mounted camera into a scene. This is partially because the participants in a scene have a less formal awareness of the mobile hand-held camera than a tripod rig, and partially because the camera operator can function with the camera as a part of him- or herself. As an operator, your body language is part of your camera movement technique. It is helpful to be aware of the established camera movement techniques and to use them as the structure on which your own sense of camera movement is built. They are the materials of experience and can be invaluable if you accept them.

If you are taping in a location where neither tripod nor brace is available, the next best thing is to rest either the camera or your body against some solid, stationary object. Doorways, chairs, walls, columns . . . anything you can find that you can lean against will take some of the weight off the camera. You'll also find after a few months that your arm muscles develop enough to make the constant weight of the camera more bearable, but don't count on this: Get a brace and tripod and pack them with your gear. Most braces come apart for transport and even the cheapest, smallest tripod will prove useful in certain situations. For instance, if you're not trying to tape your material as one continuous segment but are planning to edit various scenes together to create a whole, there is no need to keep the "action" going by constant camera motion. Each scene can be shot from a different angle and the edited series of segments will give viewpoint variety.

Often the total technique developed by video people

using only one camera and attempting to tape a half hour as one continuous event is nothing more than constant movement of the camera in an effort to vary the camera eye so the scene doesn't become static. A noble effort, since even the most loyal and dedicated audience (friends and loved ones) will tend to fidget if the camera remains stationary for too long—the cumulative effect of watching TV and going to the movies all these years.

Camera Language

Every video camera has a balance, a sense of weight and mass that you have to judge as you use it. Don't pick up the camera, smack the viewfinder into your eye, and grip the camera as if it were going to fall out of your hands any minute. Work with it in front of your face, holding it gently, allowing it to move with the elasticity of your body. The camera must breathe with you; if it doesn't, your video will look as if your nerves were shot, the picture will jerk and jitter with every breath you take.

Body to Camera

There are three points at which the camera is *steadied:* the lens, the body, and the eyepiece viewfinder. The main support comes from under the camera body, against the gravity pushing the camera into your palm. The viewfinder rests against your eye, a point of contact that steadies the camera. The lens barrel should be the other point used to steady rather than hold the camera. Your free hand should hold the barrel, gently pushing or pulling as the hand under the camera and your body move.

THE VIDEO CAMERA

Body movement is camera movement. You must understand this when you use a portable camera. If you want to move the camera to pan left, your body must pan left. If you move your body, the camera must be readjusted to keep in step with your physical flow. There's a lot of footwork involved in this.

Your feet form a two-legged "tripod" on which both you and your camera are mounted. They should be firmly planted, allowing your body to twist and turn with the same sense of balance a skier demonstrates traversing a hill. As you move your body, your balance should remain perfectly steady. When major motion is going to take place, your feet must lead the way. You won't get very far in sandals or clogs. Wear tennis sneakers or work boots; either will give you a sense of steady progress from one point to another. Often rather involved footsteps come into play, like crossing one foot over the other and then shifting weight from foot to foot as you truck the camera left or right. Work from the knees and waist, allowing the walking motion to be absorbed before it gets to your arms, shoulders, head, and camera.

The physical actions used to place and motivate the camera are linked to the aesthetic decisions you make in framing the picture. Many beginners tape what I call "wandering video." The camera slowly roams the picture, running continuously, and expressing the boredom of the operator for the chosen subject. Portable systems allow great freedom; they suggest the ability of the medium to penetrate the camera-subject barriers of traditional TV and movies. The camera is part of the action, or so the video maker thinks. This line of reasoning leads to a sort of drunken wandering. A person is talking about handicrafts: The camera focuses in on the person, then floats off to the left to examine the craft products, goes just a little too far left, then comes back, then gets the craft in the middle of the screen, then goes out of focus, then focuses, then zooms in on the craft, then refocuses. It gets so that you begin to beg for a commercial.

Part of this nontechnique is caused by the operator's inability to see where the camera is going. With both eyes open the operator would be able to anticipate changes of subject-action, but it is often difficult to keep the eye that isn't glued to the viewfinder open, although you can learn to do so with practice. Thinking something is off to the left as you pan with your camera and being able to see it out of your non-viewfinder eye are two different things. The ideal is to have both eyes open, but it's difficult to achieve. Once you've set your focus, open your free eye from time to time—especially when you're moving, so you don't run into any walls or trip off the edge of a cliff. When you move backward, put your lens-focusing arm behind you, feeling for possible obstructions.

When you can see where you're going it's much easier to get there, and with a camera movement that won't look so jerky on replay. If you can't keep track of your subject-action yourself, you might want to let someone guide you—have him take hold of your waist and pivot you around as you concentrate on working the camera. This can be especially useful in working through crowds or backing up.

I totally disagree with the aesthetics as well as the lack of polish of the handicrafts scene described above. Like the great majority of television watchers I'm not interested in the camera. I'm interested in the subject. I want it presented to me as visual information, not as home movies. TV is The Big Time, and if you're going to make TV you must learn the tools of your trade. I don't want to read a book in which every other line of type is set upside down. Sure, sure, if it were really important, brilliant writing, I'd put up with it, just as I

Eye to Camera

THE VIDEO CAMERA
_____

31

put up with the poor quality of the tape recording of Johnson being sworn in on the plane back to Washington after Kennedy was shot, with the misty black and white of Oswald being murdered, or with the flickering picture coming back from the first moon landing. But I really haven't got the time to waste to put up with the same poor quality in your program *Low-Income Housing Problems in East Minneapolis.* Life is just too short. So if you have a message, get this message: it is the medium that's in command; until you get past it by understanding it, you don't have any message.

There isn't any such thing as guerrilla television, not when it gets on the screen, anyway. When you're on the medium, you're part of it.

I became interested in video because I discovered I could make my own television. The mystery and magic of television would be mine, to alter and originate as I saw fit. But I never intended to make junk and then force other people to watch. When I was a kid I used to do magic tricks. I discovered very early on that to get my parents to watch their two hundred and fifty-second magic trick, I had to get their attention and hold it, or else they'd fall asleep just as I was pulling the rabbit out of the hat. Practice and presentation became more important than the trick itself, for without them there was no trick. The spectacle is the trip here, the event in its physical parameters rather than its intellectual implications. And whether you're playing the magician in a backyard circus or having your own television show, you've got to present what you have to say in a manner that will attract and hold people's attention.

Hot Shots

There are three major framings of a subject by the camera which serve as reference points when taping an event. The first of these is the **long shot**. A long shot is the camera view of the subject at a distance. In terms of people this means that you can see from head to toe, usually establishing not only what the total person looks like but also what the surrounding environment is. From there we move in to a **medium shot**. This encompasses head to waist. A medium shot of a person sitting at a desk is from the desk top to the top of the head. Then we have the **close-up**. This is known as being in **tight**. Television and movie close-ups are in the range of from the top of the head to the shoulders or from the top of the head to midneck.

In movies the long shot is an **establishment shot**. It's used to tell you where the action you are about to see is taking place. Often video people fail to appreciate the importance of this kind of camera view, because of the nature of the portable equipment. You begin to consider the video camera as nothing more than an electronic eye and ear and want to get close enough to what you're recording so it can be seen and heard. You'll often see video tapes that begin and end within a few feet of the subject without giving any sense of where the subject is or what the surroundings are.

A simple way to avoid this habit is to begin taping while still a few hundred yards away from the building in which your subject is waiting to be taped. Movies have stuck to the long shot to establish the scene because it works—maybe not every time, and maybe liberties can be taken because of our familiarity with it as a visual technique, but it does work. News cameramen begin their coverage of a fire with a long shot of the building burning before they cut to the firemen scurrying around with their hoses. After all, the audience has to be shown there's a fire before they'll watch it being put out. It gets right down to the old newspaper dictum: who, what, when, where, why.

The medium shot is the most commonly used shot in video, as it is on TV and in the movies. The close-up shot is a much freer technique in video than in any other medium. A medium shot is just a general view between long and close-up. But a close-up in TV and the movies is very carefully handled so that it doesn't get too close. This is because the mystique of the subject (say an actor or TV personality) can be destroyed if the camera gets close enough to see the capped teeth and the wrinkles under the make-up. People sweat in the movies or on TV only when there's a point to the sweating. Hopalong Cassidy never went to the bathroom.

With your own camera you can get right in there and revel in the quality of humanity up close. You can fill the screen with a set of lips or an ear. A hand can become a major object of interest. This can get as boring as it was initially interesting, but it is an effective technique and you should work on letting your camera flow backward and forward to absorb and discover all the aspects of your subject, rather than structuring your camera work into long, medium, and close-up views.

Proficiency with the camera develops only by using it as much as possible. Work with the camera both on and off a tripod. Experiment with the camera on a tripod to understand the basic principles behind the traditional forms of camera movement. Then duplicate the same techniques using a hand-held or body-braced camera and discover the freedom allowed. A wise combination of the two extremes will lead to the ideal camera technique—not only for your own aesthetic satisfaction but also for your audience to perceive what you're trying to say with your camera.

Avoid the wandering camera phenomenon. To do this

CLOSE-UP   MEDIUM SHOT

LONG SHOT

Shoot Now, Edit Later

you're going to have to edit, to use your equipment as a production tool rather than a magic wand waved in all directions as if casting a spell. With the video camera and deck you record information, events, life, time. You must present this recording in a coherent manner, and to do that you must readjust it, remove the excesses, trim it down, and make it snappy. People haven't got all day.

The editing option is crucial to how and what you record in the first place. If you think that what you're doing initially is it—the beginning and ending of your tape—then you'll be prone to swinging the camera back and forth in a desperate effort to get everything in. What you will actually be doing is imitating on an unconscious level the form of conventional television. If you leave the face of someone who's talking about building model airplanes to tilt the camera down, zoom in, refocus, and wobble a little till you get the airplane in his hands onto the screen, you're doing nothing more than **intercutting** to the object being discussed to explain the discussion. Real television does that too. They'll cut to a close-up shot of something being talked about. But their cut will be clean and smooth, with no time lost and with no technical disruption of the picture. You can do that too, if you think in terms of editing. If you are going to edit later, the recording can be done differently. You might leave the camera on the speaker's face while he or she talks, then later tape some **wild footage** of the model airplanes. During editing this footage can be intercut with the speaker's face to produce the final set of visuals. Or you might just want to tape a portion of the speaker's face talking, then record the rest of the dialogue on an audio cassette machine and get some footage of airplanes as well—eventually combining the whole during editing. There are any number of ways to approach the problem.

What you have to do to make good video is to break down each tape into a series of shots. Get into real detail about what you're taping. Don't just wander in someplace, turn on the portapak, and hope everything goes well. Each bit of action, each subject, has a form and substance, and you must decide how you're going to relate those variables with your camera picture and mike sound. Frame your subject, decide the angle of view, perspective, and proximity of the visual subject while you try to record the sounds being produced as audibly as possible. Set up a shot as if you were taking a prizewinning photograph, for in fact that's the game you're playing. You have to compose visuals to be viewed by others, and your composition will say a great deal about your subject matter.

Composing Your Shots

The TV screen is a powerful mode of communication, perhaps the most powerful medium yet created for people to speak-think to each other. You've got to control it, not let it control you. You must master the art of using the screen. You've got to be able to plaster what you're trying to say across it so that people understand you and your subject-message.

Evaluate the area of subject-object-action. Mentally trim off the portions of area that are not important to the communication. Study the dynamics of the subject-object within the scene. Then determine what views at what distances will most effectively present the subject-object-action as visual information.

Two people sitting in chairs talking gives you any number of camera-to-subject relationships to choose from. The camera can be set in front of or behind the subjects to frame the two of them. The camera can be set to one side or the other to get one full-face and the other over-the-shoulder. The camera can concentrate first on one person and then the other. The camera can be somewhere else entirely, perhaps on the subject the two people are discussing. Whichever one or combination of these views you choose, be as neat as possible in framing the view when you record. Visual aesthetics are crucial to visual communication.

If you are going to edit, these visual aesthetic decisions can be even more particular. Two people talking can be broken down into segments—a number of scenes to create a whole. First a long shot of both people talking to establish who and where they are and what their relationship is to their environment and to each other. Then a close-up of one talking, then a close-up of the other talking, then back to a

THE VIDEO CAMERA

long shot taken at an angle different from that of the initial long shot.

This shouldn't be done with a lot of camera wander. There's nothing wrong with asking your subjects to stop talking while you reposition the camera. Break a ten-minute discussion down into segments. Let them talk for three minutes into your long shot. Ask them to stop. Then move yourself and camera in for the first close-up. During this movement stop recording. Then start recording again and let them begin to talk again. Stop talking and recording for repositioning of second close-up. Record second close-up and conversation. Stop again, reposition, and wind up the discussion with a long shot. Later you can edit out the blips caused by starting and stopping the camera and you will have a smooth, well-framed visual flow: long shot, close-ups, long shot.

Try to keep from zooming in. If you must zoom, zoom out (zoom lenses hold their focus better when zooming out than when zooming in). When you zoom in on a portion of a scene to capture detail it is very difficult to keep the camera from getting shaky and to avoid at least a slight readjustment of the framing once you're in tight. If you set up the close-up before you begin to record and then, after it has appeared on the screen, zoom out to give a wider view, there will be little instability and you won't have to be as critical in your close-up-to-long-shot framing.

As I've pointed out, there is nothing wrong with learning about camera handling by watching the pros. You can do that any hour of the day or night by turning on your TV.

Spend a week evaluating location news footage as shot by union cameramen for local and network news shows. These guys have been at it for years and they've got their moves down. Watch how they compose shots, often at only a moment's notice; watch how they use their zoom lenses; watch how they vary their technique when working with tripods and shoulder braces. You may not appreciate their aesthetic visual decisions, but you've got to admire how they get their subject-object-area across with a minimum of effort and a maximum of visual communication.

The basic techniques of video are nothing new. The advent of easy-to-use video equipment shouldn't herald a giant step backward to prehistoric technique. Video should be television: You want your audience to appreciate what you have to say, rather than get bogged down in how you're saying it.

There are times when you must capture an event instantly, when it won't wait around for you to change positions. It's during these times that your understanding of camera technique becomes crucial. You've got to get into the swing of the action taking place and record it as lucidly as possible. When these situations occur, try to stick to long and medium shots to cover the action, rather than trying to pick up detail.

When you do pan, tilt, or truck the camera to follow action, strike a middle ground for your speed of movement. If you swing the camera too fast, your audience won't be able to follow the shift of action, and if you swing it too slowly they'll be nodding out by the time you're through. Set up a rhythm for camera movement by tapping your foot and moving the camera on the beats, not a jerky set of beat/move, beat/move, but a waltz tempo in which the camera describes its arc steadily but with a sense of rhythm.

Video taping begins with the camera, and the most important part of any camera is the operator. In and of itself the video camera does nothing more than record what it's pointed at. The aesthetic decisions of the camera operator determine the results produced by the camera. Labeling camera-handling an art is not extreme, for your use of the camera will determine, to a great extent, the kind of video you make. Your ability to work the camera in a fluid yet precise manner will develop with time, but an understanding of the basic camera techniques available to you should be your starting point. Too many video makers have simply unpacked their gear and started taping without regard for the camera as an eye. The results have been shaky at best. True, a few interesting effects have come out of these first years of nontechnique video technique, but on the whole the result has been a lot of unstable zooming in and out, continual loss of focus, and confusion as to just what/who is the subject of the scene. I'm not suggesting that you lock yourself and your camera into a certain set of standard operating procedures, but you must be at least sensitive to the camera techniques used in film and television that have been perfected over the last fifty years.

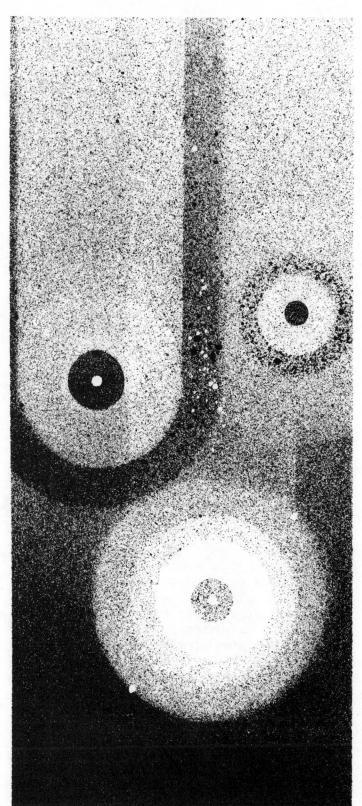

MEDIA
REPLICATIONS

VIDEO
RECORDING

The **video tape recorder** is the heart of every video tape production facility from the single portable recording deck slung over your shoulder to the block-long network studios with their million-dollar equipment. Referred to as a **VTR** by television people, the video recorder is the logical extension of the audio tape recorders that have become a major consumer entertainment item in this country, thanks to the Japanese. Whereas the tape recorder with which we are most familiar records only sound, the video tape recorder simultaneously records sound and picture.

There are times when it is not possible or desirable to immediately broadcast the scene being viewed by the video camera. A video tape recorder is used to store the video information being generated by the camera. The video tape recorder is a tape recorder capable of placing onto magnetic tape the electronic signals fed to it from the camera. The resulting tape can then be rewound and played back on the recorder or it can be stored for later playback. The video tape recorder, when used for recording a camera signal, is connected to the camera by a cable through which the electronic signals created by the camera flow to the recorder. For playback, the video tape recorder is connected by a cable to a TV set and the electronic signal stored on the video tape flows from the video tape recorder to the TV set, where it appears on the screen as a picture.

The elements of video recording are similar to those used in audio tape recording. In fact, the VTR is actually an audio-video tape recorder since it is capable of recording both sound and picture. The audio portion is collected by a microphone and played back through the speaker of a TV set; the video is collected by the camera and played back on the screen of a TV set.

The major obstacle to explaining video recording is that most people associate film with the presentation of visual material. When you show them a reel or cassette of tape and tell them that it is the medium of video, their confusion can reach profound philosophical levels. Most of this disappears when you demonstrate to them that the procedure actually works. Exactly how or why it works is another set of obstacles that is often impossible to get around. All the same, once anyone sees that video recording does work, and masters the mechanical actions involved in using a video tape recording machine, further explanations of a technical nature are not crucial unless a total relationship with the equipment is desired. Blind operation of that sort is common in audio recording as well. It is doubtful that even a small percentage of the millions of machines in use are owned by people who really understand how they work.

If you intend to take such an attitude toward your video equipment, fine; but I would strongly suggest that you

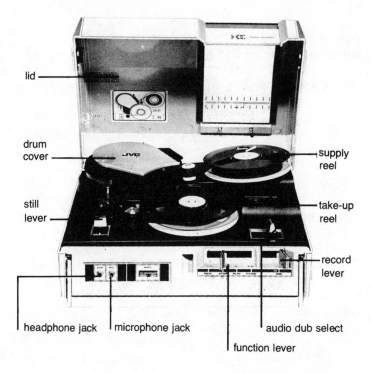

lid

drum
cover

still
lever

supply
reel

take-up
reel

record
lever

headphone jack   microphone jack

audio dub select

function lever

VIDEO TAPE RECORDERS

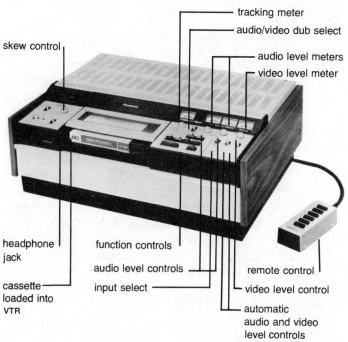

tracking meter

audio/video dub select

audio level meters

video level meter

skew control

headphone
jack

function controls

cassette
loaded into
VTR

audio level controls

input select

remote control

video level control

automatic
audio and video
level controls

memorize the instruction booklet that comes with each unit,
since there are certain quality-control procedures—such as
cleaning the video heads—which you must observe. These
procedures are grasped naturally by those who understand

VIDEO RECORDING

how their equipment functions. Those who wish to remain permanently uninitiated may not even understand what the video heads are, but they still will have to care for the heads as if they did. Personally, I think it's easier to have some appreciation of the system. It makes everyday maintenance less of a task and probably means that your equipment will last a good deal longer since you understand its capabilities and won't demand anything of it that it cannot inherently deliver.

Function

The Video Signal   The basic function of the video tape recorder is to transfer the electronic picture signal coming from the video camera onto a length of magnetic video tape and then to retrieve the signal from the tape at a future date and send the reconstructed signal to a TV receiver for display. This video tape also records the audio portion of the event being viewed by the camera at the same time as it records the picture information, so that the information stored on the tape includes the total audio-visual representation of the subject. The camera receives the light values from the scene and the microphone receives the sound waves emitted by the scene. To avoid confusion, the audio portion of recording will be dealt with separately (in Chapter 8). For the present we need to be concerned only with how the video information is stored on the video tape.

Having established that the scene viewed by the video camera is, by the action of the camera, transformed into a series of electronic impulses that represent picture information, we can now focus on the electronic signal, which is fed along a cable to the video tape recorder. What every tape recorder does is pulse these signals through components called **video heads,** onto the magnetic tape as the tape passes by the heads. On the tape are metal particles capable of forming invisible magnetic fields, which will remain in force unless a different, stronger magnetic field disturbs them. These magnetic fields correspond directly to the electronic pulses applied to the tape through the video heads. When it is necessary to reproduce the information stored on the tape, the tape is again passed by the video heads and a reverse process takes place wherein the magnetic fields on the tape induce electronic pulses across the video heads and those electronic signals are retrieved from the tape and sent out of the VTR to the TV screen. If there were no video tape recorder inserted, the electronic pulses from the camera could only go to the TV. The addition of the VTR stores them on the magnetic tape.

Once the signals are on the tape, the tape can be "played" indefinitely. In other words, the signals aren't removed from the tape to effect playback. They are "read" by

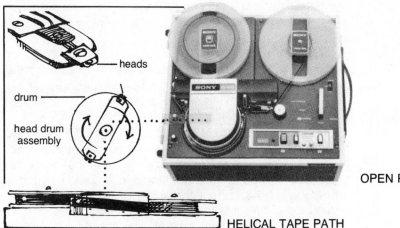

heads

drum

head drum
assembly

HELICAL TAPE PATH

OPEN REEL VTR

CASSETTE VTR

the video heads to produce a video signal, but they aren't destroyed by that reading. The tape can be played again and again, unless and until a strong magnetic field such as a bulk tape eraser or the erase head of the VTR is used to disturb the pattern.

All of the mechanical aspects of the video tape recorder—the reels, control knobs, internal motors, and the like—are intended to get the tape from the **supply reel** (the reel from which the tape originates, on the left as you face the recorder) to the **take-up reel** (the empty reel on the right where the tape is collected as it is run from the supply reel by way of the video heads). These mechanical devices and their related controls make up the greater part of the external equipment that you see when you look at the VTR, and are nothing more than governors of the tape's motion, speed, and path so that the tape contacts the heads properly. With cassette formats—U-matic, Beta, and VHS—the supply and take-up reels are enclosed in a plastic case, but

VIDEO RECORDING

the recording and playback process is identical to that of open reel recorders.

VTR Controls

Operating a video tape recorder is not complicated, but it does require the full attention of the operator, from threading the tape, through making proper cable connections, to manipulating the controls.

Open reel video tape recorders as well as video cassette machines use the recording system known as helical scan recording. Helical scan (see pp. 250–52) requires very close tolerances for tape-to-head contact to record and play back properly, and the correct threading of the tape onto the video tape recorder is crucial to this contact.

Threading Delicate handling of the tape is very important, since the tape must remain uncrinkled and unspotted by greasy fingerprints. In the cassette format it is not necessary to actually thread the tape—you simply insert the cassette— and so you should *never* touch the cassette tape. Open reel video recorders must be threaded manually, and a threading diagram illustrating the tape's path is usually attached to these machines. With open reel recorders, the best way to thread the tape is to put the supply reel on the deck with the machine turned off and unwind about two feet of tape, holding only the very end of the tape with your fingers. Then position the tape through the various guides around the heads to the empty take-up reel. Wrap the end of the tape onto the take-up reel manually, turning the supply reel by hand if you need more tape to do this. When you finish this procedure, the tape should be firmly wound through the tape path and onto the take-up reel so that there is no slack when you put the VTR into operation—if there is slack, the take-up reel will jump forward as it engages the tape on its

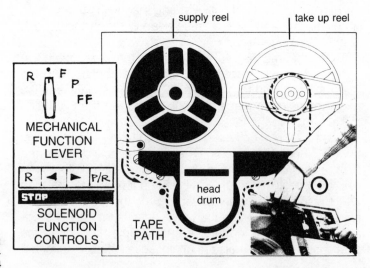

core and begins to pull it. Make sure that the take-up reel has enough tape firmly wrapped around the core so that it will start to take up the tape smoothly.

Before operating the VTR, make sure that all the connections have been made between the VTR and whatever external equipment is being used with it. There are several switches on the deck of the VTR that must be properly set for recording the signal. One switch is for the input you're using—camera, TV, or line—and must be set for the incoming signal. Another is for the mode you're in—record, play, audio dub. Finally, the meters and their settings (if any) must be adjusted. Make sure that all cables are securely connected and check the tape threading once again.

In general, operation of the video deck is not a big problem. And if you do mess up, you'll be able to tell when you play back the tape since you won't have the picture you wanted. The only really invisible problems are tape distortion resulting from jockeying the controls and the possibility that you've misthreaded the tape. If you approach the VTR calmly and move slowly through the operating procedure, you'll have few real problems.

VTR Operating Instructions

To record a signal on your VTR, first determine where the signal is coming from (camera, TV receiver, another VTR) and check that the proper cable connections have been made. Clean the video heads, audio and control head, erase head, and tape guides. Take a reel of blank tape and thread it onto the machine. Once the tape is securely on the take-up reel, set the tape counter at 0 and run the tape through the machine for about three feet—.91m—(approximately to the digit 10 on the counter). This allows enough extra footage at the head of the tape so that constant playing over the years, which may fray or crush the first few inches of the tape, won't affect the program.

When your signal is ready to be recorded, engage the record button, turn the function handle to play, and let it run, allowing about six seconds for the tape to stabilize before actual recording begins. If your recorder features meters for video and audio level, and if these meters are not equipped with automatic gain controls, you must monitor the meters throughout the recording process. Don't be constantly "riding" the meters (turning them up or down) with every change in video and audio level. Just make sure that they are in the proper range (just below the red section of the meter's dial) except for minor excursions into the red during peak signals—a very brightly lit scene in the case of video, a loud noise in the case of audio.

When the recorded signal has come to an end, do not turn the machine off immediately, but let it run a few sec-

A-V Level Meters

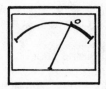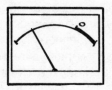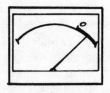

V.U. METERS   correct level   too low   too much

onds to avoid loss of stability during the final moments of the recording. If there are video and audio gain controls, you might want to let the recorded information end and then turn the video and audio level controls slowly down to zero so that the signal information fades out completely before you stop the machine.

Playback   Playing back a prerecorded tape presents no problems if you are using the same machine for playback that you used to originally record the tape. If, however, you are using a different machine, give the tape a test run before you gather your audience to show the tape. The most crucial factor is, of course, that the tape was recorded on a machine of the same standard as the playback machine. If this is not the case, you'll get a scrambled picture signal. (The audio may be fine, so don't presume that there's something wrong with your machine.)

Even if the tape is the same standard as the VTR, there may be a certain amount of distortion in the picture, such as white lines shooting through the signal. This can be corrected by adjusting the skew and tracking controls until the disturbance disappears. Turn these controls very slowly, since the least movement of either of them marks a large amount of internal adjustment. The beginning and end of the tape are the areas where some form of distortion is most likely to occur, so you should be prepared to make the necessary adjustments during the playback of those sections.

Actual playback is simple. The deck is connected to a monitor or TV receiver, the tape is loaded in the deck, and the play function is activated by turning a lever or pushing a button. If you get no signal at all—neither audio nor video—you haven't made the connections properly. If you get only audio and no video, the video heads are probably dirty. If you get video but no audio, it is possible that the audio head is dirty but more likely that you haven't made the connections properly or that the volume control is turned down on your TV set or monitor.

Rewind/Fast Forward   Both the rewind and fast forward (ff) modes of VTR operation require only one concern on your part: that you don't switch from play, record, or stop, to rewind or ff too fast. Get a firm but light grip on the handle and turn it deliberately.

Don't snap the handle from one function to the next and be very sure that you don't rush through its positions if you're going from, say, fast forward to rewind. If you do, the tape will literally jump out of the tape path. A smooth, easy, responsive turning of the handles with some sense of the mechanics behind them is essential to proper VTR operation.

More expensive video decks have **solenoid switches**—little electrically operated buttons—rather than one function handle, which must be turned from point to point to engage the deck's operations. Solenoids are much easier to use than the handle switch; they provide a smoother and surer operation, and give a more instant response. They also allow for remote control of the VTR, which is not possible with mechanical function handles. When using a solenoid-operated VTR always go from one function through stop to the next function. From rewind to stop (pause until the tape stops) to ff. To slow down the tape even more before you hit the stop button when coming from ff or rewind, you may want to hit the opposite button, then hit the other button, then the other: ff to rewind to ff to rewind, in a cradling, back-and-forth motion so that the tape slows down. Then, as the tape slows, hit the stop button. This rocking takes a little practice, but many professionals recommend it over going from ff or rewind directly to stop. If you're in play or record and want to go to ff or rewind with a solenoid-operated machine, always go to stop first, wait for the tape to stop, and then go to the next function.

Placement

Care and Handling of the VTR
Video tape recorders should be placed in an area where they aren't hit by bright sunlight, radiators, rain, and other heat or humidity producers. VTRs with solenoid buttons can be mounted either horizontally or vertically; those with mechanical function handles must be set horizontally. The only exception to this is the portable deck, which will work in any position, although when set on a horizontal it has fewer problems than when held vertically—the tape reels won't rub on the cover.

There are a number of precautions to take when using a helical scan VTR:

Care

1. Never press down on the head drum cover
2. Never thread the tape when the heads are rotating
3. Use the same size supply and take-up reels for the best skew and tracking
4. Don't block ventilation grills on the bottom or sides of the VTR
5. Keep away from dust, high humidity, and excessive heat

It's very handy to attach a sheet of paper to the back of the VTR on which to list the dates when you demagnetize the heads, oil the deck, and so forth, so you can keep track of these needs. You must demagnetize the heads after every few hundred hours of use or they'll build up a magnetic field of their own, which can ruin tapes played back on them. Use a tape head demagnetizer, which you can get at any hi-fi store; it doesn't have to be especially made for video. Be sure the power to the deck is off when demagnetizing heads. Don't ever touch the tips of the demagnetizer to the heads: If you scratch the heads you'll have to replace them. Cover the tips of the demagnetizer with plastic tape to ensure against this. Also, follow the oiling instructions that come with your VTR. If you have trouble getting the rollers

HEAD DEMAGNETIZER

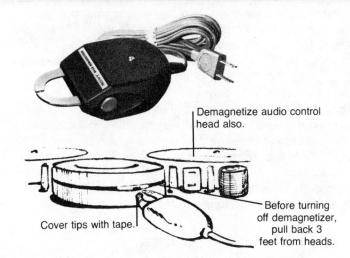

Demagnetize audio control head also.

DEMAGNETIZING THE HEADS

Cover tips with tape.

Before turning off demagnetizer, pull back 3 feet from heads.

loose (they're secured by Phillips head screws on some machines), don't pass on oiling; have your video dealer do it for you.

Maintenance   When you clean the heads of your VTR, check the rotary guides and other parts of the tape path for dirt. If you find a great deal of oxide build-up on any of these guides, it means that they're out of line and ruining your tape. Have them adjusted.

It's not within the scope of this book to provide information on major repairs and adjustments. Every piece of video equipment varies in its design according to manufacturer. Correct procedures for one might well destroy another. But it's possible for you to keep your equipment in good repair. There are manuals available from all manufacturers which explain the details of their equipment's operation and how to set things right when they get out of whack.

These repair manuals cost between five and ten dollars. Each manual deals with one particular unit—a VTR, camera,

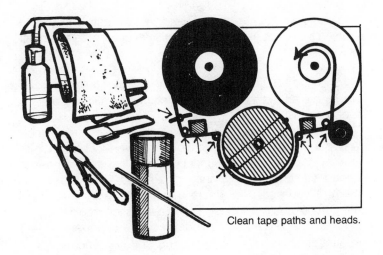

Clean tape paths and heads.

monitor, SEG, etc. Your video dealer can order them from the manufacturer for you. Even if you're not planning to do your own repairs, get manuals for the equipment you own. They take some of the mystery out of electronics and give you a more thorough understanding of the equipment, its functions, and its limitations. Written in a straightforward, paint-by-numbers fashion, they are intended for use by servicemen who haven't necessarily got engineering degrees. They tell you what can go wrong with a piece of equipment and then, step-by-step, explain how to readjust the components so everything is working again.

Every electronic circuit has a characteristic waveform. A graphic representation of that signal—be it sync, picture signal, or blanking—can be displayed on a waveform monitor or oscilloscope as a pulsating line of certain dimensions. A picture or drawing of each of these waveforms is given in the manual along with instructions as to where to monitor them on the printed circuit boards inside the piece of equipment and what to adjust until the waveform on your monitor looks exactly the same as the waveform shown in the manual. With these manuals it's possible to check and adjust all the functions of the video circuits, one by one. Even if you can't correct what's wrong, you can at least determine where the problem is and what needs to be fixed.

Repair manuals also include any modifications (such as improved circuits) that have been made on the equipment during its production run; they list the order numbers of all the parts down to the smallest screw or washer, which allows you to order replacements for faulty components.

Whether you're working with a portable or ac-operated deck, you must learn the ground rules of VTR operation and maintenance. You may never warm up a soldering gun, but you should develop a program of preventive maintenance—things like cleaning the heads properly, demagnetizing them, and being careful with the tape and the tape path. All of these elements are factors in the production of good tape and must not be ignored.

"She'll do 110 flat out." The human qualities that machines develop start in the fingertips of their owners. You begin to feel responses. "I coaxed her into high gear." Most often these riffs are ejaculated by Van Johnson types on their way back from bombing Tokyo. Or Steve McQueen types as they drive off into the sunset. Despite the romantic-sexist overtones of these comments, there is a basis of truth to them. If you can "dig where machines are at," they'll do things you won't find listed in the instruction booklet.

Machines, like human beings, respond to understanding treatment. If you're a knob-twisting button-wrencher, you'll probably find that the "damned thing broke," as if it were all a big planned-obsolescence plot with you as the ultimate victim. It doesn't have to be that way. All you have to do is learn to be considerate of the machines you use and you'll discover that they'll give you longer service with a greater degree of dependability than you might have expected.

The external buttons and knobs that activate the internal electronics and mechanics of machines are what you, the operator, must deal with. Even if you fall into the all-thumbs category, you can walk again, talk again . . . if you're willing to drop your electronic belligerence and admit that, when things break, most often it's because you break them. Of course, it may be that the thing you broke was going to break anyway, but it's usually the human factor that initiates the failure, even if it's the fault of the manufacturer for making tacky equipment.

There is one general rule to follow when operating electronic equipment that will result in its working literally years longer than it should: Go slow and be gentle. If you take your time and very carefully manipulate machines, they'll work better, longer.

Let's say you have a video cassette machine with a set of buttons for play, record, fast forward, and rewind. When you depress the particular button for the mode you want, start by placing your index finger on the button. With the tip of your finger feel the surface and the resistance set up by the mechanics behind the button. Now, very slowly push down the button, feeling the point at which the mechanism engages as you push. The same technique can be applied to VTRs, monitors, SEGs, and the like. If you have a VTR that has a lever that is turned to different positions for play, rewind, etc., hold it firmly and turn it slowly. Feel your way around. Again, you'll find that it takes much less effort and force to operate the machine than you might have thought.

The mass production of sophisticated electronics at

competitive prices means that the manufacturers have to give you something that does more than it should at the price they're asking. Since the advent of integrated circuits and clusters of circuits known as hybrid chips, most of these units are fairly stable internally. The big problems arise from the mechanical components used to activate the electronics. They get dirty and they wear down. Controls are the most sensitive to failure in terms of the life of the machine, and controls are what most of us handle roughly. Just as a racing car driver will learn to feel the gearshifting mechanisms through the stick shift, so you must learn to feel the guts of your electronic equipment through the controls that regulate its performance. This means going from play to stop to rewind with a pause at every step, rather than brutally twisting the knob from play to rewind. If there is brutality involved, the result will be that you break your toys.

So the next time you see someone talking to his equipment, don't confuse the conversation with your views on materialistic attitudes in a materialistic society. The conversation between people and their machines isn't one-sided.

We're a nation of bangers, pounders, and slammers. Somewhere along the line we began to assume that everything is made of cast iron. I haven't quite figured out if it's the last traces of our rough-and-tumble frontier spirit or just a callous grossness peculiar to our society. The first evidence that this attitude might eventually undo us came with the great media phenomenon of the fifties, the Japanese transistor radio—we called it the "cheap Jap radio." We bought 'em, used 'em, broke 'em. The truth is that the radios weren't cheap at all, they just weren't made to be handled by rough American paws. Nobody ever said a radio had to be made of sheet metal or inch-thick plastic, but somehow that's the kind of electro-mechanical device that lasts longest in our society. The heavier duty the better. When video equipment came along, the problem was even more evident. The literature of the video culture is full of mutterings about the fragility of the equipment, with knobs breaking off, covers rubbing against tape reels, and so forth.

I've produced record albums in L.A., San Francisco, New York, and London. My last production was done in London, and I was surprised to find that the British never turn anything up all the way. In a New York studio when you ask the engineer to play back a tape *loud,* he cranks it up until the speakers are popping out of their cabinets. In London, loud is about two-thirds the volume it is in New York. Not that the equipment they make can't handle it; it's just that the British think things perform better and last longer if you don't turn them up to their greatest potential.

This philosophy must be applied to your video equipment. When you first start to use a piece of equipment, treat it as if it were made of crystal. Pamper it, handle it as if it's going to fall apart if you look at it sideways. As you gain a knowledge of the ins and outs of the gear, you'll be able to use it a little more casually, but you'll also understand its tolerances.

I've been using my portable system for years without any problems, not even all those problems that are supposed to be inherent in the design of Sony's portapak. I've lugged it through Europe, across the US, and, most dangerous of all, gotten it in and out of cabs in New York. So far, not even a scratch. I'm not a compulsive saint, but I do appreciate the fact that the system isn't a baseball bat. It's a highly sophisticated piece of equipment, and just because it's made to be carried around doesn't mean it's a Mack truck. If you learn to accept this fact, your equipment will last longer and you'll make better tapes.

Portable Operation

More care and attention is required with the portable deck than with the larger nonportable models. The major areas of concern with the portable VTR are: power supply, camera-to-deck connection panel, tape loading, and controls.

Connections   The camera-to-deck connection panel is where the cable from the camera is connected to the deck. As already noted in the discussion of the camera, this connection must be securely made and checked regularly. On the panel you'll also find a TV/camera function switch. This switch determines where the input/output signal is coming from or going to. When you're using the camera for either record or playback through the camera viewfinder, the switch must be set at *camera.* When you're recording an off-the-air signal from a monitor-receiver or playing back a signal through a monitor, the switch must be set at TV. The panel will have a tracking control dial or knob as well. This shouldn't be adjusted except during playback. Follow the directions in the instruction manual for the deck you have as to where it should be set for record.

Controls   The portable VTR controls are easy to operate and require only that you don't twist them off or break them by pushing or pulling too hard. The portable VTR is a much more sensitive piece of equipment than a normal VTR—at least as far as its external controls are concerned, and you must always have the greatest respect for the fact that you've got an entire TV studio tucked under your arm.

Mounting   It has been the fashion to mount the portable deck on a backpack frame and let it ride behind during taping. I disapprove of this carrying system as it removes the deck from your control. If you carry the deck hanging from your shoulder on the strap supplied, you'll have fewer problems keeping it working. First, because you can see what you're do-

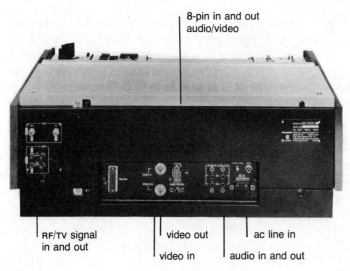

8-pin in and out
audio/video

RF/TV signal
in and out

video out

video in

ac line in

audio in and out

remote control connection

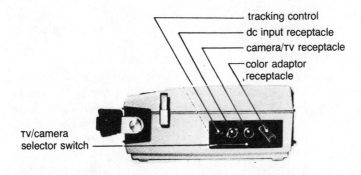

tracking control

dc input receptacle

camera/TV receptacle

color adaptor
receptacle

TV/camera
selector switch

ing; second, because you have the feel of the machine between your arm and body; third, because you will lessen the chances of damage when putting it on and taking it off.

Even with the discomfort of the shoulder strap digging into you, the advantages of having the deck at hand outweigh other considerations. The straps, by the way, can be a problem in themselves. On some portables, especially those made by Sony, the catches that hold the strap to the deck tend to come loose at their own whim, which may send your deck crashing to the ground. You can get around this, in some cases, by just turning the strap inside out so that the catches (which normally hook inwards toward each other) are hooked outwards.

With the camera connected to the deck and the deck properly loaded with tape, you're set to record. When you anticipate a situation you want to record, engage the function buttons or knobs in a record position. Don't leave the deck in a record-idling position for long periods of time—you'll run down the batteries and clog the heads. At the same time, don't jump right into recording. Your camera

Recording

VIDEO RECORDING

tube needs time to get up to its proper voltage, and your scanning sync has to lock in.

The correct procedure when using the portable camera and deck is to put the deck into record-idle, then look through the camera viewfinder until a picture appears on the screen, at which point everything is ready to begin recording. You can create an interesting effect by using the time-lag between record-engage and record-ready. First, time it—say it takes twenty seconds. When you get fifteen seconds into the process, start to record even though there is nothing on the viewfinder screen. When you replay the tape the scene will seem to burst onto the screen as the automatic light control on the vidicon comes into action—it's like keying from nothing into the scene. It's a good way to begin a tape occasionally; don't overuse it.

Power Supply   The power supply of the portable VTR can be either ac power from a wall outlet by way of the ac adaptor, or batteries, which may be installed within either the deck or a separate battery pack. The self-contained, rechargeable batteries supplied with the portables often lose their ability to store a charge. There's nothing you can do but buy a new set of batteries. When your deck's batteries do run down, I'd suggest you spend your money on an external battery pack. They cost from $100 to $300, but they'll give you several hours of use per charge rather than the thirty to forty-five minutes you get with the internal batteries. You can even adapt other battery systems to the portable VTR as long as the voltage of the battery pack matches the input voltage of the deck. It's usually 12 volts dc. Be extremely careful that the voltage is correct—anything over 12 volts will blow the fuses in your deck. Also be careful that the ground and conductor of the jack from the battery pack is of the correct polarity with the input receptacle on the deck (ground to ground, conductor to conductor). If you're not sure which is which, have your video dealer wire it for you. The most dependable and handy external battery packs are those which power movie cameras. Check movie supply-house catalogs for what's currently available. Some video groups have used motorcycle batteries. I'd say this is dangerous at best and would even recommend you stay away from anyone using such a setup. Motorcycle batteries are a fire hazard, can blow up, and can leak acid all over you. It's just not worth the small savings involved.

To keep rechargeable batteries up to full charge, recharge them for a couple of hours every day. Never store batteries in a discharged condition. Batteries tend to have a "memory," and if they're left uncharged for long periods of time, they'll literally "forget" what their full charge potential was and won't ever take a full charge again. Once batteries are recharged, they should be allowed to sit for an hour or two before use.

Some external battery packs come with rechargers that are not designed to be used when the battery pack is connected to the deck. They are *only* for recharge. Never try to run your deck off ac by plugging the recharger into an ac outlet and then plugging the battery pack into the deck. You'll ruin the batteries and burn up the transformer of the recharger. The ac power supply/battery recharger that comes with the portable VTR, however, regulates the recharge process and provides a trickle charge once the battery is up to full potential. You can adapt external battery packs to use it. They're much more dependable and do less damage to your batteries than the cheap rechargers supplied with some battery packs.

Repair Tools

The following items should be put together in a kit for use with your equipment: one regular screwdriver, one very good Phillips head screwdriver, some single-edged razor blades, a pair of long-nosed pliers, a flashlight, a pair of diagonal cutting pliers, a sharp knife, a wire stripper, a 25-watt (or less) soldering iron with a small tip, a few rolls of plastic electrician's tape, some 60/40 solder, a very small jeweler's screwdriver, and a roll of gaffer's tape (a very sticky tape used in film productions to secure lights to walls and for other temporary setups).

Make sure that your screwdrivers don't become magnetized through use; the proximity of such magnetized instruments could erase the video tape or seriously damage the heads. If they pick up metal screws with their blade, they're magnetized. To demagnetize, pass them through the space between the arms on a tape head demagnetizer, or else lay them on a bulk tape eraser, turn it on, and then lift them away slowly.

Cheap Phillips screwdrivers will strip the heads of Phillips head screws, so buy the best one you can find. If you do strip the heads, the only way to get out the screw is by drilling—have your video dealer do this, don't try it yourself.

On the Road

An awful lot of gear has been described in this section, from screwdrivers to VTRs. When you're working with video

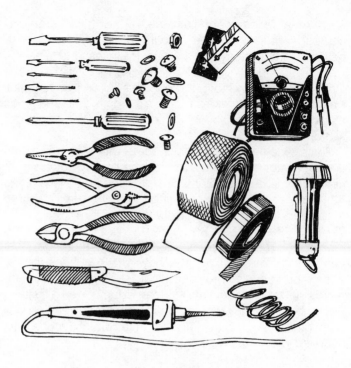

you have to be familiar with all of it, but you don't have to lug it all around with you from taping to taping. Most of us won't ever be in the position to build our own studios complete with SEGs, amps, and other processing equipment. We'll be working with portable VTRs, editing decks, and a couple of monitors at most. It's not a bad setup, but you've got to get down to the practical aspects of working with it, both on location and in your own studio situation. Most often you should divide up your equipment—the editing deck and monitors should be in one fixed spot, set up and ready to work with. The portable camera and deck is what you'll travel with and you should prepare a full set of ancillary equipment to go with it.

Portable Ensemble   The ideal portable video ensemble is made up of the following: portable camera and deck, extra external battery pack, RF adaptor, ac adaptor/recharger, audio mixer and mikes, a lighting unit, and accessories like repair tools, video tape, head cleaning spray, headphones, and, if you're going to be away from your video dealer, an extra set of video heads and an extra vidicon. These last two items will cost you about $200. You may never be in a situation in which you need them (eventually you'll have to replace both heads and vidicon, but it could be several years before that happens); on the other hand, if either malfunction when you're working on location they will be extremely difficult to purchase at a local TV shop. A manual for your camera and deck is also handy to have along as it gives instructions on the replacement of heads and vidicons and, in a pinch, would enable you to do it yourself.

Packing all this can be a problem unless you've got a small van. There are two ways of dealing with this problem. One is to pack it all in one or two big cases, the other is to break it down into bite-size pieces. At the moment, there are carrying cases available from video manufacturers that will store the VTR, camera, and related equipment (except lights) in one big padded box. This is fine if you've got space to transport it or can afford the air freight charges. Most lighting units can be purchased in sets that come complete with a carrying case, so you wind up with two luggable cases.

If you're going to be doing a lot of wandering with your equipment, however, I'd suggest that you bypass putting everything into one or two boxes. Instead, break the equipment down into sections: lighting units in their own container, VTR in the shoulder-strap bag that comes with it as standard equipment, camera in a shoulder bag you make yourself, and audio equipment, tape, repair tools, and the like in a separate case. You can build a camera case from lightweight sheet aluminum and nuts and bolts. I put one together in one evening. It was just an oblong box with a divider down the center—one side for the camera and the other for the lens. It's very handy since the deck and camera can be carried by one person and stored under the seat on a plane or in the back seat of a car with no trouble. Use plenty of foam rubber inside the case to protect against vibration. You might also want to slip a thin sheet of foam rubber into the bottom of the deck bag before you put the deck in—it will act as a cushion when you put the deck down. A small aluminum suitcase made for photographers to carry their cameras and lenses is ideal for the ac adaptor and other equipment and accessories. These suitcases come with foam rubber sections in them, ready to be cut to size to accommodate whatever you want to store in them. Keeping track of four or five small cases is more of a problem than keeping track of a couple of big ones, but you'll find the going easier when you have to put your back into carrying it all around.

The lights, camera and deck, tape, and battery pack plus recharger are fairly obvious units necessary to have. But you shouldn't neglect all the extras. Movie crews have what they call a ditty bag—a little bag or suitcase chock-full of every conceivable item, from Band-Aids to a ball of string. You never know what you're going to need and when you're going to need it. Among the items you should put in this video first-aid kit are: a couple of cans of head cleaning spray, a box of cotton tipped swabs, the repair tools and supplies outlined earlier, every conceivable patch cord and

First-Aid Kit

VIDEO RECORDING

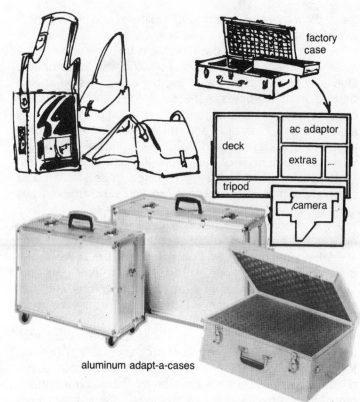

factory case

deck | ac adaptor
extras | ...
tripod
camera

EQUIPMENT CARRIERS

aluminum adapt-a-cases

jack adaptor, short lengths of audio and coaxial cable, the instruction manuals for your equipment—put them in a separate plastic bag for protection from dirt—and extra lamps for your lights, plus as many extension cords as will fit.

Europe   If you're going to be working in Europe with the system you'll also need a set of ac plug adaptors, a step-down transformer, and different voltage (220v) lamps for your lighting units. You might also want to take a small US standard TV with you for displaying tapes. For more information on European operation see pages 329–31.

3

WHITE LIGHT
WHITE HEAT

VIDEO DISPLAY

By order of the Queen, twenty-five colour TV monitors in Westminster Abbey will be switched off when her own procession and that of Princess Anne enter on Royal wedding day. For seeing themselves walk puts the Royals off their stride.

—London *Daily Mirror*, October 1973

The TV screen. Kids cluster like Indians surrounding a wagon train; construction workers in ratty t-shirts clutching cans of beer stare myopically into the bluish glow; the housewife, up to her ears in hair rollers, irons with one eye and watches the soaps with the other. Television gives us the images to describe images.

The TV screen. It is nothing more than a glorified light bulb intruding its fishbowl midriff into our lives as well as our living rooms. A new deity, replacing the cross on the wall and the embroidered "home sweet home." Certainly a new method of communication that can stereotype as instantly as it can inform. Wars have to be fought only on television. Politicians exist only on television—when was the last time you saw one in person? And that goes for football, Academy Awards, and Johnny Carson.

We all have our personal reminiscences of our TV lives. For the first television generation it was *Rootie Kazootie, Captain Video,* and *American Bandstand.* For the second, it was *Star Trek, The Monkees,* and *The Man from U.N.C.L.E.* Tv nostalgia is the most viable of all warm memories. It captures a time that never was more than imagination in the first place. Adlai Stevenson pales next to the vivid detail of Ernie Bilko. We live in a world that Hugo Gernsback, the pioneer of media technology, so aptly named "science fiction." Except that our truths as well as our fictions are replenished daily by science, with the help of a good deal of human imagination.

It all winds up on the TV screen. And the relationship we have to that screen, the sense of coexistence we've grown up with, has much to do with the emergence of video as an obvious and undeniable fact as soon as one is exposed to it. There is no reason why everyone should have a video recorder, no logic by which to explain that it will happen. But it will. Mirrors have always been a big item.

Video does the same thing, and it's with the same sense of wonder that the initial video experience takes place. We are all like little children seeing our reflections for the first time. There is some sort of primeval sense of order in the fact that this experience leads us to the conclusion that everyone will want one. Of course, it may be like a caveman sticking his fingers into the first fire he's ever seen, but so what? That's what it's all about anyway.

Getting the Picture

A television set is just the opposite of a video camera, in terms of function. The TV screen displaying the video information is a **cathode ray tube (CRT)**, which transforms the electronic signals coming from a TV camera or VTR into a series of brightness values which glow on the face of the tube to produce what we see as a picture.

There are three variations of the CRT TV set available for display purposes. The most familiar CRT display device is the home TV set. This is known as a **receiver** in video parlance, meaning it receives video transmissions and converts them into shades of gray on the **raster** or face of the TV tube (CRT).

When video signals are transmitted by a local TV station through the air or by cable, they are impressed on a **radio frequency** carrier known as an **RF** carrier. These radio waves, carrying the video, sync, and audio information, emanate from a transmitter and are then received on the home TV set by an antenna, which is tuned to the various TV channel frequency ranges (the pole pieces of antennas are of varying lengths in accordance with the various TV channel frequencies). Once these RF signals are fed into the TV set through the antenna input, they are processed to again become the composite video and audio signals as they were put together at the TV station from the signals derived from the video camera and associated audio equipment.

The progress of the signal through the TV set is as follows: VHF and UHF tuners are used to locate the RF broadcast of any particular channel frequency range. The channel selected is then fed into an IF or intermediate frequency amplifier where the RF carrier signal is strengthened. The signal then travels to a video detector, which separates the video and the audio signals from the RF carrier, converting the video into the original composite video signal. The video signal is sent to a video amplifier and to a separator, where the horizontal and vertical sync signals are removed and directed to their own separate amplifiers from which they control the display of the video signal on the CRT. The audio portion of the signal goes to other amplifiers, where it is amplified and then routed to the speaker for reproduction.

The process is a relatively logical one: The total TV broadcast signal enters the TV receiver through the antenna input, is separated into audio and video, and the two signals then follow their own separate paths to the display components—the CRT and the speaker.

A video **monitor** is a variation of the TV receiver. This set has no VHF or UHF tuners to select a particular frequency channel, no IF amplifier, and no video detector. The signal coming out of a video camera or VTR is not an RF broadcast

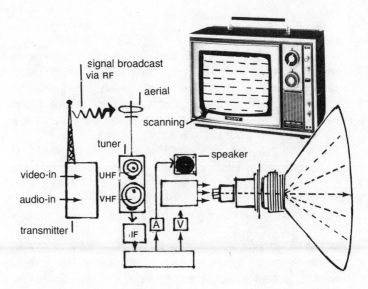

signal broadcast
via RF

aerial

scanning

tuner

speaker

video-in → UHF

audio-in → VHF

transmitter

IF A V

COMMERCIAL
TV RECEIVER

signal but a **composite video signal,** which, to be broadcast, would have to be impressed on an RF carrier. If the signal is to be displayed locally only and not broadcast, it can be displayed as a composite video signal. The composite video signal is fed directly to the video amplifiers and control circuits while the audio goes directly to the audio amps.

The advantage of the video monitor over the receiver is the maintenance of the quality of the composite video signal. Every time a signal, be it audio or video, is impressed on a carrier frequency, broadcast, amplified, and then reproduced, there is a loss of picture and sound quality—possibly not noticeable on the home screen but, in terms of the original quality, a reduction in values.

The pure video monitor, such as that used during production to watch signals coming from several cameras, does not have an audio system. It displays video only. There are also video monitors with both audio and video facilities which are used to display taped programs. The third type of monitor is the **monitor/receiver** which has both an antenna input for RF TV signals and a set of inputs for audio and composite video signals. A monitor/receiver is essential if you want to tape off-the-air broadcasts, since the broadcast signal must come in through the tuner assembly, be stripped of RF, and then be fed out of the video outputs after it has been converted to composite video and pure audio. Monitors and monitor/receivers are considerably more expensive than normal TV sets because they are not yet a mass-produced item. It is also possible to do this using a specially constructed tuner without a TV set. Many video cassette recorders have such a tuner built in, and can thus be used to record an off-the-air signal. But these tuners have no "monitoring" facilities, so the result can be a less-

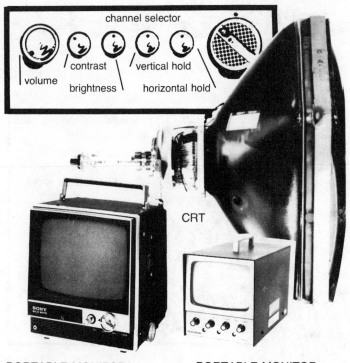

channel selector

contrast

vertical hold

volume

brightness

horizontal hold

CRT

PORTABLE MONITOR/
TV RECEIVER

PORTABLE MONITOR

than-perfect recording. It's advisable to use a monitor/ receiver when taping a TV broadcast.

Operation

The operation of a standard TV receiver is familiar: You turn it on, tune in the particular channel you want to watch, and then adjust the contrast, brightness, color, volume, and fine tuning controls until you get the optimum picture quality. The video monitor works the same way except that it has no channel selector or fine tuning since the particular signal you want is fed directly into the works of the TV. In a way, it's similar to a hi-fi setup in which an FM broadcast is tuned in on an FM tuner and then fed into an amp and then into a set of speakers. But a record played through the same system requires that the signal from the record be fed directly into the amp and then to the speakers—you don't tune in the record, you just adjust the volume and tone controls to suit your listening pleasure. The same differences exist between a receiver (tuner-amp-display) and a monitor (amp-display).

Located on the back or side of every monitor are sets of input and output jacks, which let you plug directly in or out of the set. These jacks are used to send video and audio signals in or out of the monitor. There are three types of input-output jacks in general use. The most standard is a

VIDEO DISPLAY

UHF CONNECTOR

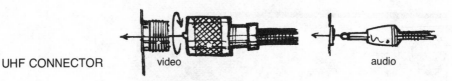

video   audio

MONITOR/RECEIVER
PATCH PANEL

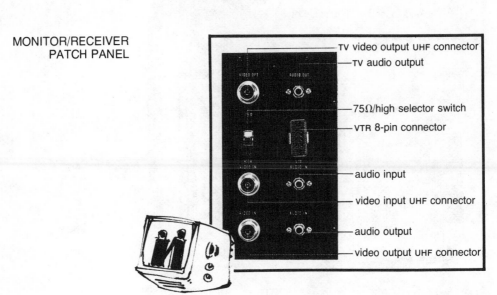

- TV video output UHF connector
- TV audio output
- 75Ω/high selector switch
- VTR 8-pin connector
- audio input
- video input UHF connector
- audio output
- video output UHF connector

UHF connector for video-in and another UHF connector for video-out. These are also called line-in and line-out. UHF connectors deliver only composite video (no audio) in and out of the monitor and are found on professional monitors, especially those which have no audio facilities and are for viewing the video picture only. The second set of connection jacks on many monitors are eight-pin connectors. These connectors are used between VTRs and monitors. When connected, they provide a full set of video and audio connections in and out of both the VTR and monitor (two connections per function—one ground, one lead—equals eight connections). When an eight-pin connection is made, you have video-in, video-out, audio-in, and audio-out, whether you're going from VTR to monitor or from monitor to VTR.

The final set of input and output jacks on monitors equipped with sound facilities are audio-in and audio-out jacks. These may be in any configuration from mini plugs (favored by Sony and other Japanese manufactures) to Cannon connectors, phone, or phono plugs. (See pp. 169–73.) The audio-in and out are used in conjunction with the UHF line connectors to send both audio and video in or out of the monitor. If there's an eight-pin connector, one connection takes care of both audio and video; with audio and UHF lines, the audio and video signals are separate and each must run on its own cable from the VTR or other equipment to the monitor.

THE VIDEO PRIMER

For video display purposes, the best monitor to buy is one that has all the connections listed above, so that you can have a full set of options: UHF line-in and out, audio-in and out, and eight-pin.

When buying a monitor you have to consider its screen size, video quality, audio quality, and the jacks supplied. The controls will vary with the quality of the monitor. Most have the standard controls found on any TV set: contrast, brightness, horizontal hold, vertical hold, volume, and on/off. These controls are sufficient for most purposes. If you're into video production on a multicamera level, combining a variety of signals to make a final master tape, you might want to invest in one high quality monitor. Most TV studios have at least one such **master monitor,** which is laden with extra controls and is more sophisticated than normal monitors or TV sets. Such a monitor provides the production crew with a picture that shows all the imperfections and instability in the video signal during recording so that the equipment can be readjusted to correct those deficiencies. Extra controls on a master monitor include focus for the picture-tube deflection coils; internal and external sync, so that you can run a non-composite video signal into the monitor and get your sync signals from the same sync generator that is supplying the signals to the cameras and other equipment; the ability to adjust the horizontal picture detail at various frequencies; vertical linearity controls; vertical height adjustment controls; and scanning controls.

Master monitors also have a much higher **horizontal** and **vertical resolution** than normal TV sets or less expensive monitors. Of course, the resolution quality of the monitor will not improve on the resolution of the camera or VTR, but it does permit you to check just how much resolution the total system is reproducing.

Display Techniques

The Sony showroom on the Champs Elysées in Paris has twenty video monitors filling an entire wall of the first floor of the display area. It's a knockout. You walk in off the street and there is a mosaic of beautiful Trinitron color, very impressionistic; a checkerboard effect is created by the display of two different tapes on alternating screens. This little bit of Sony showmanship provides an example of the potential and the problems that you'll encounter when it comes time to show your tapes.

If you have a number of monitors, you may want to try "looping" them—that is, sending the video signal into a monitor and then running it out again to another monitor until a string of monitors are connected together, all showing the same video picture. The first monitor is connected to the VTR through its UHF-in jack. The second monitor is then con-

Multiple Screen

VIDEO DISPLAY

nected to the first by running a cable from the UHF-out jack of monitor 1 to the UHF-in jack of monitor 2. This connecting process can be repeated on up to five monitors without any loss in picture signal strength. The addition of a **video distribution amplifier** makes it possible to supply even more monitors. The video signal is first amplified in and out of the **VDA** and then the monitors are connected. It's also possible to supply the same picture signal to up to forty regular TV receiver sets by running and RF signal out of the VTR's **RF adaptor** to an **RF amplifier** and then connecting each of the sets in series to a coaxial cable from the amp. This, by the way, is the basis of cable TV—what you're doing is creating a small cable TV network. The RF adaptor is a circuit that can be purchased separately and plugged into your VTR. It costs about $50 for b&w and about $100 for color. The composite video signal is fed to it and an RF signal is produced, which can then be fed on a cable to the antenna terminals of regular TV receivers, where it is treated as a broadcast signal.

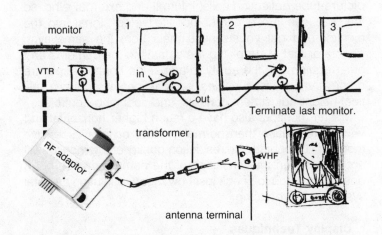

The ability to show a tape on more than one screen has to be handled wisely. The first time I set up two televisions to simultaneously display the same tape to a group of people, I ran into a problem that the video addict has trouble understanding. Most people are accustomed to watching television on one screen. When they're bombarded by the same images coming from a number of sources, the impact of the visual message can often be destroyed or at least be made less effective. People are used to focusing on that one screen and then passing into the medium to get the message. When they are given a number of screens to watch the message on, the medium becomes too much for them; they are distracted by the very novelty of multiset display. The chances that they'll appreciate what your tape is all about are less than if you put them in their normal TV-watching environment: one set, one group.

There are, nonetheless, methods of using more than one set that can add to your presentation. Some video groups have created tapes that take advantage of the multiset situation by programming various portions of the tape program onto different monitors à la Warhol's *Chelsea Girls*. If you have a large audience to present a tape to, you may want to use more than one screen. This is still consistent with the one screen to one group situation since the use of more than one set here simply ensures that everyone can see properly. All the same, such a multiset display can be distracting since the audience watching one screen will often have a peripheral awareness of the flicker from the other screen. If you want to get into larger group presentations, another option is to use a video projector.

Display techniques, even with one screen, must take into account the audience that has gathered to watch your tapes. People are used to watching television in the scratch-when-I-want-to privacy of their homes. They're geared to commercial interruptions, talking on the phone or to others, flicking the dial if they get bored, and so forth. You will probably be asking them to sit down to watch a minimum of a half hour of uninterrupted video. I've seen video makers squirm as their audiences murmured through their tapes or seemed to strain to maintain attention. All that can be said about that is: You're not making movies, you're making television, and you have to accept and accommodate your presentations to the *form* of television. This means allowing for an amount of audience unrest that would be unconscionable in a movie theater and attempting to create as comfortable and dramatic a setting as possible for the viewing area.

An effective enhancement to video display is excellent sound. By running the audio signal out of your VTR into a separate hi-fi amplifier and playing the sound through one or two speakers, you can augment your video presentation with an unusual and pleasing break from normal TV. If you have two speakers available, place one on each side of the monitor screen. If you're using your RF adaptor, you can run the audio out of the audio output jack on the VTR and turn down the sound on the set. People like good sound. By supplying it you'll find that your visuals will take on an extra dimension. Paul Morrissey once told me that TV is more than fifty percent talk. Make the talk as audible as possible. You may find that good speakers tend to accent the bottom, bass frequencies of your audio and be lacking in the more trebly sounds that usually come out of the speakers built into TV sets. You can overcome this by also running the audio signal into the set you're using (either RF or audio-in)

Sound

VIDEO DISPLAY

and then adjusting the volume on the set so it blends with the auxiliary speakers to create a hybrid, fuller, and more dynamic TV sound.

The Screen   Another important consideration in presenting the best possible display is to use as large a screen as possible. There is a rule used in setting up TV control-room monitors which holds that the proper viewing distance from the monitor should be at least three times the height of the screen. If the screen is a foot high, the ideal viewing distance should be at least three feet. The eye has trouble focusing on a screen that size if it's any closer. In monitor situations, this rule has led to the use of very small screens for displaying pictures coming from the various cameras, since the director and engineers are usually seated very close to the screens. Often these screens are only two or three inches high. Conversely, if your audience is going to be six to ten feet from the TV screen, you must use a large screen, approximately 19 to 23 inches (48.3–58.4cm). Small monitors are great for editing and checking footage, but you can't expect any real impact from a presentation made on a very small screen. I've seen video people get hysterical with glee the first time they've watched their tapes on a large screen. It does make a difference, especially for an audience that expects to see something on the screen and isn't impressed solely with the fact that you, not the networks, have turned out the show they're watching. If you can't afford a large monitor, get a used large-screen TV set for a few dollars and have it jeeped. **Jeep** is the video term for turning a regular receiver into a monitor. If you don't want to invest the money (about $100) in having that done, use it on RF. The picture may be a little fuzzy, because of RF amplification, but the size of the picture will compensate for any lack of quality when using an old, big-screen TV.

Perhaps the most effective method of displaying video is to go to your audience and show the tape on *their* TV set, using your RF adaptor. They may not understand how you can show video on their TV, but chances are they'll be much more comfortable and more likely to absorb the content of what's on the screen. This is the ultimate extension of video—the audience, a large but separate group, watching video tapes or discs at home.

Cable TV is also available as a display medium. Bring your tape to the local cable outlet and they play it on their VTR into their RF amplifiers out through the cable network they've set up to the receivers in the homes of their subscribers. Regulations for cable display vary from community to community. Get in touch with your local cable outlet for details on their display procedures—it's usually just a matter of delivering your tape to them and not expecting to get paid when it's broadcast.

Monitor use has certain technical limitations. First, the more monitors you use, the less cable you'll be able to run between each of them unless you use a video distribution amp to boost the signal (see p. 104). Try to keep the cable length between monitors about the same—five to ten feet. If you're using a monitor to determine the contrast ratio of a scene, remember that the monitor controls affect those values. The proper brightness and contrast for a scene is a subjective evaluation once you get out of the too-light, too-dark areas. But you shouldn't allow the monitor to one-up you because you forgot that the controls were set for full contrast and full brightness. Finally, the last monitor in any series must be terminated so that the video signal doesn't bounce back along the cable. Most monitors have a 75-ohm termination switch built in. It should be in the "on" position for the last monitor in a series. If it doesn't have such a switch, a 75-ohm resistor connected across the last monitor's line-out leads will work in a pinch.

Cable Connections

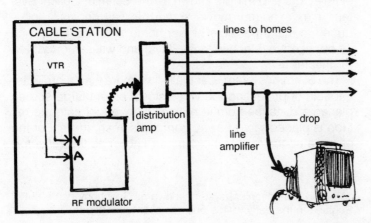

CABLE STATION

VTR

lines to homes

distribution amp

line amplifier

drop

RF modulator

VIDEO DISPLAY

Video Projections

A TV set with a screen that fills one whole wall of your living room—the omnipresent future-toy of science fiction settings. Cinemascope TV screens, hanging like huge technicolor canvases, which allow the viewer to have a private movie theater. *Fahrenheit 451* aside, the possibility of large-screen color television has always been a goal for those involved in the development of TV. The main obstacles have been the limitations imposed by the present TV system. A television picture consists of 525 horizontal lines stacked vertically on the face of the screen. The larger the screen, the farther apart the lines are; the farther apart the lines, the worse the resolution of the picture. On a 17-inch (43.2cm) TV screen the lines are close together, providing a mass of picture information in a confined area. On a 23-inch (58.4cm) screen, the lines have begun to spread, making it necessary for the viewer to move farther away from the screen to see a clear and in-focus picture. On a 30-inch (76.2cm) screen you'd probably have to be seated at the other end of a long room before the resolution looked right. In addition, it's technically difficult to construct a cathode ray tube to such dimensions (the largest to date is a 30-inch experimental model); the cost would range in the thousands of dollars and the voltage needed to make it glow would be excessive.

It is necessary to eliminate the CRT as the display device before large-screen television will become a reality. For almost fifty years, attempts have been made to do this, some of them relying on the existing principles of television, others attempting to provide a new alternative. To date, only two* methods of large-screen television have proved practical, both developed in the forties.

Large-Screen Systems

The earliest interest in large screens was prompted by the possibility of installing them in movie theaters to present special events, usually sports, by closed circuit to a paying audience. The first method of generating such a large picture relied on the use of a very bright cathode ray tube, a set of reflectors and mirrors known as the **Schmidt Optical System,** plus a normal movie screen. It was originally introduced to make possible bigger home TV screens (12-inch as opposed to 8-inch in those days) and was later adapted for theater use.

The Schmidt system is actually little more than a projector capable of projecting the video signal onto a large screen placed a distance from it. A small, very bright cathode ray tube is placed facing away from the screen. In front of the CRT is a concave, parabolic mirror known as the *reflector.*

*An electro-mechanical system known as Scophony came out of England in the early 1940s but was soon abandoned.

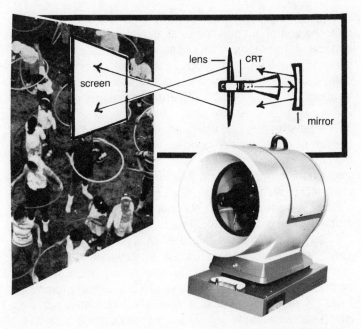

lens — | CRT

screen

mirror

GBC

The image produced by the CRT, which has been reversed so that the scanning is in the opposite direction of normal TV, is projected onto the reflector in the same way the light from a match would reflect when placed in front of a mirror. The light beams strike the reflector and are reflected back toward the CRT and past it. Around the neck of the CRT, and thus parallel to the face of the CRT and the reflector, is a plastic lens called a *corrector plate*. This lens focuses the light bouncing back from the reflector into a TV picture, curved slightly to fill in the hole in the picture area which the CRT has blocked. The light then travels through a second, stronger lens and out to the screen.

This system requires a small but excellent quality cathode ray tube running at a very high voltage so that it produces an extremely bright picture signal. Nonetheless, there is still an unavoidable loss of light between the projector and the screen, which gives a washed-out picture. This problem has been solved to a certain extent by using very bright CRTs, as well as by limiting the size of the screen and using screens with surfaces that reflect light images more efficiently than normal movie screens.

The other method of video projection is the **Eidophor System,** which also projects a picture onto a screen, but through a vastly more complicated technique. A series of steps involving oil, light, and a CRT produces a much larger, brighter picture than obtainable with the optical system. Unfortunately, the cost of an Eidophor projector is thousands of dollars more than the most expensive optical projector, and the problems involved in operating it are extensive.

VIDEO DISPLAY

Both of these systems have been refined and improved over the years, but only the optical system has come down in price. At the moment it's possible to buy an optical video projector for $1000 to $3000, the more expensive being color-capable. In this price range the screen size is limited to about 40″ × 60″ (10.2 × 15.2m) at most, generally in the 30″ × 40″ (7.6 × 10.2m) area.

Video Projectors   The least expensive video projectors are b&w only and are not of exceptionally good quality. They can be rented, but the lenses, reflector, and CRT become disaligned through the rough handling they tend to get and a fuzzy and distorted picture results.

Sony has introduced a number of video projection formats, all based on an optical system. The basic principle of the Sony system has been copied by a number of manufacturers in the US and in Japan to produce the current plethora of "TV projectors."* Using one very bright Trinitron color cathode ray tube, the Sonys are capable of projecting a 30″ × 40″ picture onto a specially formulated screen, which comes with the unit. The drawbacks of the Sony system are similar to those of all the other projectors available

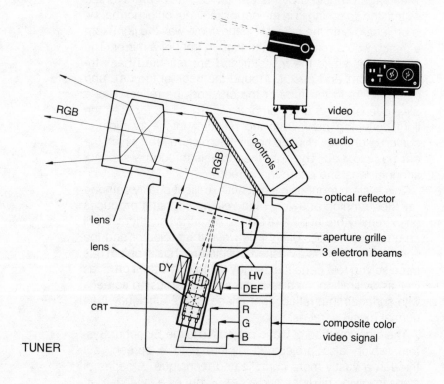

*All video projectors will accept a composite video signal from any source, including a camera, VTR, or VHF/UHF tuner. This means you can watch "regular" TV broadcasts on a video projection system with the addition of a VHF/UHF tuner. Sony sells such a tuner as an extra for its system.

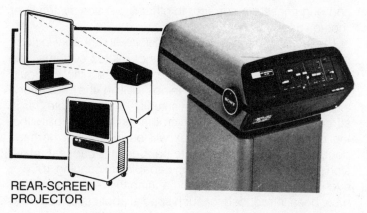

REAR-SCREEN
PROJECTOR

in that the brightest possible picture can be achieved only when viewed in a totally dark room; there is a certain loss of resolution when the picture is enlarged that much; the screen is slightly concave and so requires straight-on viewing; and the projector must sit exactly five feet in front of the screen—the focus of the picture having been set by the distance from projection unit to screen.

ADVENT VIDEOBEAM

Variations on the video projector have been developed also, one of the most interesting being a color system invented by Henry Kloss for the Advent Corporation in Cambridge, Massachusetts. It produces a 3′ × 4′ (.91 × 1.2m) picture and has a three-CRT projection system using specially designed CRTs that have reflector/lens systems incorporated into their construction. Each CRT produces one of the three primary colors. The picture resolution, contrast,

VIDEO DISPLAY

and brightness are superior to other projection systems for "home" use.

The future of large-screen television will probably include projectors that don't require optics or related systems. The use of lasers to project an image, the possibilities of liquid crystal screens, and other things still in the laboratory/ experimental stage may eventually make simple large screens possible. There is even the remote possibility that the number of scanning lines will be increased, perhaps doubled, but that is unlikely to happen in broadcast television. High-resolution cameras, capable of scanning 1000 or even 2000 lines and thus approaching the resolution of film, have been introduced for such special applications as making movies on video and then transferring them to film for projection in movie theaters, through the use of lasers, projecting a high resolution image directly from special video tape.

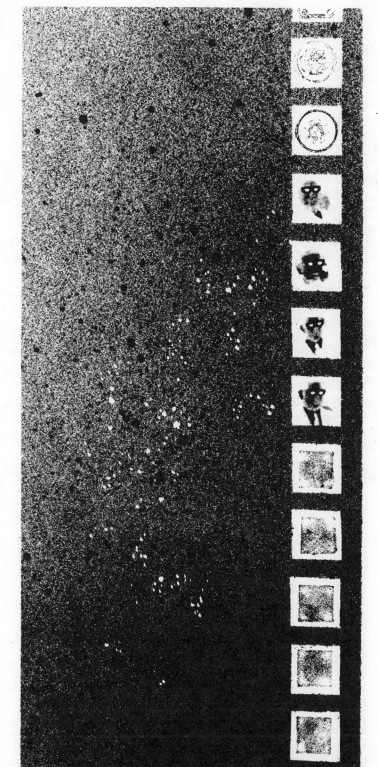

4

ELECTRIC
EDITS

EDITING

Video taping an event usually results in a certain amount of overshooting—you wind up with more video tape than the subject warrants. This is often unavoidable, especially when you're taping an impromptu occurrence and have to keep your camera and deck running to make sure that you don't miss anything important. When you get home you've got three or four reels of tape, but a lot of it is wasted footage; you can present your event in a coherent, interesting manner with just fifteen minutes of selected scenes.

When you first make tapes you may get a little dismayed at the amount of tape you record compared with the really important moments you capture; and you'll have to strap your audience to their chairs to get them to watch thirty minutes of glorious camera work if only five or ten of those minutes really say anything. You must, therefore, eliminate some of the sequences you've taped in order to make the story you're trying to tell presentable. This elimination of footage is **editing**. Most professionals see editing as the assembly of various scenes into a continuous whole. But it's largely a process of elimination. The aim of editing is to build a new whole; the action is to pare off what is not needed.

The cliché of finely sharpened scissors slicing through film to leave the best parts on the editing room floor is part of the romance of film. The political accusations that television news reporting is a matter more of carefully framed opinion than it is of honestly recorded fact is a reminder that there are people who make TV as well as people who watch it. The physical grace of finely edited film is considered the genius of some directors and their crews. The story of the sixties is the sense of non-edit that Warhol and other film makers developed. They were eventually to see this reflected back at them from a culture that demanded that things be played out to their natural conclusion. If it was going to take Jimi Hendrix all night to play his guitar solo, a tape of the same would run all night, as the gods had ordained. All this is part of editing.

Editing is a metaphysical as well as a physical process. There is a deep sense of the spiritual about it; it is media power on a very intense level. When you first discover that you can combine electronic elements of the world view you've captured on tape, it's likely that you'll be overwhelmed by the potential of the button you're pushing. You can reconstruct the universe, after a fashion, through editing. People can be made to appear, speak their piece, and disappear at your whim. A sequence of events can be reordered to imply that the exact opposite has happened.

If you want to maintain the integrity of the occurrences you've taped, editing can be used as a packaging technique to give your presentation a beginning and an end. Through editing, you can also eliminate the blips that are recorded when the VTR is started and stopped during taping. Even if it is only on this level of visual polish, editing must be performed.

When professionals make movies they have what is called a **shooting ratio.** It can be 5:1, 10:1, or 2:1. This ration indicates how much film is shot in relation to how much footage is used in the finished movie. It's not unusual for a film crew to shoot five times as much film as will eventually appear on the screen. Video taping an event leads to a similar ratio—you have more than the audience really wants to see, more than enough to tell the story. Most taping ratios run in the 2:1 to 3:1 area, but even if you have twenty minutes of tape and ten minutes of story, that's a lot of extra tape/time. Of course with video, the tape can be erased and reused, so we are speaking of a time-information factor here, rather than the sometimes prohibitive cost factor of overshooting film—a medium that is not reusable.

There are those video purists who believe that video, to be an alternative, must maintain its original electronic integrity, must present events as they happened, and if things drag for minutes on end between real content, well that's just atmosphere being established. Without getting into a long-winded confrontation with that view, I'd like to point out that the video maker makes a conscious decision as to what to tape in the first place, and if he or she can't then admit that some of the taping was unnecessary and should be eliminated after the fact, just as certain events were eliminated at the time by the decision not to tape them, then there is every likelihood that these good people will have to watch each other's tapes . . . because no one else will.

Editing is just one of the many post-production techniques that can be applied to the preparation of video programs for presentation. In a later chapter (pp. 191–94) I'll discuss other aspects of post-production as a process. Right now, let's concentrate on *how* to edit, not why.

Functions
The function of video editing is to assemble the various portions of recorded tape into a new whole. In the early days of video recording, the editing process was adapted from film editing techniques. To put a video program together from recorded tape, the desired segments for the end product were physically cut out of the total tape footage and **spliced** together. Splicing is a matter of putting the length of tape that is scene two after the length of tape that is scene one and joining the end of one to the beginning of two with a piece of splicing tape—a sticky tape similar to common cellophane tape but with better adhesive properties.

Because of the nature of the signal on the tape and the basic fragility of the tape itself, problems occur when the

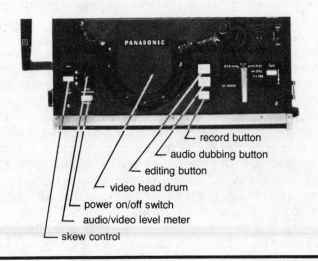

record button
audio dubbing button
editing button
video head drum
power on/off switch
audio/video level meter
skew control

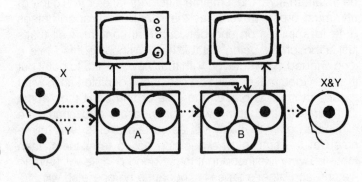

X

Y

X&Y

A

B

ELECTRONIC EDITING

video tape is physically edited. Even when the tape is cut and spliced perfectly, the results on the screen do not appear as cleanly cut. Rather, the first scene is replaced by the second in a wipe fashion; the second scene appears at the bottom of the screen and rises to push the first scene out of the picture. There are some instances when physical editing of tape is necessary. These instances and the technique involved are described on pages 100–01.

Electronic editing was developed as an alternative method when it became obvious that video tape was unmanageable as a cut-and-splice medium. Electronically edited tape is never cut; instead, the video signals on one tape are transferred to a second tape. Two video tape recorders are connected and the editing process takes place between the recorders. If, for example, you have two different scenes on two different reels of tape which you want to put together, one after the other, to create your final program, you transfer scene one and then scene two from their original tapes onto a new reel of tape. One video deck is used as a playback deck and the other as a record deck. The playback signal from VTR A is taken out along a cable from that deck and run into VTR B. The tapes of the scenes you want to put together are played on VTR A. Blank tape is placed on VTR B to record the signals as they come out of VTR A.

Say you've recorded a man running. Subsequently, and on another tape, you've recorded a woman running. You feel those two sequences would make a good program and you want to put them together, first the man running and then the woman running. Load the tape with the man running on playback VTR A, locate the sequence on the tape, and then stop the tape just as the sequence is about to begin. Load a blank tape on record VTR B. Start VTR B and push the record button, then put VTR A into play. The video signal of the man running will be transferred from VTR A to VTR B. Rewind the tape on VTR A and remove it. In its place, load the tape with the running woman sequence and locate the beginning of that sequence. Once this is done, put VTR B into record and play the woman running on VTR A. VTR B records the video signal of the woman running on the blank tape right after the sequence of the man running. You have assembled a new tape with two scenes on it, one following the other, first a man running and then a woman running. This is known as **assembly editing,** and it is the basic proce- dure for the great majority of editing done in video.

Assembly Editing

There is a terminology used to label the various tapes that take part in the editing process. The original tapes on which are recorded the sequences you want to transfer to a new tape in a new order are called original or **master** tapes. These are **first generation** tapes, meaning that they are the tapes on which the signals were originally recorded. The blank tape on which the program is to be assembled is known as the **composite master,** since it too is a master, or a new original version of the scenes you've taped. The com- posite master is considered a **second generation** tape, since the signals have already been recorded on tape and are being rerecorded on a second tape. Each time a signal is rerecorded the resultant tape is said to be another gener- ation away from the original recording. Once you have your composite master and you make copies of it, the copies are called **third generation** tapes.

Generations

Some confusion is introduced into all this by the fact that the edited, composite master tape is often called the master tape since it is the fully assembled program from which copies are made. It's likely that you'll get into calling your edited tapes masters, but be sure you understand that such masters are second generation tapes and not the original recordings of the sequences. The importance of this distinc- tion has to do with the change in quality from one generation to the next, a factor discussed in Chapter 5.

Components

As outlined above, two VTRs are necessary to assemble sequences from various tapes to create a new, composite whole. It is also important to have a monitor attached to each of the VTRs so that you can view the master tapes and the signal as it is recorded onto the composite master tape. Ideally, VTR B, on which you are recording the composite master, will be what is known as an **editing deck**. But before we discuss the design and function of the editing deck, let me say that it is possible to edit without one. You will not get smooth transitions between scenes (the purpose of an editing deck), but you can assemble sequences from various tapes onto a new tape using any two VTRs.

Editing without an Editing Deck

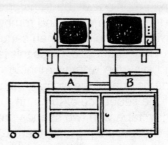

EDITING SETUP

To do this, record sequence 1 onto the composite master tape on VTR B. Just as sequence 1 ends, put VTR A into the pause position so that the picture is frozen. Stop recording on VTR B and rewind the tape slightly. Now, when you feed the signal of the second sequence from VTR A to VTR B, do it as follows: Find the beginning of sequence 2 and start to play it, going into the pause position on VTR A. With the beginning of sequence 2 frozen, engage the recording mechanism of VTR B and, as it starts to record, release the pause control of VTR A (slowly) into play. What you'll see on the screen when you play back the composite master on VTR B will be sequence 1 ending, seeming to freeze, and then sequence 2 displacing it, unfreezing, and the action continuing. The displacement of picture signal 1 with picture signal 2 will be a little messy, but the frozen action at the point of transition will tend to make it seem more like a special effect than like a total disruption of the picture. This technique is especially useful if you don't have an editing deck and want to join together various independent segments, each of which expresses a whole thought—if you'd taped a rock group, for instance, and wanted to select certain of their songs to assemble on a composite master. The fact that each song ends with a freeze and then the tape comes out of the freeze when a new song begins makes this **rough edit** acceptable.

But rough editing won't do if you're trying to create a continuous flow on your composite master. For that you'll need an editing deck in place of VTR B.. The deck will allow you to go from recording sequence 1 to recording sequence 2 on the composite master without any loss of **vertical sync**, which appears on the screen as disruptions of the video signal. In other words, there will be a clean cut from sequence 1 to sequence 2, one picture will replace the other without any of the noise or distortion on the screen that you normally get on a helical scan VTR when you stop recording one signal and then start up the deck again to record another signal.

Editing decks are readily available, usually the most expensive of any particular series of VTRs a manufacturer produces. They fulfill all the normal functions of a VTR—record, playback, etc.—but they have additional electronic components which make electronic editing possible.

There are three kinds of electronic editing decks avail-

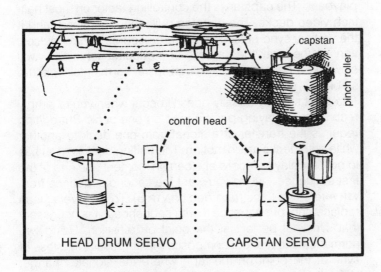

capstan

pinch roller

control head

HEAD DRUM SERVO   CAPSTAN SERVO

able: those capable of **simple editing, capstan servo editing,** and **head drum servo editing.**

Simple editing is more or less a misnomer, since the deck really does nothing more than any non-editing VTR. Beware of this kind of editing deck: The manufacturer is trying to pull a fast one, since there's no such thing as simple editing. The electronic process by which clean edits take place can be performed only by equipment that is designed to do the job at hand. I've never been able to figure out exactly what a "simple" editing deck is, but I gather that it attempts to replace smoothly the signal of sequence 1 with the signal of sequence 2 without providing any electronic stability to the process—just a few prayers in Japanese.

Capstan servo editing decks are the most popular and the most common editing decks; head drum servo editing decks (also called **head override** editing decks) are less common, though they perform pretty much the same function. The *servo* is the key. Electronic editing is electronic for two reasons: First, the tape is not cut; instead, video signals are played from one tape and recorded onto another. Second, electronic controls are applied to the recording process so that one signal replaces the other in a manner that gives a change of signal without any distortion or disruption visible on the screen.

To understand electronic editing we must understand the controls applied to the tape as it passes along the tape path from supply to take-up reel (pp. 43–44). The rotation of the video heads and of the capstan as it pulls the tape through the path past the head are controlled by control pulses derived from the incoming composite video signal. These pulses are fed to the tape by the control head during recording and read off the tape through the control head during playback. The capstan is the controlling factor on most half-inch video decks. The capstan pulls the tape along, and if the vertical sync pulses, which are used as control pulses, vary on the tape, the capstan of a capstan servo deck will slow the tape or pull it through faster to maintain picture stability.

Loss of stability usually doesn't occur when you're simply recording or playing back a tape on one deck. But editing requires the transfer of a signal from one deck to another while both decks are in motion. The editing deck (VTR B) has to go from play to record at a certain moment (the end of the first sequence—when you want the second sequence from VTR A to begin to be recorded on VTR B). To achieve a clean replacement of sequence 1 with sequence 2 it is necessary that sync not be lost at the point of transition. Using two normal VTRs to make a composite master results in a loss of sync at the point on the tape where the signals change, since the sync coming in with the rest of the composite video signal from the tape on VTR A is not the same as the sync that is controlling the running of the tape on VTR B. It takes several seconds for the capstan to stop sensing the sync control pulses coming from the playback of sequence 1 on VTR B and begin to lock into the sync control pulses coming from the new signal of sequence 2 from VTR A being recorded on VTR B.

Sync Control

Capstan Servo

With capstan servo editing, you push the edit button when you want to stop playing back sequence 1 on editing deck VTR B and want to start recording the next sequence from VTR A. What the editing button does is to lock the capstan of VTR B to the sync of the signal on VTR A. This locking takes place ahead of the edit point so that there is no loss of sync when the record mechanism engages, because the source of the sync signal has changed from the signal on the composite master tape on VTR B to the signal coming in from VTR A. If there is any variation in sync signal between sequence 1 (already on VTR B and providing sync up to the point of edit-record) and sequence 2, the capstan will slow down or speed up the tape on VTR B to compensate for that variation. That prevents any **roll** or **break-up** of the picture on the screen when the signals change.

A **clean edit** is when all this works and the vertical sync is not lost at the point of picture change. The picture simply switches from scene 1 to scene 2.

Editing Techniques

Two kinds of editing can be done with an editing deck. One is assembly editing, the process described above. You begin with a blank tape, which is to become your composite master, and assemble a series of sequences, which you record on that tape, one after the other, until you have a new whole on the previously blank tape.

XXYYYZZZZ ASSEMBLY EDITING

EEEEDDDEEEE INSERT EDITING

The second type, called **insert editing**, is done when you have a program already assembled and wish to insert a sequence in place of another sequence that's already on the tape. (Insert edits work only if both the old and the new sequences are of the same length.) With assembly editing your concern is that the transition between the end of sequence 1 and the beginning of sequence 2 appears smooth on the screen. Suppose, however, that you have a composite master tape of sequences 1–5 and you decide you want to replace sequence 3 with another sequence not on the master. You can begin all over again, assembling the sequences on a new composite master tape, starting with the first and adding each other one in turn. It would be better, however, if you could just pop a new sequence 3 onto the tape to replace old sequence 3. This is the purpose of insert editing. It is done in the same manner as assembly editing . . . up to a point. You put the original master tape of your new sequence 3 on VTR A and your composite master on editing deck VTR B. You find the beginning of new sequence 3 on VTR A and find the point at which you want

Insert Editing

it to begin (the end of sequence 2) on VTR B. You then start the machines, push your edit button, lock into the incoming signal, and proceed to record new sequence 3 over old sequence 3.

It is at the end of the recording that the insert edit function again becomes important. You have to **edit out** of the composite master just as you **edited in,** since you already have sequence 4 on the composite master. This is accomplished by pushing the editing button again to unlock the signal coming from VTR A (there are various ways of doing this, depending on the make of VTR) and then stopping the recording process by turning the function level on VTR B to stop. Unfortunately, many VTRs don't produce a clean insert edit-out. When you edit out, the deck has to compare the sync from the signal coming from VTR A with the sync that already exists on the composite master in sequence 4, which is to follow the end of the insert. This is asking for a lot of technology, basically a small computer evaluation of syncs and some sort of compromise sync. So, although insert editing is offered on most video editing decks, the more honest manufacturers will alert you that there will be a loss of picture stability during the edit-out of insert editing.

Sound Lag   There is really very little need to do insert edits. If you stick with assembly editing, you'll get good, clean edits, and since editing is usually used for sequential assembly onto one tape to create a program, assembly editing serves the purpose admirably. Insert edits should be used only when it is crucial to replace a segment of an edited tape with new information, and even then the edit quality will not be as good as the edits achieved with assembly editing.

One other problem that you'll run into with certain editing decks is the sound. Most editing decks transfer the audio portion of the sequence along with the video so that the sound changes from scene to scene at the same point as the video does. Some older decks have a lag in the sound beginning (caused by the distance between the rotating video heads and the audio head, which is situated farther down the tape path), and other decks may produce an audible popping when the audio portion changes. These are problems that can be corrected by your video dealer. Most manufacturers have licked sound lag with their newer model decks, and the adjustments can be made on older decks to solve the problem.

Editing Setup/Operation
Equipment   Editing can be a tiring, time-consuming task. To make it as easy as possible, set up your editing equipment in a convenient manner.

Ideal editing equipment consists of two VTRS, one of which is an editing deck, plus two monitors. The monitors don't have to be very large, but they do have to be large enough and close enough to be seen clearly, since you'll be making visual decisions at the edit points. Two 9- to 12-inch monitors are ideal. If the monitors are not the same size, use the larger one to monitor the signal from VTR B and the smaller for VTR A. Place the equipment on a table so that VTR A and B are about a foot apart, with the monitors behind and slightly above their respective VTRS. You should be able instantly to reach the controls of either VTR without stretching or fumbling, and you should be able to view either monitor without having to turn your head.

Connect VTR A to VTR B in the following manner: take the video signal from VTR A out on the line-out/video-out to the line-in/video-in jack on VTR B. Take the audio out of VTR A on the audio-out jack and connect it to the audio-in jack on VTR

Connections

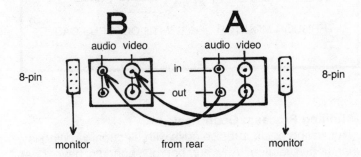

B. Connect the monitors by way of the eight-pin connecting cables on each VTR.

The ideal is to use decks of the same manufacturer and series so that the controls are identical on each. At times, however, this is not possible. You may have access to only a portable deck and an editing deck. In that case, the portable unit should be used as VTR A. Most portables don't have a video line output and an eight-pin connection. They have only one output for the video signal—usually a ten-pin out. You will therefore need to do one of two things: Either take the audio and video signal out of the portable deck and run it into a monitor on a ten-pin to eight-pin adaptor cable, then take the signal out of the monitor on the monitor's audio and video line-out jacks, and finally run the signal to the editing deck; or, run the signal out of the portable deck directly to the editing deck using a ten-pin to eight-pin (if you can run VTR B's monitor via the line and audio cables) or a ten-pin to UHF and audio plug adaptor cable. Use your RF adaptor connected to a receiver/monitor or regular TV set as your method of monitoring the signal on portable VTR A. This latter method is preferred, as there will be a certain amount of signal loss and instability if you run the audio and video signals through a monitor before you run them into VTR B.

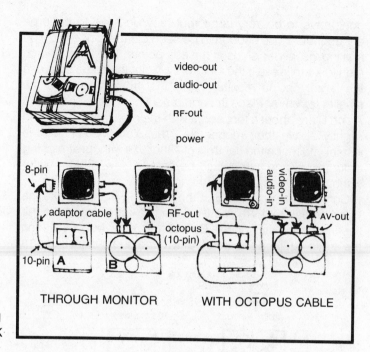

video-out

audio-out

RF-out

power

8-pin

adaptor cable

RF-out

octopus
(10-pin)

video-in
audio-in

AV-out

10-pin **A**

B

THROUGH MONITOR

WITH OCTOPUS CABLE

EDITING WITH
PORTABLE DECK

Editing Process: Open Reel

Achieving clean, precise edits with electronic editing is a matter of practice. Any number of shortcuts have been passed around to make editing easier, such as marking the tapes at the edit points with a felt pen or marking off the reels with scales to set reference points. If you are willing to put some time into working with your editing equipment, you'll discover that these shortcuts take more time in the long run than does learning proper editing technique.

To begin editing, load the master reel with sequence 1 on VTR A. Then load the blank reel of tape, which will record the composite master, on VTR B.*

On VTR A run the tape until you find the beginning of the sequence you want to transfer to VTR B. Set the number-counter on the VTR to 000. Now rewind the tape back from 000 to 990, ten digits before the point where you want the sequence to begin.

On VTR B run the tape for a few seconds to allow for a length of leader on the take-up reel prior to the beginning of the program, then stop the VTR and set the counter at 000 as well. Rewind the tape on VTR B back to 985, five digits farther back than the setting on VTR A.

Both decks are now in editing position. Start VTR B running forward and as soon as the counter gets to 989—on

*The decision about what to edit onto the master and the preparation of an editing sequence will be covered in Chapter 10.

the way to 990—start VTR A. The two counters will register 000 at about the same instant, give or take a second.

While both tapes are running toward 000, engage the edit function button on VTR B in preparation for pushing it as the counter reaches 000. With your finger poised over the edit button, watch the signal on the two monitors (in the case of the first sequence edited onto the tape there will be no signal on monitor B) and push the edit button if everything on the screens appears to be correct.

Let the sequence run on VTR A as it records on VTR B, and when the sequence ends let both VTRs run another couple of seconds. This is important; don't stop the VTRs exactly at the end of the sequence, but leave a certain amount of breathing space (about four digits).

Rewind VTR B back to 000 and check to see that the edit is clean, with no roll-overs or distortions. If the edit isn't clean, do it again. Now move VTR B to fast forward and find the end of the edit sequence. Play VTR B and note at what point you want the sequence to be replaced by sequence 2. At that point, push the counter on VTR B to 000. Rewind VTR B so the counter again reads 985.

Remove the first tape from VTR A and replace it with the tape containing sequence 2. Play the tape until you find the point where you want sequence 2 to begin, push the counter to 000, and then rewind the tape until the counter reads 990.

You are now ready to edit sequence 2 onto the composite tape on VTR B. Start VTR B running and when VTR B's counter is about to turn 990, start VTR A running. Watch both screens to see that the first sequence—which is now on monitor B—is ending as the second sequence is about to begin on VTR A's monitor. If this is the case, engage the edit button and push it when the counters reach 000. If they aren't lining up, run either VTR A or B ahead until they do.

Repeat this procedure to edit each new sequence onto the composite master.

As your proficiency in this technique improves, you'll find that you don't have to rewind a full 15 digits on VTR B and 10 digits on VTR A. Instead, you can rewind VTR A from 000 to about 993 and VTR B from 000 to about 990. Any closer to 000 will not give the tape enough time to achieve a stable running speed from stop to play positions and your edits will be messy.

The reason for letting the tape run a little beyond the end point of the sequence is that if you're off in your edit by a second or two, you'll suddenly be confronted with blank tape on VTR B. Also, the audio portions of the tapes will blend together more smoothly if one is being erased from the tape as the other is beginning to be recorded over it.

The editing technique described above assumes that you

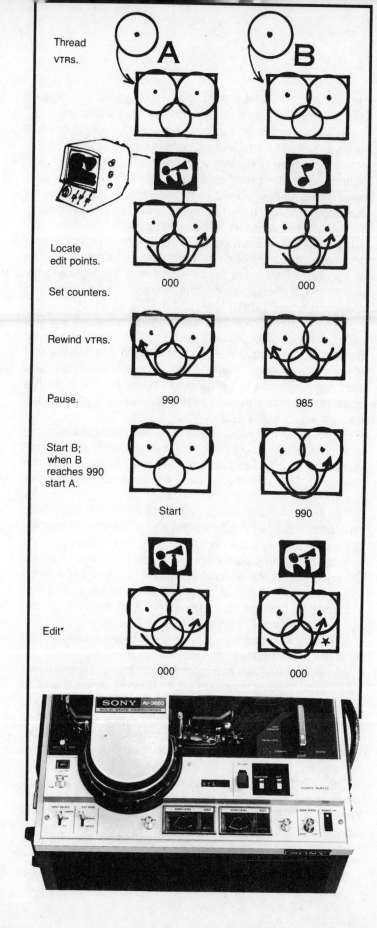

are using two VTRs from the same manufacturer that have the same general requirements for getting up to speed and the same method of counting footage. If you're using two different machines, such as a Sony and a Panasonic, you can still use this technique; you'll just have to determine what compensations you have to make for the digit-counting variations between machines. For instance, one counter may register two numbers in place of one on the other counter. If the two speeds are slightly different, this can be compensated for by starting one deck a second or two sooner than the other. As you begin to get the feel of this technique, you'll find that you won't spend too much time looking at the counters and running the tape back into position for the edit will become more or less subconscious.

A trick you may want to try if you feel that this technique takes a little too long: When you rewind the tape on VTR A past the point where you want the sequence to begin, put it in play and then into the pause position. Less time—about two digits less on most counters—is usually required for the VTR to get up to speed and stability when you move the control from pause to play, rather than from stop to play. As a result you have to rewind VTR B back less and the editing process doesn't take as long. Be careful that you do achieve stability before the edit point if you're going from pause to play on VTR A, and also that the heads don't clog and get dirty while they're rotating against the tape in the pause position.

Editing is both a video *and* audio process. Although your main concern is exchanging one video picture for another, you'll find that the audio portion of the tape provides the most problems in editing—you'll need to edit at points where the necessary audio of sequence 1 has ended and the necessary audio of sequence 2 has yet to begin. Say you have a person talking in sequence 1 and another person talking in sequence 2. The important edit point is after the first person has finished talking and before the second person has begun. You will find that you often edit in these situations using **audio cues**: If the first person has stopped talking and the second person hasn't started, this is the time to push the edit button. Working with audio cues and virtually ignoring the video can often be the fastest and most effective method of editing if your program material lends itself to that technique.

Audio Cues

When you begin editing, don't try to achieve perfection the first time out. If you miss the precise edit point you want by a second or two, let it go. Your audience will never know the difference and it will save hours of frustration as you learn the fluid actions of editing technique. You'll find that the more time you spend editing, the better, cleaner, and tighter your edits will become.

The editing process described above is one I have developed and used myself for some time now. Although the counting done by the digital foot counter isn't strictly accurate, I have found that one develops a feel for the equipment and the editing procedure that makes that lack of precision rather unimportant. Eventually, light-emitting diode digital readouts may be available on VTRs, in which case the drawbacks will be eliminated.

If you can't get the feel of this editing method, there are three others you can try. The first was presented by H. Allen Frederiksen in his book *Community Access Video,* although I've seen variations on it which have been developed independently. It relies on editing cues in the form of markings made on the tape.

Alternate Editing Methods:
Tape Marking

Using two similar VTRs, run the master tape on VTR A until you reach the point at which you want the new segment to begin. Put the deck into pause. With a felt-tip marker, make an X on the tape just to the left of the audio/control track head, rewind the tape until the X is just to the left of the skew bar, and mark a second X on the tape next to the audio head. Repeat this process with the composite master on VTR B. First, mark an X just as the previous segment is about to end, then rewind and make another X to mark the distance between the skew bar and the audio-head.

With both tapes rewound so that the first set of Xs are wound back onto the supply reel and the second set of Xs are at the same point on their respective machines—just to the left of the skew bar or erase head—start both VTRs at the same time. When you see the first set of Xs pass the audio head, push the edit button on VTR B. When you see the second set of Xs pass the audio heads, push the record button to engage the edit function.

There are two major setbacks to this system. One of them, which is shared by the second alternative editing method I will describe, is that the decks have to be started at *exactly* the same time. That isn't easy to do, and, after a few frustrating attempts, you'll wind up using your sense of feel (as in the editing by digits I described above) rather than strictly watching your X markings. The second drawback is that marks should never be made on the tape—even a felt-tip marker will make an impression on the tape, which, though slight, could mess up the signal. If you use this system, mark as lightly as possible and make sure the ink has dried before you rewind the tape onto the reel.

Alternate Editing Methods:
Scales

The other editing system uses a set of scales attached to the take-up reels of both editing decks. (Why doesn't somebody make a reel with scales printed right on them?) The scales allow you to rewind the tapes on VTR A and VTR B equally. Determine the point on VTR B's tape at which you want to edit in, then rewind the tape (first noting where the

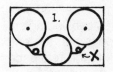

Mark first x.

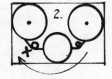

Rewind.

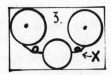

Mark second x.

Do this on both VTRs,
then start simultaneously.

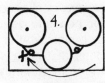

Rewind.

EDITING WITH MARKER

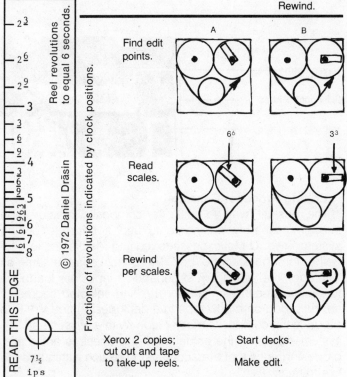

READ THIS EDGE

Reel revolutions
to equal 6 seconds.

Fractions of revolutions indicated by clock positions.

© 1972 Daniel Dräsin

2 3/6
2 6/6
2 9/6
3
3/6
6/6
9/6
4
3/6 6/6 9/6
5
9/6 6/6 3/6
6
6/6
7
6/6
8

7½
i p s

A   B

Find edit
points.

6⁶   3³

Read
scales.

Rewind
per scales.

Xerox 2 copies;
cut out and tape
to take-up reels.

Start decks.

Make edit.

EDITING WITH SCALES

outer wrap of tape on the supply reel is on the scale) a set
number of steps on the scale. Find the point at which the
new segment begins on VTR A, and rewind the exact
number of steps on the scale as you did on VTR B. The
result: the edit points on both VTR A and B will be the same
distance away from the video heads and record/edit position. You have to be especially careful when using this system that you view the scales on both take-up reels from
exactly the same angle. If you're off, your edit will be off.

The most accurate backtiming is done with a stop watch.
None of the aforementioned methods really measures the
amount of time between the cue and edit points. Choose a
point on tape A (first generation) that is at least ten seconds

*Alternate Editing Methods:
Stop Watch*

EDITING

91

before your edit point and has a recognizable audio or video cue, a point that you are confident you can find again. Time the space between this reference point and the edit point. Do the same with tape B. Go back to the reference point on both tapes. Let's say backtiming works out to twelve seconds on tape A and ten seconds on tape B. As always, decks are started from the pause mode in order for the heads to get up to speed. Start VTR A and click on the stop watch. After two seconds start VTR B. Your edit happens at twelve seconds. I would suggest watching the monitors rather than the clock once the decks are rolling.

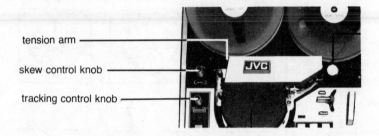

tension arm

skew control knob

tracking control knob

Editing Process: Cassette

The easiest way to make a composite master tape through assembly editing is to use a cassette editing system—two **U-Matic** cassette editing decks locked together through a control box. The principle is the same as with open reel editing. First generation tapes are loaded in turn into a deck designated VTR A and selected segments are recorded on a new tape in a deck designated VTR B to produce the composite master tape. With a cassette editing system, however, the actual editing process is automatic, producing edits that are accurate in position within three to five frames.

The U-Matic cassette recorders used in the editing system have solenoid-operated controls, which makes remote control of the decks possible. (Remote control of functions is not possible with mechanical controls.) The decks are connected to a special remote control console, which operates both decks simultaneously. This black box lets you advance and rewind the tapes on each deck—in slow motion if you want—to determine the edit points. When the points are located, you push the edit button on the remote control box and the two decks automatically roll in unison to produce the edit.

Cassette editing systems offer a variety of user controls to make editing as easy as punching up 2+2 on a pocket calculator. These include preroll rewinds, which automatically take the tapes back to a predetermined position before

the edit points; slow motion frame search, which moves the picture forward in slow motion so you can find the exact frames on which you want to edit; preview facilities, which permit you to see what the edit will look like before you go ahead and make it; and digital tape counters of highest accuracy with search and memory facilities to let you find and mark specific edit points.

Cassette editing systems are expensive ($10,000 to $15,000 for the two editing decks and control box), but editing time on these machines is $50 to $100 an hour at video studios.

Perfect Edits

Even with cassette editing, achieving perfect edits with no visual jitter or bounce at the edit point requires a combination of technique and the use of ancillary equipment. **Dropout compensators, time-base correctors, processing amplifiers,** and the like* are often used between editing decks to correct the deficiencies of the original recordings at the moment they are transferred during editing to the composite master tape. One close to indispensible unit for editing is a **cross-pulse generator.**

Cross-Pulse Generator

The video signal from VTR A is run through the cross-pulse generator on the way to VTR B. The cross-pulse generator displays the horizontal and vertical sync information not normally seen in the TV picture when displayed on the screen. (In a pinch, a monitor set to underscan the picture will serve in place of a cross-pulse display.) Observing the cross pulse on your monitor, you can adjust the stability of the picture using the skew and tracking controls on the VTR until the horizontal and vertical sync bars are as straight and coherent as possible.

---

*The application and function of these pieces of equipment will be explored in greater depth in Part II, when we talk about the technical background of video recording and editing. Be patient, or, if you want to know about it now, flip ahead to Chapter 12.

EDITING

Broadcast Quality

Almost all video formats can record tapes for commercial broadcast if the original format recording is transferred to U-Matic cassette for editing and if the editing is done with maximum picture stability using whatever electronic process is needed. Once the program material is assembled on U-Matic it can be broadcast directly from the cassette or transferred to 2-inch **quadraplex** tape for broadcast. Both will require a time-base corrector to insure the coherency of the picture on the home screen. Transfer from U-Matic to 2-inch quad with time-base correction is about $50 per fifteen minutes of program, plus the cost of the 2-inch tape.

5

MYLAR
IS THE
MESSAGE

VIDEO
TAPE

Whenever I demonstrate video equipment to people who've never seen it before, they seem to be more amazed by the capabilities of the actual video recording tape than by any other part of the process. They've seen cameras before, and the deck just looks like some bizarre form of tape recorder, but the fact that the tape is capable of storing both sound and picture at the same time—well, that's just incredible.

Cameras are things that take pictures; tape recorders are things that record sounds. And video tape just doesn't fit into any preconceived notions about recording and playing back a picture. When I explain that it's just like audio tape, but instead of recording sound only, it records both sound and picture, they shake their heads and glance at me to see if I'm kidding.

I continue by saying that video tape records and stores the electronic signals produced by the camera and the microphone. That the tape is able to record the sight and sound of people talking is comprehensible to them. Then I say that all you have to do to see the people and hear what they were saying is rewind the tape and it can immediately be replayed through any television set. The fact that it can be played through a TV set doesn't faze them, but that you don't have to send the tape to Eastman-Kodak to be developed is a mind-blower. I decimate what's left of their already overloaded ability to deal with these realities by saying that, just like audio tape, this tape can be played over and over again, can be erased and a new sound and picture put on, and can be stored indefinitely.

"You mean you can erase it when you're through with it and use it again?"

"Just like audio tape," I reply, but the connection still isn't made—audio tape can't record a picture! That is the key to the confusion—the concept that somehow the tape is what does the recording. And while that may be all right for sound, everyone knows that you need a roll of film to take pictures.

To pep them up, I say that the same tape can be used to record color or black-and-white pictures. The going really gets rough at this point. To demonstrate, I record a segment of a color TV show off the air. Once again a sense of disbelief fills the air. I didn't tell them that it would record TV shows too! Even if they got as far as understanding that it would record pictures taken with a camera, it doesn't logically follow that it should record TV shows as well.

The tape really is the hang-up here, for the tape symbolizes the whole video process. If they could only understand how the picture got on the tape and why the tape didn't need developing and how the tape could also record a TV show . . . then everything would be straightened out.

If you don't understand that a video camera changes light into electronic signals or that the VTR pulses those signals to the tape or that those signals are reconstructed into a picture on a TV screen, it would seem that the tape should be the least of your troubles. But people are able to accept the video camera, video tape recorder, and television set as having a function that is defined and explained by what they're called. A camera really is just a camera, no matter how it works.

As you get more involved with video you become concerned with the real electronic hardware necessary to produce the picture and you begin to take the tape for granted. Yet it is the video tape, and its ability to store electronic signals, that makes video possible. Television did all right without video tape, but only as a live broadcast medium, and use of that medium has traditionally been available to only a very few. So the advent and success of video has depended on the ability of the equipment to record, store, and then reproduce a scene at some future time when an audience would be available. The tape, then, is vitally important, much more important than the rest of the equipment, and so it deserves the attention that those having their first video experience give it.

Tape Composition/Function

An understanding of what video tape is and how it works is essential to a total appreciation of the video recording process. First and foremost, video tape is *not* wide audio tape. It is formulated especially for use in video and can't be replaced by audio, computer, or any other tape you happen to find in the dollar-bin at your local hi-fi store. It is vital to understand that if you wish to prevent damage to the heads of your video tape recorder.

Video tape is a plastic ribbon coated with magnetic iron oxide particles. The plastic ribbon is usually made of mylar, the same material used for audio tapes, and is known as the tape's **backing**. Spread on the ribbon is a **binder**, which is used to hold the magnetic particles to the backing. The magnetic particles are microscopic splinters of iron oxide arranged lengthwise in a pattern running in the direction of the path of the video heads across the tape when it's played on the VTR. As video tape has become more sophisticated, the oxide coating has been polished to reduce the tendency toward abrasion as tape passes and contacts the video heads. In addition, a back-coating has been applied to the backing to make it more supple and to reduce the possibility that dust or other particles will stick to the backing. Other oxide formulations have been experimented with, but at the moment iron oxide is the most commonly used.

During recording, the oxide particles are magnetized and

VIDEO TAPE

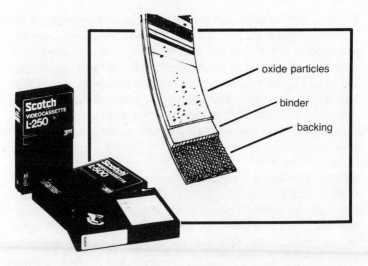

oxide particles

binder

backing

arranged into patterns corresponding to the frequencies produced by the electromagnetic gap of the video heads on the VTR. The signal remains on the tape and can be continuously replayed by the video heads, inducing the patterns off the tape, as long as the magnetic patterns are not disturbed by a magnetic field, such as introduced by the erase head on the VTR when you record over a tape. A more detailed explanation of recording and playing back a tape is included in Chapters 2 and 3.

After the tape has been erased, it still has a series of magnetic signals on it but they are arranged at random rather than in any coherent order. You can see this series of magnetic fields producing various frequencies through their random orientation by putting a reel of erased tape on your VTR and playing it. A salt-and-pepper field, what is known as tape **noise,** will appear on the screen. That is a signal being produced by the video tape and corresponding to the random patterns of the oxide.

Tape Speed   Video tape comes in a number of widths: quarter-inch, half-inch, three-quarter-inch, one-inch, and two-inch, in both open reels and cassettes. Tape is often referred to in terms of its length—"half-hour tape," "an hour reel." The length of time for which tape can record depends on two factors: how long the tape is physically and the speed at which the tape is run on the VTR. A one-hour cassette is capable of recording an hour of information only because the ratio of VTR speed to tape length allows it.

Thus the Betamax and VHS cassette formats use approximately the same lengths of tape to record one, two, four, even six hours, depending on the speed at which the tape runs by the heads. The Betamax cassette runs at 1.59 ips to record one hour and at 0.79 ips to record two hours; the VHS runs at 1.31 ips to record two hours and at 0.66 ips to record four hours. The slower the tape speed (the longer the recording time), the lower will the overall picture quality be, so there is a trade-off between how long you can record and how good your picture will be.

Tape Problems

Tape can sometimes cause problems in the recording and playback process. If the tape has been used too many times, has gotten dirty or greasy from being touched, or has become wrinkled, the result can be **drop-out** and **head clogging,** which can ruin your tapes and eventually damage your machine. With proper care, the tape will not produce either of these problems. Drop-out is caused by spots of oxide missing from the binder or by dust or other particles stuck on the tape and covering the oxide. Drop-out appears on the TV screen as little black or white lines darting across the picture, and it is one of the main reasons for keeping your greasy fingers off the tape and for storing it in a plastic bag when not in use. Head clogging results when oxide comes off the tape and gets caught in the video heads' gaps. If the tape clogs your heads, you can't play back or record, the latter being more of a problem since there is no way to see if you are recording (except for presuming that you are because the record button's engaged) without stopping the tape, rewinding, and checking.

You'll occasionally experience head clogging during very humid weather or if you're showing tapes in a hot, stuffy room full of people. The only way to solve the problem is to put the VTR in front of an air conditioner. It is normal for the heads to clog after recording or playing back half a dozen tapes. It is therefore a good idea to clean the heads between each record or play so that oxide doesn't build up in the gaps. The best product to use for cleaning the heads is a freon spray. It comes in pressurized cans and is more effective than the fluid cleaners supplied with most VTRs.

Tape Handling

There are a number of precautions you can take to maintain the quality of your tape. When you first unwrap it, keep the protective plastic bag usually supplied, as well as the

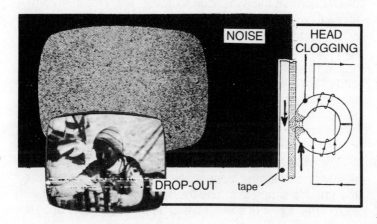

NOISE

HEAD CLOGGING

DROP-OUT   tape

VIDEO TAPE

99

the outer case. A plastic sandwich bag will work as a replacement. Whenever the tape isn't in use, store it wrapped in its plastic bag in the box in an upright position. Don't put things on top of the reels or cassettes—any pressure could damage the tape. The same goes for handling the tape when you take it out of the box. Don't squeeze together the reel flanges (rims), since the edges of the tape are most susceptible to damage.

It's a good idea to run the tape through the VTR—fast forward/rewind—when you first unseal it to give the tape a little extra "polish" and to remove any loose oxide. Having done this, clean the video heads. You don't have to prerun the tape if you don't want to—it is a pain and does wear the heads a bit—but it will ensure better quality control of your recordings.

Touch only the very beginning of the tape with your fingers when you thread it onto the VTR and through the tape path. Grease and dirt from your fingers causes drop-out, and handling the tape tends to bend and crease it—both of which do your video heads no good, besides messing up the signal.

If you do manage to step on the end of the tape or if it gets eaten by your dog, you should have allowed enough leader so that you can cut off the ruined section. I run the tape in about 10 digits on the VTR counter before I start to record. The longer you intend to keep the tape, the longer your leader should be, since some chewing of the end does result every time you thread it on the take-up reel.

The tape may break occasionally. More likely you'll stretch it by flipping too quickly from fast forward to play, or from rewind to play. The tape will jump off the VTR or get wrapped around the capstan. Stretched tape is worse than broken tape and, thanks to the tensile strength of mylar, a tape will stretch to a fine string before it breaks. If you try to play a section of stretched tape, your video heads will sound as if they're changing gears and the picture will get all scrambled, because the sync, control track, and picture information will have been distorted and won't be where it should be on the tape. A stretched section of tape must be removed and the remaining sections of tape to the left and right of it spliced together. If you stretch the tape at the beginning or end of the reel, it's easier just to cut off the whole length from the stretch to the end of the tape. If you stretch it in the middle, where it is possible to make two shorter reels of tape out of the longer reel, don't splice, just cut the tape in two and make two new reels.

Physical Splicing

Physical splicing is a problem because it isn't easy to do neatly and even the best splice is messy when played on the VTR. It's simplest to use the splicing tape provided with most VTRS, a pair of scissors, and a pair of gloves. Put on the

THE VIDEO PRIMER

gloves, trim the two ends of tape at a diagonal to approximate the path of the field signals across the tape, and butt the ends of the tape together. Now apply the mylar-backed splicing tape across the joint and trim the edges of the tape so there is no splicing tape hanging over on either side. This method ignores two points: 1) If the joint between the tapes is not flush, it will expose a bit of the sticky side of the splicing tape to the video heads, and that will muck up the heads; 2) No consideration has been made for the flow of sync/control track signals and field information along the length of the tape. It is more than likely that the field before the joint and the field after the joint are not going to be in sync.

This last problem can be overcome by using sync-developing chemicals and a tape splicing block, a block of aluminum with a groove cut down the length along which the tape lies. Slits cut across it at the proper angle correspond to the angle of the fields across the tape. This method is, of course, old fashioned and can be both messy and imperfect, but it does work when you need to splice together tape broken at a crucial point in the program or master tape. The first step is to put an eyedropperful of **magnetic tape developer** on the upper edge of the tape. The developer will make the sync and field signals visible on the

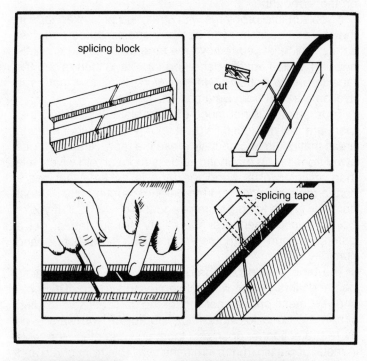

PHYSICAL SPLICING

audio
video

EIAJ TAPE

tape. Place the tape in the splicing block and cut between the sync/field lines. Repeat this at both ends of the break or stretch so that you wind up with two ends of tape, each with a final sync/field line before the very edge. After cleaning the tape ends with freon, replace them in the splicing block groove and slide the ends toward each other until they butt together. Now place a piece of splicing tape over the joint and smooth it down. Remove the tape from the splicing block and you will have as perfect a physical tape splice as is possible.

When you play the splice it will appear as a horizontal wipe as one picture signal moves up from the bottom of the screen to push the first picture off the top. You may hear a little bit of a whir from the heads, which is unavoidable. The segment of tape that has stretched or broken is lost forever. Splicing cannot repair damage to the program flow; it can only minimize the effect.

Splicing kits, including block and developer, are available for sale but it may take some searching to find one, since Sony and other video companies who used to have them in their catalogs no longer carry them. It's not possible to use a splicing block made for half-inch audio tape to splice video because the angles at which the cutting slits are set across the block do not correspond to the angle of the field signal on the video tape.

If you run into problems with stretched or broken tape in the U-Matic cassette, you can buy special take-apart cassettes, which let you remove the tape from the old cassette housing, splice out the damaged tape, and then reload the tape in a new housing. No similar replacement feature is available as yet from Betamax or VHS.

Storing Tape   Tape storage is not much of a problem. It's unlikely that there are any magnetic fields in your home or other storage areas that could accidentally erase the tape. Keeping tapes away from strong magnetic fields is more an old wives' tale than anything else. Keep the tape in an area that has a comfortable humidity and temperature, around fifty percent humidity and a temperature of about 70°F. (21°C.). If you're going to lose sleep over the safety of your tapes, get a barometer and thermometer to check temperature and humidity levels.

The tape should be stored upright in its original plastic bag and storage box, with nothing resting on top of it. Bookshelves, metal prefab shelving, or the cabinets sold to keep film in are good for video tape storage. If you want even more protection for open reel tape, you can place the reels in metal or plastic 16mm film cans available from photo and movie supply houses, but this isn't really necessary. One extra precaution might be to keep a CO_2 fire extinguisher near your storage area. Although tapes are not likely to

ONLY A PHILLIPS-TYPE SCREWDRIVER AND A PENCIL ARE NEEDED TO INSERT NEW TAPE INTO A U-MATIC VIDEOCASSETTE USING A SCOTCH BRAND "U-DO-IT" RELOAD KIT.

catch fire or burn, they would be damaged by water used to put out a fire. Carbon dioxide will not hurt the tapes. At all costs, try to keep from spilling soft drinks, coffee, and other liquids on tapes, VTRS, and the rest of your equipment.

Self-stick labels on the box—inside and on the spine—and on the reel or cassette housing are a good idea if you want to keep track of what's on which tape. Include the date taped, by whom, the standard, and any other information you think necessary. I use a color coding system on the spines of my tape boxes to indicate the subject matter, whether the tape is color or b&w, and when the tape was recorded. The self-stick labels that are supplied with Sony tape are no good; they self-destruct after about six months. Use Dennison, Avery, or other self-sticking labels you can buy at any stationery store.

Tape Care

It's a good idea to run any tapes you haven't played for six months or so through your VTR when you get the chance. It keeps the tapes from aging. It's also advisable to store your tapes **tails out** rather than **heads out.** Heads out means that the end of the tape extending from the reel is the beginning of the program. A tails-out tape is one that has been played and not rewound. By storing tapes tails out you have to rewind them before playing, thus "limbering up" the tape before play.

One final consideration regarding tape care: If a tape is mailed or otherwise exposed to hot or cold weather, don't play the tape for about twenty-four hours after it has been returned to room temperature. This gives the tape binder a chance to stabilize.

VIDEO TAPE

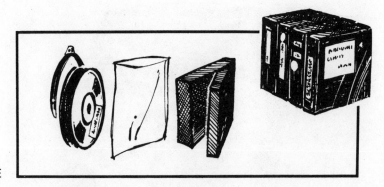

STORAGE

Tape Copies

Making duplicates of your tapes is a time-consuming, but simple process. To make one copy, connect two decks; the one with the original tape is the **master VTR**, the one with the blank tape to record the copy is the **slave VTR**. The connections can be eight-pin to eight-pin or line to line. Keep the cables connecting the two decks short so that no extraneous noise is picked up and no frequency degradation takes place due to unduly long cables. You'll need a video distribution amplifier if you want to make more than one copy at a time. The VDA accepts the video signal from the master VTR, amplifies it up to its peak level—usually one volt of signal consisting of 0.7v video and 0.3v sync—and then has four outputs from which four slave VTRs can be fed. The quality of the tape copies can be improved by inserting a processing amplifier between master and slave VTR or between master and VDA.

The real problem with making copies is that a half-hour tape takes a half hour to copy, an hour tape an hour. Of

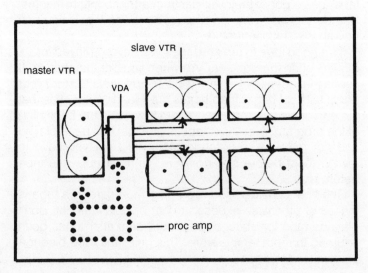

course, you don't have to stand over the vtrs once the recording process has started, but it still takes real time. If you need a great number of copies, you should investigate some of the professional copying services. They use other methods of tape duplication which are very speedy, but they're also more expensive.

Accessories
A few hints on using tape:

1. Get stick-on letters to mark *rewind* on your take-up reels for easy visual identification. When you're working with a number of reels at once, you'll know at a glance that all the tapes on the rewind reels have to be rewound.

2. Label your tapes directly on the reels or cassette housings as soon as you record them, to avoid confusion.

3. Keep your empty take-up reels in boxes when they're not in use so they won't get warped or broken.

4. When you get a bulk tape eraser to erase tapes, make sure that it's UL approved, that it will erase half- or three-quarter-inch tape, and that it will take up to a 10½-inch (26.3cm) diameter reel.

5. As you erase, do one side of the reel, then flip it over and do the other side, even though the instructions don't tell you to. Slowly withdraw the tape as far from the eraser as possible before turning off the power.

Tape and Tape Reels

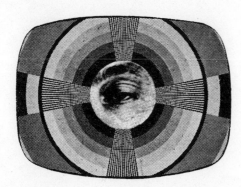

A test pattern can be used to adjust your camera and monitor to peak operating condition. You can order professional tv test patterns from the Electronic Industries Association, 2001 I Street, N.W., Washington, D.C. 20006. They cost a few dollars each. You'll have to order a set of gray scales separately and glue them onto the test pattern (the printing process is different for the pattern and scales). Set up the test pattern exactly parallel to the plane of the front of the camera lens. Use the horizontal and vertical lines to test resolution, the circles to test distortion, and the gray scale to

Test Patterns and Test Tapes

VIDEO TAPE

test the horizontal resolution and contrast ratio of your equipment. The manual for each camera and monitor gives the exact specifications to which the unit should conform when using a test pattern. The EIA chart comes with detailed instructions for checking shading, streaking, interlace, ringing, and frequency responses.

Test tapes are also available. You can get them from Sony and other manufacturers. They're expensive—$50 or more each—but they're first generation recordings made to the exact tolerances of EIAJ equipment, and they let you test your VTR and monitors to see that they're properly processing the signal. Again, the manual for each unit will describe the various test procedures for that unit.

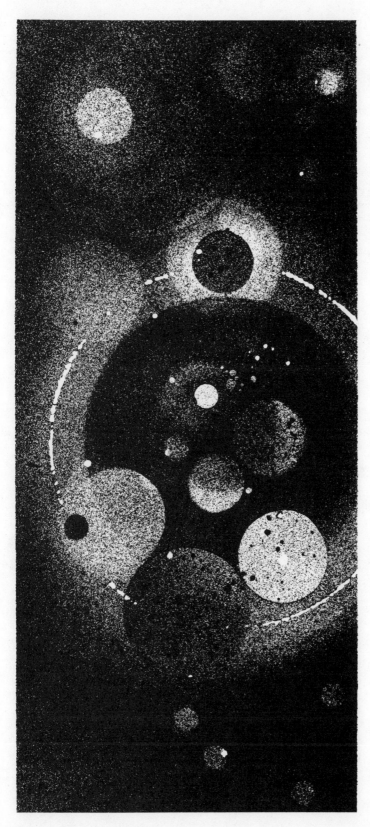

6

GALILEO
WORKED
FOR SONY

LENSES

The initial contact most people have with lenses when they start working in video is gripping the **zoom lens** mounted on the front end of their cameras. It has become an almost universal practice for video manufacturers to supply a zoom lens as standard camera equipment, especially on portable cameras. This is in a way unfortunate, because it allows the novice video maker such a wide latitude of lens capabilities that the question of just what a lens is and does may not occur for months. The zoom lens has become a sort of catchall piece of equipment, capable of giving a close-up picture, a wide-angle picture, and other views of the scene being taped, without requiring the least understanding of what function the lens is performing or how the particular angle of view in the viewfinder is being achieved.

An immediate, and unfortunate, result of the ease with which a zoom lens can be manipulated is the urge to constantly "zoom" in and out, from detail to overview and back to detail, in a manner that those watching the final tape find disturbing. This zoom syndrome gives the horrors to professional cameramen, who are convinced that such camerawork is somehow representative of the potential of video as a communications and entertainment medium.

In actual practice, the zoom lens should be used very rarely to zoom. If you are about to embark on your first series of video experiences and want to eventually learn the full potential of your equipment, remove the zoom lens and replace it with a fixed lens, which demands that you and your camera move in order to capture the various dimensions of the action taking place in the scene you're taping.

More Than You Want to Know about Lenses

Before we get into the theoretical and practical aspects of the lenses available to you and the techniques that can be employed in using them, let's first consider the lens, what it is, and what it does.

A lens is a piece of clear glass with one or both sides curved either in (**concave**) or out (**convex**) and which, because of its inherent optical properties, is capable of collecting and focusing rays of light into a coherent image. Most of us experimented with the basic principles of the lens as children when we found that we could use a magnifying glass (a **bi-convex** lens) to collect and focus rays of light from the sun onto a piece of paper, creating enough heat to set the paper on fire. To do this we inserted the magnifying lens between the sun and the paper and moved the lens back and forth between the two until the image of the sun, in the form of a very bright circle of light, appeared on the paper.

The lenses used in still photography, motion picture pho-

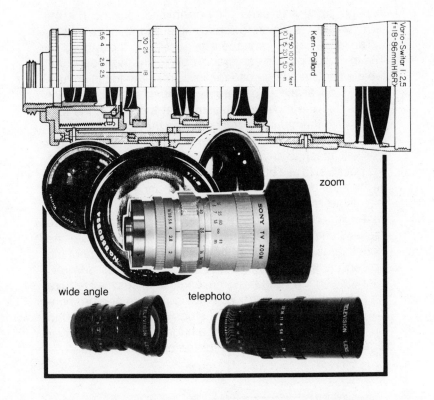

zoom

wide angle

telephoto

tography, and video are more complex versions of that simple magnifying lens. They have more than one piece of glass or element included in their construction. A photographic or video lens consists of a metal cylinder or barrel, inside of which are mounted two or more lens elements, each element varying in the shape* it has been ground to, and set at certain distances parallel to each other along the length of the barrel. The distance between the elements and their shapes determine the ability of the lens to collect and focus light rays. You're probably familiar with the long, or telephoto, lens, which has a long barrel with the elements placed far apart as in a telescope, and the fish-eye lens, in which the elements are placed close together and the front element is very convex, seeming almost to wrap around in a half globe. You should become familiar with the effects created by various combinations of elements and barrel lengths and the names given to such combinations. But there is no real need to understand the shape of the actual elements inside each lens. In fact, if you do get curious and decide to take the lens apart to see what the elements look like, you might as well throw out the pieces when you get

*Lens element shapes include: **plano-convex, bi-convex, converging meniscus, plano-concave, bi-concave,** and **diverging meniscus.**

through since it is almost impossible for anyone but the most skilled craftsman to reassemble and adjust a lens.

The lenses used for video are identical to those used on 16mm motion picture cameras. Technically, this means that the elements are large enough in diameter to focus an area that will cover the target area of the vidicon tube. It also means that the mounting assembly at the end of the lens mates with the mounting assembly on the front of the camera. This assembly is known as a *C-mount*. Any C-mount lens will attach to any camera having a C-mount assembly. It is possible to use other lenses on your video camera, such as those made for 35mm still cameras, since the diameter of the elements is again large enough to cover, and in this case exceeds the diameter of, the vidicon target area. To use a 35mm lens you must get an adaptor to mate the mounting assembly of the 35mm lens to the C-mount of the vidicon camera. In most cases this is known as a P-to-C adaptor, although certain 35mm camera lenses may require variations of this adaptor. When using 35mm or other non-C-mount lenses, note that the f-stops do not correspond to their normal settings. This is not a huge problem, but it does require that you compensate for the change.

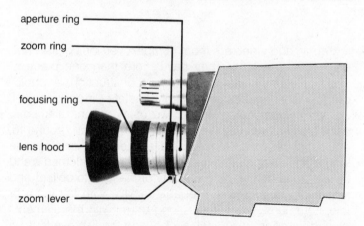

aperture ring

zoom ring

focusing ring

lens hood

zoom lever

Lenses used on vidicon cameras with two-thirds-inch vidicon tubes are C-mountable, can be adjusted to **standard focusing,** and have an adjustable **iris.** The opening at the far end of each lens barrel, which allows the light collected by the lens elements to be passed onto the vidicon target area, is known as the **aperture.** In this aperture is mounted an iris diaphragm, usually just called an iris. The iris is a circle of interleaved metal plates, which can be opened and closed to describe various diameter holes by twisting a ring surrounding the outside circumference of the barrel. The iris allows the camera operator to control the amount of light that passes out of the aperture to the vidicon. The range of light to which the vidicon is sensitive is limited and, as you'll learn when you read about lighting (Chapter 7), can be determined in terms of footcandles. It is possible for the light reflected by a scene and collected by the lens to be too bright when the iris is at its widest opening. If so, you must close the iris slightly, cutting down the amount of light that passes through the lens and so controlling the brightness of the image that is cast on the vidicon target area until it is within acceptable limits. If there is not enough light when the iris is closed down to a very small aperture, you can open up the iris, enlarging the aperture, to allow more light to pass through the lens to the vidicon.

lens mount

focus

iris

F-Stops

In still and movie photography it's customary to calibrate the openings of the iris as **f-stops** or **T-stops**. It's not really necessary in video, since you are viewing the picture produced by various iris openings directly on the camera's electronic viewfinder and can see if the setting is right. But since most video lenses are adapted from 16mm, the f-stops have remained. T-stops are rarely found on these lenses. F-stops begin at a certain point and then continue to another point according to a standard range of f-stops. This range starts with the point at which the iris is open its widest, allowing the most light to pass through the lens. This f-stop is known as the **lens speed**. It could be f 1.4, f 1.9, f 2, or higher, depending on the quality of the lens. This first f-stop, the speed of the lens, is determined by a set of optical calibrations of the particular lens where f is equal to the **focal**

length of the lens divided by the diameter of the lens. A lens with a first f-stop number that is very low—between f 1.2 and f 2—is known as a fast lens; that is, it will admit a great deal of light at that f-stop and allow the camera to be used at low light levels. Lenses opened to their largest iris position, the lowest f-stop number, are said to be **wide open**.

The lowest f-stop number is an important item to note when you're buying a lens because it determines the minimum amount of light you will need to use the lens. An f 1.4 lens will operate in lower light levels than an f 2 lens, whereas an f 2 lens will operate in lower light levels than an f 3.5 lens, and so on. If the vidicon tube is sensitive down to 2 ft-c (footcandles) and the room in which you're taping is at 5 ft-c, but your lens will admit only 10 ft-c or more of light, then the lens is hampering recording in that light. A vidicon will accept light from an f 0.9 lens on up, so you should buy the fastest lenses you can afford.

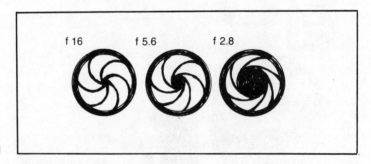

IRIS DIAPHRAGM

Once the lowest f-stop number has been determined for a particular lens, other f-stop numbers are calibrated up to f 16 or f 22, depending on what is the smallest opening of the iris in relation to the widest opening. Each succeeding f-stop means the iris is closed down twice as much as the stop before it. The amount of light going through the lens is therefore halved with each f-stop closing. An f 1.4 lens at f 1.9 (the next f-stop) permits half as much light to hit the target area of the vidicon as it did at f 1.4; at the next stop, a fourth as much light is allowed through; at the stop after that, an eighth as much light, and so on. Going in the opposite direction—starting with the lens closed down to the smallest possible iris opening, say f 16—the next f-stop would allow twice as much light in, the following four times as much light, then eight times as much, sixteen times as much, and so on up to the widest iris opening, which would equal the first or lowest f-stop. You'll find that the majority of lenses opened to the same f-stop will allow just about the same amount of light through the aperture to the vidicon. The differences occur only at the extremes: Some lenses have lower f-stop numbers (allowing more light through) and some have higher f-stop numbers (allowing less light through).

All this can get very confusing, especially when you hear someone talking about **stopping down** or **opening up** a lens. The confusion lies in the fact that there are two starting points when you work with a lens: the lowest f-stop and the highest f-stop. Video terminology has confused this even more because, whereas the filmmaker usually has to cut down the amount of light passing through the lens, the video maker is usually forced to open up the lens to allow through as much light as possible. So for the video camera operator, the starting point is the lowest f-stop number, that f-stop which opens up the iris widest and allows the greatest amount of light through the aperture to the vidicon. From there, if more light is available, the camera operator will "stop down" the lens by moving to a higher f-stop until the picture in the viewfinder is acceptable. The reference point in video is the smallest f-stop/widest iris opening. It is most common for the video camera operator to use the camera with the lens wide open, closing down only in bright light areas. The filmmaker, on the other hand, starts with higher f-stops and is more concerned with closing down the iris as the situation demands it. This is due to the fact that video work is usually done with available light only, which makes it necessary to admit as much light as possible to get an acceptable picture.

The other control you'll find on lenses is the **focus ring**. *Focus* There may be a focusing scale marked off in feet or meters, which tells you the proper setting of the focus control in relation to the distance from the camera to the subject. Since you achieve focus by looking through the viewfinder of your camera and then turning the focus ring until the image is sharp, it's unlikely that you'll need to pay much attention to the calibrations on the focusing ring.

Three factors—**focal length, depth of focus,** and **depth of field**—go into determining lens use and type. Lenses are designated as such-and-such millimeter: 10mm, 12.5mm, 16mm, 25mm, 50mm, 75mm, 100mm, and so on. These numbers refer to the focal length of the lens.

The focal length is the distance, measured in millimeters, *Focal Length* from the **optical center** of the lens to the **image plane.** The optical center is the point at which the rays of light entering the lens converge and begin to spread out again. The image plane is the point at which this convergence and spreading-out forms an image of the scene viewed by the lens. The image plane in video is the target area of the vidicon tube, for it is at this point that the rays of light coming through the lens must be focused so their transformation to an electronic signal can take place.

The focus of the image on the image plane is established **LENSES**

113

by moving the vidicon tube back and forth until the image thrown on the target area by the lens is in focus. Once this is set, the lens is generally "in focus" for the camera, although it is necessary to adjust the focus of the lens itself with the focusing ring, depending on the distance of the subject from the lens. The internal lens-to-vidicon focus is adjusted with the focusing ring set at its farthest point—so that it is focusing on the farthest possible object that can be re-solved. Then, as the subject moves closer to the camera, the focusing ring is used to readjust the focus of the lens, to maintain the focus of the image cast by the lens onto the vidicon target area. If the internal lens-to-vidicon focus is adjusted properly, the calibrations on the focusing ring should work; if the focusing ring is set to ten meters, objects ten meters away should be in focus. If this is way off, have the adjustment corrected by your video dealer.

Angle of View   Now the going gets tricky. The focal length of the lens determines the area that the lens can view at any given distance. The smaller the focal length, the more area viewed by the lens; the larger the focal length, the less area viewed by the lens. A lens with a focal length of 12mm (expressed as a "12mm lens"), focused on an object twenty feet away (6.1m), will view more of that object than a lens with a focal length of 50mm. A person is standing ten feet (3.05m) from the camera: with an 8mm lens, you would see the entire person, from head to toe; with a 50mm lens, you would see the person from shoulders to top of head; with a 100mm lens, you would see from the bottom of the nose to the top of

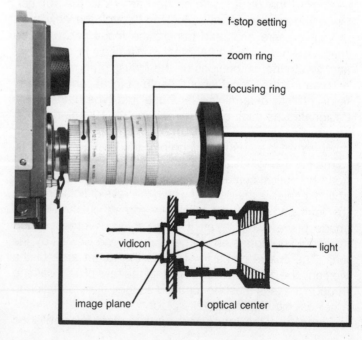

f-stop setting

zoom ring

focusing ring

vidicon

light

image plane

optical center

the eyebrows. The *greater* the focal length, the *less* area viewed by the lens at any given lens-to-subject distance, although the image size of that area as it is cast on the vidicon remains the same.

Focal length is the determining factor in the angle of view of the lens and of the **perspective** the lens creates of the subject. The focal length in millimeters is used to determine the angle of view in degrees of the lens. Lenses with small focal lengths (6mm to 12mm) are said to be **wide angle lenses,** lenses with large focal lengths (75mm to 135mm) are narrow or **telephoto lenses,** and those in between (16mm to 50mm) are considered **normal lenses.** A zoom lens is a lens whose focal length can be varied from wide angle to telephoto. The determination, except at the extremes (6mm and 135mm), is relative, the result of the individual lens user's judgment.

When dealing with the angle of view and the corresponding focal length, you must be aware that the smaller the focal length number, the greater the area viewed, at any distance. You've probably seen the fishbowl effect in which everything seems distorted in a wraparound panorama. This is the result of using a lens with a short focal length and thus a very wide angle of view. The word *angle* in "angle of view" is an expression of the horizontal width of the picture viewed by the lens.

As you work with lenses of different focal lengths you'll discover that two lenses of different millimeters can be used to cover the same angle of view, but the resulting picture will not be the same. You could use a 75mm lens twenty feet (6.1m) away from a subject and a 25mm lens six feet (1.8m) from the subject and come up with a view of the same area. Only the perspective is different. Perspective is a response from that part of our brain which tells us that things are the right size in relation to the things near them; that things farther away than other things *look* farther away. Our sense of perspective is related to our depth perception and our learned sense of proportion.

Different lenses create different senses of depth and proportion, and there are times when we will find, upon viewing a picture, that the depth and proportion seem all wrong. A wide angle lens used too close to the subject will tend to exaggerate depth as we know it; a telephoto lens will seem to remove depth if used too far from the subject. A person standing ten feet (3.05m) in front of a tree will appear through a telephoto lens as though he were right under the tree, since that lens tends to squeeze things together. The same view taken by a wide angle lens—at a closer distance to get the same angle of view—will make the person look like he's standing on the lens while the tree will appear to be miles away in the background. These distortions won't arise

LENSES

if you learn how to use your lenses and realize that the distance between the lens and the subject must be varied to accommodate the ability of the lens to view perspective and depth.

Before we finish with the subject of angle of view and focal length, I'd like to indicate the ability of various lenses to cover a subject. At a distance of 10 feet (3.05m) from lens to subject, a 100mm lens will have an angle of view of about 10 inches (.25m); that's 5 inches (.127m) to the left and 5 inches to the right of an imaginary line stretching from the lens straight to the center of the subject. At the same distance, a 75mm lens will cover about 22 inches (.56m); a 50mm lens about 33 inches (.84m); a 25mm lens about 70 inches (1.8m); and a 12.5mm lens about 138 inches (3.5m) of subject area. To get a 75mm lens to cover the same angle of view as the 100mm lens, you'd have to move the 75mm lens to a distance of about 5 feet (1.5m) from the subject. The difference between the two identical angles of view—75mm at 5 feet *vs.* 100mm at 10 feet—would be the perspective and depth proportions created by these two lenses. Objects would seem closer together with less sense of depth in the 100mm/10-foot view than in the 75mm/5-foot view.

In addition to focal length, two other factors are at work in focusing. Most lenses are able to focus from a certain minimal object-to-lens distance out to infinity. Three feet (.91m) is typically the nearest point at which a lens can focus. This limitation can be corrected by the use of special lenses or special lens adaptors. But within the range of distance from lens to subject that the lens can focus, two factors decide the quality and quantity of that focus: depth of focus and depth of field, the latter being the more important of the two.

Depth of Focus

The two terms are often confused, even by those who work in professional television and photography. In video, the depth of focus is the distance toward and away from the lens that the vidicon tube can be moved while the image on the target area of the vidicon remains in focus. Depth of focus is an optical factor, preset at the factory. It is not used as an operating control in focusing in on a particular picture, but it is a variable that can be used to adjust focus under extreme conditions—if you drop the camera, the lens-to-vidicon position may be altered and adjustment of depth of focus necessary. You should understand that there is a certain latitude of vidicon-end to lens-end distance allowable within which the image still stays in focus and that this distance indicates a certain depth of focus in the scene being viewed by the particular lens—in other words, there is an area within which everything viewed by the camera will be in focus. This area is more properly dealt with as depth of field, and it is this term that we'll use to discuss focus in relation to lens, light, and vidicon.

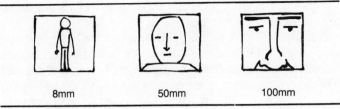

8mm 50mm 100mm

CAMERA TEN FEET (3.05m)
FROM SUBJECT

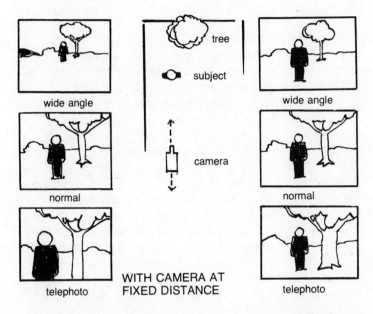

wide angle | wide angle
normal | normal
telephoto | telephoto

tree
subject
camera

WITH CAMERA AT
FIXED DISTANCE

CAMERA MOVED TO KEEP
SUBJECT SAME SIZE

Depth of Field

Depth of field is defined as the distance from the nearest object in sharp focus to the farthest object in sharp focus. In other words, the area between two particular points—one closer to the camera, the other farther away—in which everything appears through the viewfinder to be in focus. An example of depth of field: A 25mm lens is set at f 2 and focused on a man standing ten feet (3.05m) away. There is a certain distance—three feet or .91m—in which the subject can move either toward or away from the camera before he will go out of focus. The depth of field is six feet (1.8m). Everything from seven to thirteen feet (2.1m–4m) away from the lens is in focus when the camera with that lens using that f-stop is focused at ten feet (3.05m).

There are three elements that determine the depth of field and so can be used to alter it: the focal length of the lens, the f-stop, and the distance from lens to subject. With the 25mm lens at f 2 and the subject at ten feet, we have one depth of field. If we closed the lens down to f 11, the depth of field would be increased. If the subject moved back to twenty feet and the lens were refocused on it at that point, the depth of field around the subject would be increased. If

LENSES

117

the 25mm lens were replaced by a 12.5mm lens the depth of field would be increased.

Because, in video, the speed at which the light from the lens is converted to electrical signals is invariable, and we have only the variations in focal length, f-stop, and subject distance to play with, the depth of field in video can never be as great as it is in film, which has as yet another variable, film speed. You will probably find yourself limited to depths of field in the area of four to six feet (1.2–1.8m) unless you use very bright lighting and are able to stop the lens down to f 11 or f 22.

Perhaps the easiest way to understand depth of field is with a zoom lens. Because a zoom lens has a variety of focal lengths—usually ranging from 12.5mm to 50mm or 75mm— it can be changed from a short focal length lens to a long focal length lens by zooming in or out. Set the zoom lens at the shortest focal length, stop down the iris to the largest possible f-stop number, and focus on an object a considerable distance away. Now observe the picture you get. The depth of field is at its greatest point. As you zoom in on the object you'll see that objects in front of and behind the selected object will go out of focus—the depth of field has been reduced by the increased focal length of the lens. If you were to increase the lighting on the scene and stop the lens down more, you'd be able to regain some of the depth of field, but not all of it. The reason for this is that the amount of light collected by a short focal length lens and the resolving power of such a lens is greater than the corresponding values of a long focal length lens. Resolution combined with light detail are the elements that most affect depth of field.

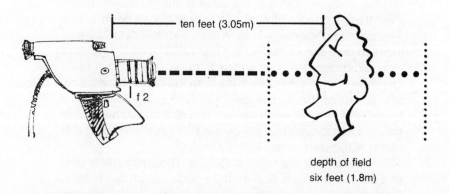

ten feet (3.05m)

f 2

depth of field
six feet (1.8m)

If you're attempting to attain the greatest depth of field, place the lens as far away as possible from the subject, stop down the lens as far as possible, and use the shortest focal length lens possible without losing the sense of your scene.

Lens-o-rama

Let's now consider the advantages and disadvantages of the lens types. The wide angle lens has a focal length up to about 16mm. It's very good for giving an impression of space and area, and it increases the depth and scope of your scene. Use a wide angle lens if you're working in a small room and want to make it appear larger than it really is. You will run into problems, though. First, you'll notice that there is an exaggeration of the perspective and the depth of the scene. The distance between things close to the camera and things a few feet farther back will seem like miles; objects at a distance will seem too small in relation to things closer up. With a wide angle, be very careful when focusing on the faces of people close up, as they'll look more like they're being reflected in a fun house hall of mirrors than anything else. You'll also notice some **barrel distortion:** things near the edges of the screen will tend to curve off the horizontal and vertical. One more problem with a wide angle lens is that people moving toward or away from it will seem to be taking giant steps at twice normal walking speed.

All of the peculiarities of a wide angle lens can be used to create interesting effects, but don't overdo it. Like the zoom lens, the wide angle can be so much fun to play with that you'll destroy its effectiveness by overuse.

Wide Angle Lenses

FISHEYE LENS

A normal lens is one whose focal length is between that of the wide angle and telephoto lens. For video cameras, the designation of a normal lens varies from 16mm to 30mm. You'll have to make your own decisions. I favor 25mm as a normal lens, since it doesn't produce a wide angle distortion effect when used at reasonable distances and still doesn't magnify the image as a telephoto does. A lens with such characteristics should be used for most of your standard taping, when you plan to cover a scene whose action permits you and your camera to move around.

Normal Lens

LENSES

119

Telephoto Lens Telephoto lenses are lenses that magnify the area of view so that it seems closer than it really is. These are lenses in the 50mm to 135mm range. A 75mm to 100mm telephoto will do for most applications. Not all 75mm lenses with high focal lengths are telephotos; you can buy a 75mm lens that has a five-inch-long (12.7cm) barrel and one that has a ten-inch-long (25.4cm) barrel. The shorter one is a telephoto. The difference between a **long lens** and a telephoto lens is that the telephoto has extra optical elements in it that allow a shorter barrel length but still have the same focal length. The advantage over a long lens of the same focal length is its smaller size—which is really important since a long lens often can't be mounted on a vidicon camera without some form of ancillary support to keep it from stripping the threads of the camera mount and falling off. But a 75mm long lens and a 75mm telephoto give the same picture; they are not different in their angle of view.

Because telephoto lenses have very shallow depth of field, they are principally used for what is called **selective focus**. You can only focus on one object or one plane at a time. Use this to your advantage by employing telephotos to pick out an object in a field of action and turn the rest of the action into a blur. A person surrounded by thousands of other people in a crowd can be singled out using a telephoto lens.

One of the other telephoto lens effects is compression of space, bringing objects in the foreground and background closer together and creating a more compact sense of perspective and depth than really exists. The farther away the subjects are from the lens, the more compression of space results.

Zoom Lens A zoom lens is a combination lens, having the advantages and drawbacks of all other lenses. You'll often find that zoom lenses are described in terms of a **zoom ratio**—the ratio between the longest and shortest focal lengths—such as 10:1 or 3:1. A 12mm to 120mm lens would be a 10:1 zoom while a 12.5mm to 50mm lens would be a 4:1 zoom.

As I've noted above, longer focal length lenses have less depth of field than shorter focal lengths, so when you're focusing a zoom lens it is best to zoom to the longest focal length and to focus at that point. When the focus has been set, moving the zoom to any shorter focal length will increase the depth of field, and the lens should stay in focus.

There are two basic ways to use a zoom lens. One is to use it as a **variable focal length lens,** changing its focal length between shots. The other is to actually use it to zoom in or out on a subject during the shot. The first use of the lens simply makes it a very versatile, convenient piece of equipment, which replaces half a dozen lenses of different focal lengths. The second use requires more care. If the

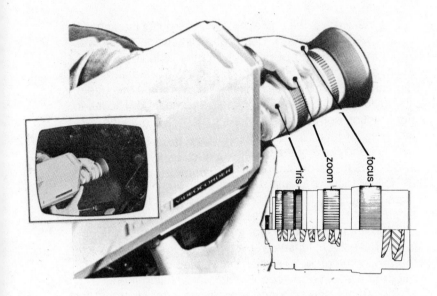

iris zoom focus

zooming motion is not smooth it will show up on the screen as an erratic and unstable picture, even though at the time of zooming, you think it's real smooth. As you use the zoom lens to zoom, you'll notice that the angle of view changes with the zoom action but the perspective stays the same. Even though the angle of view changes—becomes narrower as you zoom in on a subject—the position of the camera remains the same and the corresponding perspective from which the scene is viewed also remains the same. But since the angle of view has shifted, the perspective won't *look* the same.

If you were twenty feet (6.1m) away from a person and viewing that person with a 25mm lens, you'd have a certain angle of view and perspective. Now, if the lens were a zoom and you zoomed in on the person by increasing the focal length of the lens to 75mm, the angle of view would become narrower as you zoomed in. But the position of the person in relation to his surroundings would become distorted, compressed, and unreal because of the maintenance of the same subject-to-camera position during the zoom. Luckily, the depth of field would also decrease and so the area around the subject would go out of focus as the zoom is effected, cutting down somewhat on the distortion of the perspective. If you were to dolly in on the subject with your 25mm lens, moving in until the angle of view was reduced and the subject enlarged on the screen, there would be no loss of perspective even though the perspective would change with the movement of the camera. When you dolly in, a scene spreads out to fill the screen; when you zoom in, certain portions of the scene just get bigger to keep the screen full. Also, as you dolly in on a subject, parts of the

LENSES

scene that were in the angle of view move out of the angle of view, giving a feeling of depth and dimension. When you zoom in, this feeling of depth is not there; instead, everything seems to get flatter and a portion of a flat, two-dimensional surface is magnified. To overcome this effect, it's advisable to pan and zoom at the same time. The left to right motion compensates for the lack of depth as the subject gets larger.

One handy use of the zoom is to maintain the focus on a subject as the subject or the camera moves. This is known as **follow focus** because the focus is adjusted while the scene is actually taking place. This must be done carefully since it's an "on the air" adjustment.

It's interesting that most of the cameras used by the networks have zoom lenses, yet you'd never know it from viewing any commercial television production. They realize the effect of a zoom and keep it in its place, using it only as a special effect and only occasionally as a replacement for repositioning the camera to change the angle of view.

Lens Accessories

Most lenses have a minimum focusing distance of between one and three feet (.3–.9m). It's sometimes desirable to focus on an object only inches away from the lens. **Plus diopters** are used to turn your lens into a close-up lens. They're mounted on the front of the lens and let you get right up to the object and still be in focus. Diopters come in gradations of +1, +2, +3, and so forth; the higher the number, the closer you can get and the larger you can make the object. It's also possible to reduce the minimum focusing distance through the use of **extenders**. These are extra

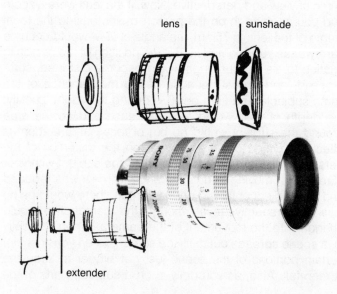

lens   sunshade

extender

lens barrel lengths, which are mounted between the lens and the camera lens mount. With extenders, however, the focal length of the lens is changed and the lowest f-stop, or lens speed, is raised. If you want to invest in a separate lens for such close-up work as magnifying a penny so it fills the screen, you can get **macro lenses,** which will focus down to a couple of inches.

It's important to note that the longer the focal length of a lens, the greater the minimum focusing distance. A zoom lens will rarely focus in less than five or six feet (1.5–1.8m). But diopters, extension tubes, and the like are relatively inexpensive and can easily be mounted on your lens.

Occasionally you will want to use **filters** on your lenses. A filter is a piece of glass or plastic mounted in front of the lens, usually in a filter carrier or holder, which screws onto the end of the lens. Filters are useful in various situations to exclude light or to create impressions with light. If you're taping in bright sunlight, for instance, and can't close down the lens enough to get a good picture—say your lens only goes to f 11 and you need f 22—using a **neutral density filter** will help. These filters cut down the light to varying degrees; you choose the filter that fits your needs at any specific time. Filters are rated in terms of their **transmission ability,** the amount of light they transmit through their density. For instance, a ninety-percent transmission filter will eliminate ten percent of the light passing through it, while a ten-percent transmission filter will eliminate ninety percent of the light passing through it. It's a good idea to have a selection of neutral density filters if you plan to tape outside a studio, where you have less control over light levels.

There are also other filters available for creating special effects—such as **fog filters,** which give a sense of fogginess in the scene, with halos spreading out from the highlights and a blurring image. Fog filters are available in various gradations from heavy to light fog. And there are **polarizing filters,** which, when rotated on their axes, can darken a brightly lit sky to make it look like night without changing the light values of the other objects in the scene. There are also filters available in any number of colors, which can be used to create effects. A red filter, for instance, will darken its complementary color—the green of grass—while lightening its own color—the red of lips. Other colored filters will do the same for their complementary colors. You can cut through haze in an outdoor situation by using a red or orange filter and you can darken a background—especially the sky— with a yellow filter. Another useful lens accessory is a **sun shade,** a metal extender that screws onto the front of the lens and has a depth of about two inches (5cm). It was designed to keep random reflections of bright lights from hitting the lens, but it is also a very good protective unit for

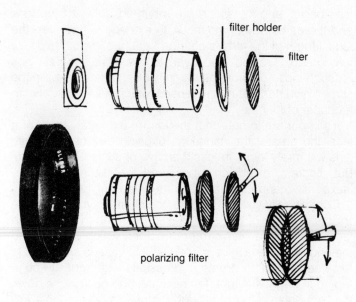

filter holder

filter

polarizing filter

your lens. And it makes your lens look longer than it really is. A sun shade is often supplied with portable cameras.

Buying a Lens

Lenses are very expensive items that must be chosen with care and an eye toward the use to which they're going to be put. There is no sense in buying a cheap lens, but at the same time it's usually necessary to compromise somewhat on the quality of the lenses you get unless you have a lot of money. A really fabulous lens will set you back between $500 and $1500. A lens is a precision piece of equipment as complex and delicate in its own way as any electronic circuit. You can buy a cheap tape recorder for $20 or a good one for $300; the difference is not in their *capability* to record sound—they both can do it—but that the *quality* of the reproduction will improve with the quality of the manufacture. By the same token, you can get a 50mm f 2.2 lens for $40 or for $400; both will gather light and throw the image onto your vidicon, but the higher quality (more expensive) lens will ensure a much sharper, more defined picture. This doesn't mean you shouldn't buy cheap lenses. But if you're considering getting one, test it first to be sure that it will perform up to your expectations.

The first consideration of most video people when they buy a lens is what the speed is. Since a lot of taping is done in available light only, a lens speed of f 1.4 is ideal. At the very least the lens speed should not be greater than f 2.2. Wide angle and normal lenses are available with these speeds for $50 to $100. But telephoto and zoom lenses with high speeds are more expensive, the principle being that the longer the focal length, the more care must be taken in lens construction and element quality.

The sharpness of the lens is the next thing to consider. The more expensive a lens, the better its resolving power and sharpness. You should always check out a lens before you pay for it. Mount it on your camera and check for sharpness at the lowest and highest f-stops—it's at these two points that a lens is usually least sharp. This test is especially important when buying used lenses.

Zoom lenses are the most difficult to choose because the best zooms are very expensive and the rest have some defect—you may notice that at the farthest point in the zoom it will not hold its focus. A zoom with a large zoom ratio is more expensive than a zoom with a smaller ratio, and a zoom with a large ratio and a very fast speed—say 10:1, f 1.4—is going to cost you as much as the rest of your equipment combined.

When you're working with a studio camera rather than a portable camera, you may want to get a zoom with a manual or automatic zoom and focus control, which can be operated from behind the camera. Such units are expensive, the automatics being a great deal more than the manuals. The lens on this kind of zoom is enclosed in a boxlike affair that contains the gears to move it in and out and to adjust the focus. The entire box mounts on the front of the camera, getting some support from the tripod. With a manual unit, a rod with a knob on the end extends back along the length of the camera either on the side or under the camera, or the knob is located at the far end of the camera with the other controls and monitor. With an automatic unit, the controls are often electric—a set of buttons are pushed for zoom and focus. By turning the knob or pushing the buttons you can zoom while monitoring the effect on the viewfinder.

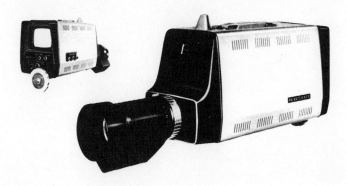

REAR CONTROL
ZOOM LENS

Lenses are not as terrifying or complicated as they may sound. As you work with your camera you'll discover the parameters of the lens as a precision instrument and you'll

LENSES

get a feel for what different lenses can do in the way of getting you the picture you want. Portable camera owners will probably find themselves working with the standardly equipped zoom lens as their first point of reference. Unfortunately, the latitude that zooms provide can often keep you from fully appreciating the function of a good lens. You might want to develop a small lens "kit" as soon as you can afford it. Added to the zoom you've already got should be a wide angle lens, say 8mm to 16mm; a telephoto lens, 80mm to 100mm; plus a set of filters, sunshades, and perhaps a rotating prism or split field lens if those effects appeal to you. The cost of such a kit will be a minimum of $200 to $250, but having a variety of lenses to choose from will give you new ways of framing your subject-action. The most important thing to keep in mind when choosing and using lenses is that you first have to decide what you want to see through the viewfinder then find a lens to match that vision. The lenses are there; it's up to you to use them creatively.

7
ILLUMINATIONS

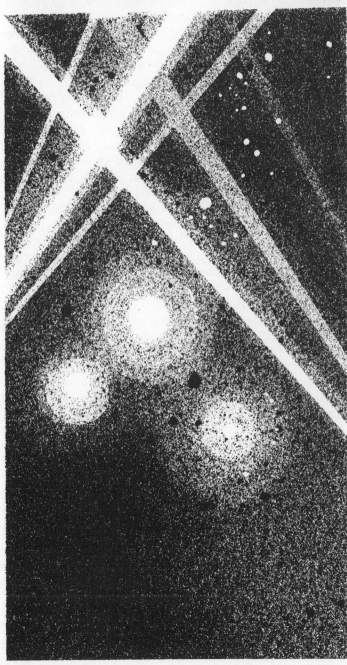

LIGHTING

Some people work very hard
But still they never get it right
Well I'm beginning to see the light.

— Lou Reed

The quality and quantity of light available while taping is of primary importance. Light and the absence of it are the basic factors that must be adjusted and controlled to ensure that the video you're making will be visible to your audience. The sensitivity of the vidicon tube permits recording with very little light, often with just available room or outdoor lighting, but to achieve the highest possible technical quality it's necessary to introduce some artificial lighting, even if only to augment an already bright light source.

Light in a scene is evidenced by the relative brightness or darkness of objects in that scene. All objects absorb and reflect a certain amount of the light that strikes them and can be considered modifiers of the light directed at them. Since the lens-vidicon assembly of the video camera is sensitive to relative values of the light striking objects in the scene the camera is pointed at, and since it performs the process of collecting the light reflected back at it and turning these bright and dark values into a series of corresponding electronic pulses, it is important to understand how various amounts and types of light will affect the picture as it is recorded by the video camera and VTR for eventual display.

Measuring Light

The measurement of light intensity used most often in video is the **footcandle** or **ft-c**. A footcandle is equal to one **lumen** per square foot, and you'll occasionally see the light-sensitivity of a video camera expressed in lumens rather than footcandles. Even more rarely the term **lux** will be used to describe the light-sensitivity of a camera. One footcandle equals 10.76 lux. What all this means is that a standard method of measuring the intensity of light has been established which allows manufacturers to describe the sensitivity of their equipment to various light levels. These measurements are all based on the lumen. It's the common denominator and signifies the total amount of light given off by one candle, lumens per square foot representing the number of lumens given off, or reflected, by a light source which falls on a square foot (.84 sq. cm) of space. If a light gives off ten lumens per square foot when it casts its light on an area three feet (.91m) away, it will give off fewer lumens per square foot if it has to cast its light on an area six feet (1.8m) away. One lumen per square foot equals one footcandle, so a footcandle is a measurement of the light intensity as it affects the area or surface being hit by that light.

Any light meter can be used to determine the footcandle rating of the lighting available. Set the light meter to 200 ASA. Take a 12-inch-square (.84 sq. cm) sheet of white paper and place it in the area you want to measure in such a way that the light illuminates the paper. Now point the light meter at the paper, holding the meter close enough so that it reads only the value of the light reflected from the paper. You will get an f-stop rating from the meter that corresponds to the following footcandle values: f 2 or below is less than 20 ft-c; f 4 = 40 ft-c; f 5.6 = 75 ft-c; f 8 = 150 ft-c; f 11 = 300 ft-c. This is only a rough reading and does not correspond to the f-stop setting of your vidicon camera.

When selecting cameras you must consider the resolution the camera provides at certain ft-c levels. At times the manufacturer will indicate only the minimum amount of light needed for the vidicon to be excited, such as "minimum illumination: more than 2 ft-c." At other times the ideal illumination levels will be given as well, such as "optimum illumination: 200–300 ft-c." And when the vidicon has an automatic gain control (standard in most portable cameras), which adjusts the vidicon sensitivity to the light levels of the scene, thus preventing overexposure, the range of that control will be given in footcandles, such as "automatic gain control range: 90–800 ft-c."

With b&w video, the ideal illumination range in footcandles is roughly between 75 and 150. The problem you'll encounter with most vidicon video cameras is that when you get below a certain footcandle level the automatic gain control on the camera will compensate for the lack of light so that the resulting picture—while it may have visible and distinguishable objects in it—will be so full of noise that it will be unviewable. This noise is simply the electronics of the vidicon and recording process showing through the picture, and is evidenced as a very grainy, gray picture with no real blacks or whites. If this happens, you need more light.

There are a number of light variables to consider. You must determine where the light is coming from, if there is enough light to properly illuminate the scene, and what kind of light it is. These considerations are most important when working with available light.

Available Light

Available light is of two types: outdoor sunlight and indoor artificial light. Both kinds of light have certain similarities, but, at the risk of some duplication, let's take them separately.

Outdoors   The sun is usually the only source of available light in outdoor video work. It provides what is known as **general lighting** because it illuminates the overall scene. In theory, it does not provide **specific lighting,** in that you can't adjust the sun to shine on one particular object in the scene, although it is possible to move objects so they are directly in the sun's path and so obtain a certain amount of specific lighting. The quality of the light coming from the sun affects the definition and resolution of the video picture. If the day is hazy and overcast, the shadows cast by objects struck by the sun's rays will not be as sharp and clearly defined as when the day is brilliant and clear. As a result, bright sunlight gives a better, sharper, more defined image than overcast sunlight. There are drawbacks to sunlight, however, specifically as regards the **brightness ratio,** which you must contend with when using a vidicon camera.

The brightness ratio is a relative measurement of light levels, a comparison between the brightest and the darkest objects in a given scene. You can determine the brightness ratio using the f-stops on your camera lens. Begin stopping down, one f-stop at a time, until the brightly lit areas appear in the viewfinder to have the same light value as the shaded areas had before you began stopping down. If it took one f-stop, the brightness ratio is 2:1; if it took one and a half to two f-stops, depending on your f-stop spread, the ratio is approximately 3:1.

This brightness ratio can also be expressed in terms of the contrast of the resulting video picture, the range from black to white on your monitor screen. The **contrast range** of a b&w receiver is usually 20:1 (this corresponds to a brightness ratio of 2:1), meaning that the most illuminated/brightest object is twenty times as bright as the darkest object. Taking brightness and contrast into consideration at the camera end, it's advisable to work in a 2:1 or 3:1 relationship between the brightest and darkest values reflected by the scene through the camera.

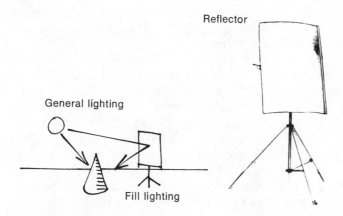

In between the brightest and darkest areas of any scene are the shaded areas. Shaded areas are considered the **fill-light** areas—parts of the scene that aren't receiving direct light but also aren't totally black shadow. When you gauge the brightness ratio at the camera (or the contrast ratio at the monitor) be sure that you work from the blackest to the whitest portions. Also remember that this is a relative set of values. Most important is that you understand the general idea involved.

With direct sunlight as the only source of illumination, the vidicon picks up very bright areas and very dark areas with little tonal value in between. This causes a contrast ratio that can exceed 20:1 on the receiver and a lighting or brightness ratio that may be more than 3:1, possibly 5:1, 8:1, or 10:1. The result when this type of scene is recorded without any modification of lighting is a very black and white picture with no grays between, an effect that looks like the picture has been burned into the display tube.

When working in sunlight, some kind of fill-light is usually needed to provide a general lighting of the area to bring the level of the indirectly lit portions of the area up near the level of the sunlit portions. Fill-light may be provided by a floodlight or by sunlight reflected off some aluminum foil mounted on a piece of cardboard. Ideally, the fill-light should provide illumination that is one f-stop less bright than the sunlight; this will give you a lighting ratio of 2:1 between sunlight (the major illuminator of the key objects in the scene) and the fill-light (the general illuminator, which establishes the darkness levels of the scene).

Working with available light indoors presents similar problems. If you are planning to tape in an area without any auxiliary lighting equipment, you must work with the lighting that is available. You must determine if the lighting level is high enough to produce sufficient footcandle illumination for taping. If there is enough light for this, the general lighting of the area is acceptable. You then have to determine the brightness or lighting ratio between your subject and the rest of the scene's area.

If the general lighting is such that your subject is hit by the same amount of light as are the rest of the objects in the scene, your subject will not be differentiated and will appear flat and uninteresting. There are several ways to get around this when working with available light. One is to put the subject as near as possible to the source of the available light—illumination increases as you get closer to the light source so your subject will be more brightly lit than the rest of the scene. Another solution is to put your subject against a background that provides the greatest possible contrast. If neither of these results in an acceptable picture in the viewfinder, you must introduce some form of specific lighting to "spotlight" your subject and provide a suitable contrast and brightness ratio to give your subject dimension and importance in relation to the rest of the scene.

Artificial Light

This brings us to the use of artificial lighting. I recommend that you include a set of lights on your basic equipment list. You'll be surprised how the quality of your tapes will improve once you start to use the right lighting. Some video makers object to using artificial light because, they say, it destroys the "naturalness" of video and the intimate nature it can have as a recording medium. I agree with this to a degree, but I think it's vastly more important for your audience to be able to see what's happening on the screen than for your subjects to groove on the fact that your camera *almost* works in the dark. Adequate lighting is especially important when preparing tapes for cable broadcast, because the inevitable loss in signal quality of an already low-light tape makes what does get through look drab. There is no need to blind your subjects with high intensity lights, but you must provide adequate lighting to bring the tape signal above the noise level of the equipment and to provide sufficient contrast and brightness ratios.

Standard reference points have been established in television to describe the type of lighting needed for various applications. Most of this terminology was originally borrowed from the movies and the·theater, but it has been successfully adopted by TV, so it can be considered the TV standard. First there is **base-light,** the general lighting given to the total area of the scene to ensure that the picture signal is above the camera noise level. Base-lights are usually placed as high above the scene as possible, often hung from the ceiling in studios, so that they create an artificial sun to bathe the area in light. Then there is the **key-light,** often called the **spotlight.** The key-light can be either a spotlight or floodlight. In either configuration it is the brightest light used and is directed at the action or subject of importance. It illuminates the subject, casts shadows, and brings the subject into the center of attention. As a spotlight, the key-light pinpoints the subject and sets it off dramatically from the rest of the area. As a floodlight, the key-light is more diffuse, just shining on the front of the subject and the area around the subject and creating a flatter, less distinct area of light concentration than the spotlight does.

Although it is often considered the same as the base-light, the fill-light should be singled out at least for purposes of definition. The fill-light is a soft, diffuse light used to illuminate the dark shadow areas that are created by the key-light. It is not meant to destroy the shadows completely, simply to bring the lighting ratio to the point of a suitable brightness and contrast range.

The two final forms of lighting are the **back-light** and the **eye-light.** Back-lighting is accomplished with a spotlight, similar to the key-light, placed behind and pointed at the back of the subject so that a rim of light appears around the outer edge of the subject to give depth and definition from the rest of the scene. The eye-light is a small, pencil-beam spotlight used in professional productions. It is directed at the subject's eyes and causes a certain glitter or sparkle, which makes the eyes seem more alive and natural on the screen.

Equipment

A wide range of lighting equipment is available, costing from $10 to $1,000 and more. You have to determine the power limitations in the areas in which you want to set up the lights before you decide on what lighting equipment to use. TV studios and sound stages have special electrical power provisions for running large lighting units. But home, loft, school, and other taping environments usually have only wall outlets with a limited supply of electricity. Electrical wiring is divided into circuits, each circuit servicing a certain number of outlets with electricity. No single normal circuit can supply more than 15 amps or 1,500 watts of power.

Power Before you plug in your lighting equipment, find out two things: the **amperage (amp) rating** of the electrical circuit you're using, and the total **wattage** of all the lights you're plugging into that circuit.

If you have three lights, each containing a 500-watt bulb, you can safely plug all three of them into a 15-amp circuit. If you need more wattage than that, you will have to divide the lights between two circuits. And remember to take into account the wattage used by your VTR, monitor, and other equipment if you're plugging them into the same circuit as the lights. The wattage requirements are listed on the back of the equipment on the plate containing the serial number. A VTR using 80 watts, a monitor using 180 watts, a camera using 50 watts, all combined with a lighting system using 1,500 watts will most likely blow the fuse on the circuit they're plugged into, and possibly blow the fuses in the equipment, if not the equipment itself.

There is a procedure to ensure that you don't blow fuses or equipment when you're working up in this wattage range. Determine the total number of watts used by the equipment (the wattage for lights is printed on the tops of the bulbs). Then find out the amps of the electrical outlet where you're working—household circuits are usually 15 amps *per circuit.* Finally, find out the **line voltage**, which in the US varies between 110 and 120 volts. If you're unsure of the voltage, figure it at 110 volts.

Locate the fuse box or circuit breakers for the electrical outlets in the area in which you want to work. Get a small lamp and plug it into one of the outlets. Break the circuits or unscrew the fuses one at a time until the lamp goes out. That circuit breaker or fuse is the one governing the circuit the lamp is on. Now, with all the circuit breakers or fuses in place except the one that turned off the lamp, unplug the lamp from that outlet and plug it into the other outlets in the area, noting which outlets also supply no electricity to light the lamp—all these outlets are on the same circuit. Mark them with tape or a china pencil. Repeat this process to find out how many circuits are running into the area and which outlets go with which circuits (the total number of possible circuits equals the total number of fuses or circuit breakers). If you're working with 1,500 to 2,500 watts, you need identify only two outlets on different circuits. With two different circuits you can draw up to 30 amps of power—15 from each outlet—or a total of 3,000 watts. Divide your equipment evenly between the two outlets (which means you're dividing equally between two circuits), and plug it in.

The formula this is based on is amps = watts ÷ volts. If you know the line voltage (110v to 120v) and the number of watts your equipment draws, you can determine the number of amps needed. Similarly, knowing the amps and the line voltage will give you the amount of available wattage (15 amps times 110 volts equals 1,650 watts, or 1,500, to be on the safe side).

It's not advisable to buy a lighting unit that will draw any more than 15 amps since you'll often be faced with situations in which only one outlet is available, and with it, crossing your fingers, you'll get enough power to run the lights plus a VTR and camera.

A **grounded outlet tester** is useful for checking each outlet before you plug in your equipment. This little unit, which you can pick up for $5 or $10 at a hardware store, indicates if the wiring of a particular outlet is as it should be—that the polarity is not reversed, that the ground or neutral wire is connected correctly (if it is not, the outlet isn't safe), and that power is available at the outlet. Extra 15 amp **fuses** are also handy for location work.

Bulbs

There are two types of inexpensive, low-power-drain lights that can be used for video. The first is the good old light bulb, known as incandescent light. Light bulbs are available in wattages up to 500 (the more watts, the more light). They can be mounted in lightweight aluminum **scoops,** which come with stands or other mounting apparatus and can be purchased at photo supply stores. A 250-watt bulb with scoop and stand will cost $8 to $10. These lights have what are known as **hot spots**—points where they cast more light— so they don't provide totally *even* illumination, nor can they be used as spotlights. But they can be used for fill-lighting or base-lighting and will provide a considerable saving over other types of lighting equipment. Although they can't be made into spotlights, it's possible to buy **barn doors,** door- like panels attached to the rim of the scoop which cut off the area of light throw, allowing you to specify the lighting to some degree.

The next step up in lighting is the **photo-floodlight** which you've seen used with home movie cameras. These lights have an internal coating of silver or other reflective material which acts as the bulb's own reflector. They are available in both flood and semi-spotlight configurations and are very effective. On the minus side, they use up a lot of wattage, have very poor life (they last from two to eight hours), and *must* be mounted in porcelain sockets since they generate a lot of heat and would otherwise melt the casing of regular light bulb sockets such as those built into scoops. Photo-floods were used extensively in early television because of the high level of light they gave off, but you'd do well to avoid them.

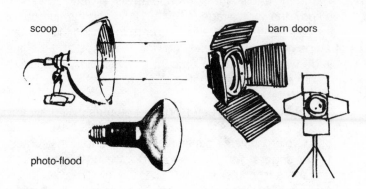

scoop

barn doors

photo-flood

You'll occasionally hear people suggesting that fluorescent lighting be used to illuminate an area. Fluorescent lights are great; they run at low temperatures, use little wattage, and give off lots of light. However, fluorescent lights don't work with vidicon cameras because they create RF interference. Use them if you don't mind a black interference bar. If you find yourself in a situation that requires taping under fluorescent lighting try wrapping your camera in aluminum foil and/or running a grounding wire between the camera lens assembly and the built-in mike (if you're using a portable camera). It may help reduce interference. With color video, fluorescent lighting is further undesirable because it makes skin tones greenish. If you are forced to tape in fluorescent light, investigate color gels, available in sheets from film or theatrical supply houses, which you can put over fluorescent fixtures to correct their color balance.

One other form of small-time lighting is the sun guns produced by Sylvania and other companies. These are portable lights that run on batteries and produce about twenty minutes to an hour's worth of light, depending on the wattage of the bulb (150 watts/fifty minutes; 250 watts/thirty minutes; 350 watts/twenty minutes). Sun guns are great if you have to tape in an area that has no other light, but be warned that you'll need more than one of them to create any kind of balanced lighting, that it takes forever to recharge the batteries, that bulb life is only ten to twelve hours, and that for the cost of two of them you could almost buy a set of professional lights.

The most professional lighting available, and the most

compact, is the lighting unit using **quartz bulb** lamps sold by movie supply houses. A complete set of three quartz bulb lamps (one spot, two floods), with lamp housings, stands, and carrying case, will cost between $300 and $400 and is worth the investment. Any number of different bulbs (they're called *lamps* in movie jargon) are available for these lighting units with wattages that range from 350 to a couple of thousand watts. A 500-watt quartz lamp (remember, you're going to need three) will last about 2,000 hours. Quartz lights are the most viable with video once you've made the investment in the lighting hardware. The units come in both spotlight and floodlight models; some are convertible from one to the other. They also have lost of accessories, including barn doors—**dichroic daylight conversion filters** to change the light to the same color temperature as sunlight; scrims, which are screen-door arrangements to diffuse the light's intensity; and **fresnel lenses,** which provide a spotlight beam with soft, indistinct edges. Particularly useful in this range are the portable lights marketed by the Lowell Company. These **Lowell lights** can be hung from any conceivable object and can even be taped to a wall.

scrim

quartz light

Sylvania Sun Gun

Lighting Setups

One-Light Setup

Even if you have only one spotlight or floodlight, you can still improve the lighting on your tape. First, find out what available light is present in the area where you're taping; then determine if the artificial light can be used as a key-light with the available light creating the fill-light. If this is possible, set the artificial light next to the camera, pointing into the scene. Raise the light as high as possible to keep the shadows to a minimum.

If you have no available light and must work with one artificial light alone, try to use it as a base-light, putting it near the camera and, if it's a spot, set it at its widest flood position. It will be garish, but it'll work.

Many TV news teams get along very well with just two lights, a fill-light and a key-light. Set the fill-light to the right

Two-Light Setup

or left of the area, midway between the rear of the scene and the camera. Raise it up as high as possible and point it down and across the area so that the shadows are cast to the right (or left) by the light. Then place the key-light next to the camera, again as high up as possible, and point it into the center of the area in which your action is taking place. The effect won't be bad, and the lighting will seem fairly balanced.

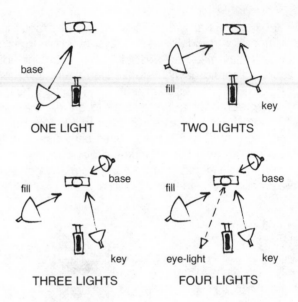

ONE LIGHT   TWO LIGHTS

THREE LIGHTS   FOUR LIGHTS

Three-Light Setup   With three lights you can follow the standard key-/fill-/back-light procedure. Set up the fill-light to one side of the area and away from the camera. Raise it so it floods the area. Then set the key-light next to the camera and adjust it into a spot to pinpoint the action or subject. Finally, adjust the back-light to its lowest possible position. Set it diagonally across the area from the fill-light, pointing up so that it can't be seen but provides a rim of light around the central area and subject of the scene.

Four-Light Setup   With a good three-light setup, the fourth light can be used as an extra fill- or base-light to provide even more general illumination of the scene. Or it can be a small spotlight mounted directly on or above the camera to point wherever the camera points, giving a little extra illumination and also functioning as an eye-light. Finally, it can be an eye-light, used specifically to get the glinting-eyes effect described earlier.

Working with Light
Once you have established the base-/fill-/and key-lighting of the scene, you'll be working with shadows more than with

light. You must learn to control shadows so they don't get in the way of the action of the scene but seem to fall naturally where they should. Exaggerated shadows are to be avoided except to create dramatic effects. The complete absence of shadows can also seem very weird. There should be a certain amount of shadowed area in your scenes, or else nothing will stand out and the scene will lack depth. Be carefull in setting up and balancing your lights, however, so you don't wind up with a multitude of shadows thrown in all directions.

The biggest shadow problem comes when lighting the human face. Incorrectly placed and focused lights will catch eye sockets, noses, and other facial features and make the subjects look grotesque. To avoid this, place the fill-lights as high as possible above the subject and point them down so that the shadows cast are as short as possible. The key-light should be slightly above head level, pointing in and slightly down to fill in any shadowy areas still present.

Proper lighting isn't difficult, but it does require trial and error and plenty of practice. It is best to start by setting up the fill-, then the key-, and finally the back-light, in that order, adjusting each as the next is set up.

The variations in lighting *between scenes* is a knotty problem you may run into if you're taping a number of scenes to be edited together later on a composite master tape. Once you've recorded a number of scenes with totally different lighting, there is little you can do to correct the problem. Your light values will change between indoor and outdoor scenes, but that can enhance the program. If your light values change from interior scene to interior scene, however, you must pay better attention to matching your footage.

Matching Footage

Make a record of the lights and f-stops used during the taping of each scene and try to duplicate the brightness and contrast ratios in the scenes that follow, either by using the same lighting and f-stops or by varying the f-stops to compensate for a change in lighting. Try recording a standard test pattern at the beginning of each take. Rewind to it and view it on your monitor at the end of the first take, then mark the monitor brightness and contrast controls as they are set. When you begin the second take, check the test pattern on the second scene against the prerecorded tape test pattern from the first take, using the same monitor and settings, and then adjust your lights or f-stops until the two test patterns correspond.

A continuous visual flow will be hampered if you constantly have to adjust the controls on your monitor during playback because one scene is too dark and the next too light. So try to match the lighting from scene to scene as you're recording.

LIGHTING

Lighting and Make-up

Bright quartz light placed close to the subject washes out skin tones, especially when you are taping in color. It's helpful to apply a light coat of pancake or stick make-up to the face, hands, and other areas of exposed skin to darken it somewhat. Basic TV make-up kits are available from many theatrical supply houses. In a pinch, use a light dusting of talcum powder on faces to reduce shine, or at least have your subjects wipe their faces briskly with a thick towel to take down shine and bring a flush to their skin.

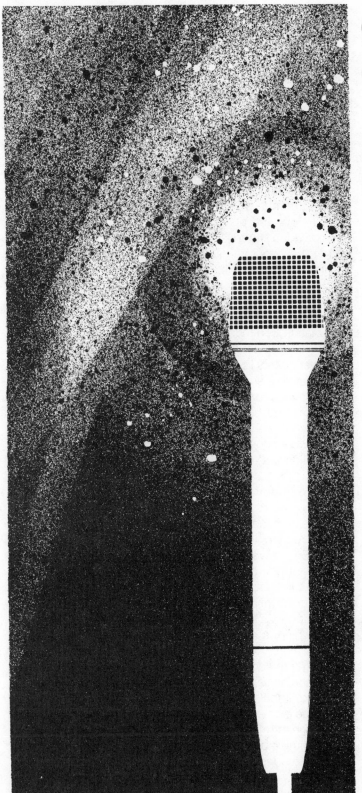

8

AUDIO
FIDELITY

SOUND

The audio portion of your video tape is as important as the visual. Sound reinforces the reality of the picture and is crucial to the total you're trying to communicate. Yet many video people mimic network television's disregard for the basic principles of good sound recording. Your tapes will be much more effective if you understand and use the techniques and equipment involved in attaining the highest level of sound quality.

Most video equipment is equipped with monophonic sound recording and reproduction facilities. There is only one sound track, corresponding to the single audio head discussed on pages 247–48. An exception to this is the U-Matic format, which has two audio tracks and can, therefore, record and play back stereo sound.

Poor video sound is the result of poor recording practices rather than equipment limitations. Think of your VTR as an audio recorder as well as a video recorder, and make the same careful preparations for recording the audio portion of your program as you do the video. If you do this, your tapes will have a great deal more impact. After all, most of the real information communicated by television, as by people, is transmitted as sound, not picture.

The speakers in most TV sets don't contribute much to the fidelity of your sound, whether that sound is good or not; all the more reason to make your audio as professional as possible. You can use auxiliary speakers, and eventually TVs may be made with better speakers. In all events, poor sound work shouldn't be excused by assuming that the reproduction system is going to be low quality anyway.

Audio Functions

When a noise is made a **sound wave** is produced, which radiates from the source of the sound in much the same way ripples radiate when a pebble is dropped into a still pool of water. Sound waves can be considered as energy, and the amount of energy that a particular sound wave produces can be measured in terms of its strength (**amplitude**) and speed of vibration (**frequency**).

Microphones Microphones are devices containing a **generating element**, which is affected by the strength and frequency of sound waves. The element is sensitive to the variations in air pressure created by the sound waves, which radiate toward it from the noise source. It transforms the sound waves into electrical impulses by vibrating according to the air pressure. The element is located at the head of the mike behind a wire mesh grill, which protects it from damage. A household example of a generating element can be found in your telephone. Unscrew the plastic cap over the mouthpiece and examine the round disc with the holes in it. That's a generating element. Behind it are two metal contact tabs—

one is the conductor and the other is the ground. The sound waves produced by your voice enter the element on one side, and on the other side the electrical impulses leave it.

The element actually consists of two components that work together to produce the electric pulses. First, the **diaphragm,** which is a sensitive membrane that does the vibrating when the sound wave strikes it; second, the **transducer,** which senses diaphragm vibrations and produces an electrical voltage. This transducer varies with the type of mike being used. There are **piezoelectric** mikes, which have a crystal or ceramic element that, when stressed or bent by the movement of the diaphragm, produces a signal voltage. Then there are **condenser** mikes, in which the diaphragm has a movable, charged conductor that forms one plate of a **capacitor;** as the conductor vibrates, a varying voltage is produced. The charge is maintained by current from a battery or plug-in power supply. An **electret condenser** is similar to a regular condenser except that the element is permanently charged. Because all condenser mikes produce very faint electrical signals, they contain bulit-in amplifiers and require a battery or other power supply for operation. There are also **dynamic mikes** and **ribbon mikes.** A dynamic mike has a diaphragm, which is attached to a coil of wire known as the **voice coil.** This coil is located near a magnet and, as the diaphragm vibrates, the voice coil moves, producing a voltage output by the coil-magnet assembly. A ribbon mike uses a metal foil ribbon instead of a diaphragm. The edges of the ribbon are exposed to the incoming sound waves and generate voltage as they vibrate.

Mike Types

Once the voltage signals have been produced by the generating element in the mike, they are sent along a cable to the audio mixer or amplifier for processing. In video, when only one mike is used, the signals are sent through the audio input into the VTR where they are amplified, equalized, and sent to the audio recording head. They are then pulsed onto the tape as it passes the audio head to form a magnetic pattern on the tape corresponding to the electrical voltages of the signal. During playback, the signals are induced across the head as the tape passes it, reamplified, and sent out of the VTR to the television speaker. The recording of audio on a VTR is the same as recording audio on a normal audio tape recorder.

Audio Components

The first and most important component needed for good audio is a good microphone. There are many types available, each with its own characteristic good points and drawbacks.

Piezoelectric mikes, called **crystal mikes** or **ceramic mikes,** are generally the cheapest on the market. They respond to a very limited range of frequencies and are sensitive to temperature changes. Avoid using them.

Ribbon mikes are excellent, especially for recording the human voice, as they give a warm, personal sound to the voice that other mikes have never been able to duplicate. But ribbon mikes are big and bulky, and are sensitive to vibration, so they can't be hand held. The ribbon element gets frayed with time and must be trimmed, and if you breathe hard into a ribbon mike you can ruin it. There are some very high quality, costly ribbon mikes on the market, but you should spend your money on a more practical, less expensive mike.

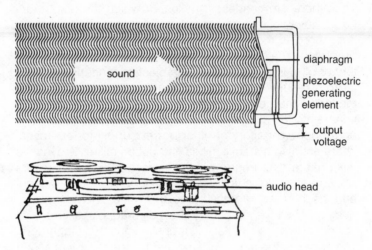

CERAMIC/CRYSTAL GENERATING ELEMENT

sound

diaphragm

piezoelectric generating element

output voltage

audio head

Dynamic mikes are one of the two varieties worthy of consideration for video. They are relatively inexpensive—$40–$80. They're dependable and have a good, smooth frequency response range.

The condenser mike is the other standard mike of the broadcast industry, but it is expensive and not portable, since it requires an external dc power supply or a large battery needing frequent replacement. Electret condensers do not use up power for charging the plate, so they need only a small internal battery, an "AA" cell right in the mike housing that will run for about 1000 hours before you have to replace it. Electrets have good frequency response, they are reliable up to a point, and cost from $50 to $100. You do have to be careful that you don't leave them in the sun, car trunk, or other hot place, and that they aren't exposed to humid conditions. If you have to choose between a moderately priced electret and a dynamic, you'd be better off with the dynamic unless you take exceptionally good care of your equipment. Personally, I like the electret better than the dynamic because of its higher output level and greater sensitivity, especially to the human voice. On the other hand, when I'm recording in the field I need a mike that can take a lot of bouncing around and people shouting into it, so I use a dynamic.

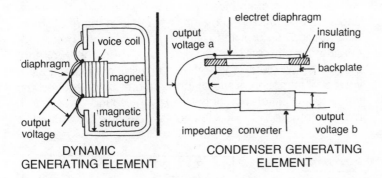

DYNAMIC
GENERATING ELEMENT

CONDENSER GENERATING
ELEMENT

Mike Models

Once you've decided on the type of mike you're going to use, you have to pick a particular model. A mike can be either **omni-directional** or **uni-directional**. These terms refer to the **pickup pattern** of the mike—the directions from which sound waves can approach the mike and vibrate the diaphragm. Omni-directional mikes are sensitive to sound coming from all directions: above, below, in front, or behind the mike. An omni-directional mike is great if you're recording in a quiet area where the sound you want recorded is the only sound being made, and if you want a total sound made up of any echo or reverberation produced by the combination of the sound and the place. Got that?

Say you're in an area where there is more than one person talking, there are street noises, or other sounds that are not the object of your audio recording and will interfere with the audience's ability to hear what your subject is saying. In that case, you want a directional mike that rejects sounds coming from the rear and sides and separates the sound you want to record from the extraneous sounds. Directional mikes have different directional abilities. Some reject all sounds except those coming straight at the front of the mike. Others vary in their degree of sound rejection from the sides and rear. To describe what sounds will be rejected by a directional mike, manufacturers have introduced **polar response patterns**. The polar pattern is a graph of the pickup area of the mike in which the front, sides, and rear polarity of the mike are established in relation to a circle around the mike. The front of the mike is 0°, the rear 180°, and the midpoint of each side is 90° and 270°, right and left as you face the mike. A line plotted within this circle describes the area in which the mike will pick up sound. This plotting results in a graph pattern termed **cardioid, super-cardioid,** or **bi-directional**.

SOUND

A cardioid mike rejects sound from the rear of the mike, is less sensitive to sound coming from the sides, and is very sensitive to sound coming in within a 30° arc of the front. It is called cardioid because of the heart-shaped pattern this pickup area describes.

The rejection of sounds approaching the mike at certain angles is accomplished by making a hole in the rear of the mike casing, so that the diaphragm is vibrated by sound coming from the mike in two directions—from the front and the rear of the diaphragm—thus canceling out those sound waves. This motion of the diaphragm can be described as the pressure difference between various sound waves, and has led to cardioid mikes being called **single-D** cardioids.

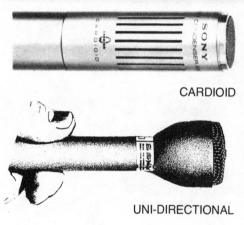

CARDIOID

UNI-DIRECTIONAL

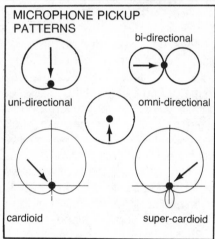

MICROPHONE PICKUP PATTERNS

uni-directional

bi-directional

omni-directional

cardioid

super-cardioid

There are problems with the cardioid mike. First, it is prone to a bass **proximity effect**; that is, the closer you hold the mike to the source of the sound, the more bottomy and bassy the sound is when you play it back. If you hold a cardioid close to your mouth when you talk, your voice will sound bottom-heavy, and you may not want this effect. Second, the cardioid is very sensitive to vocal breathing and sibilance. You have to learn how to work with a cardioid when recording people talking so that they don't constantly pop their p's and b's when they speak. Third, the cardioid mike tends to be mechanically noisy, meaning that shifting it from hand to hand while recording can sound like you dropped it out of a second-story window.

The super-cardioid mike was invented to solve some of these problems. It is also called a **variable-D** mike (trademark of Electro-Voice) in that it has a number of holes or **ports** in its case, which further adjust the frequency ranges exposed to the diaphragm so that the bass proximity effect is practically eliminated. The most common variable-D mike is the shot-gun mike—a long shafted mike that is very directional in its pickup pattern because it has high side- and rear-signal rejection. Variable-D and similar super-cardioids are expensive; they are usually used for recording a sound surrounded by other loud noises. Super-cardioids are sensitive to mechanical shocks and are very susceptible to vocal sibilance when used in close proximity to the voice.

The last directional-type mike is the bi-directional, and its polar pattern describes a figure eight. It picks up sounds from the front and rear but rejects sounds coming from the sides. There is little likelihood that you'll ever need one of these mikes.

Mike types and pickup patterns are made in various combinations. There are omni- and uni-directional dynamics, cardioid dynamics, omni- and uni-directional electret condensers, cardioid electret condensers, and so forth. Choosing among them requires an understanding of what the particular mike is made to do in relation to what you intend to use it for. It also requires an appreciation of **frequency response.**   *Frequency Response*

Frequency response is the range of different sound wave lengths to which a mike is sensitive: from bass sounds to treble sounds. Each sound wave vibrates at a certain number of cycles per second, or Hz, starting with a very slow vibration for a bassy sound to a very fast vibration for a treble sound. Your ear can hear from about 20Hz to 16,000Hz (this is written 20–16,000Hz or 20Hz–16KHz). The human voice produces sound waves in the neighborhood of 2,000Hz, meaning that they vibrate at a rate of about 2,000 cycles per second. The lowest note on a piano is 27.5Hz, middle C is 261.63Hz, and the highest note is 4,186Hz. Sounds are divided into three categories for convenient reference: low frequencies, midrange frequencies, and high frequencies. Low frequencies extend from below hearing level to about 250 cycles per second. Midrange frequencies go from 250Hz to just over 2,000 cycles per second. High frequencies (in the sense of audible sounds) go from 2,000Hz to 20,000Hz—although you can't really hear anything above about 16,000Hz.

Mikes are described as having a particular frequency response. Cheap mikes will usually only record from 100 to 8,000Hz. They're all right for voice recording, but don't try to record bass guitar or violin with them. And don't expect your voice recordings to have too much bottom feel or high-end color. (The voice and other sound makers have overtones at frequencies higher than the actual sound, which provide a sense of depth and life to the sound produced.) More expensive mikes may have a frequency range of 80–15,000Hz, and the most expensive will have a range close to 20–20,000Hz—a nominal exaggeration developed by electronics people to show that their mikes and their other equipment can deal with all audible sound.

Check the frequency response that your VTR is capable of recording before buying a mike. If it can only handle 100–10,000Hz, then you won't need to get a mike that can do a better job, although there is no need to get a mike that exactly matches that response. If you get a mike that has a more limited frequency response, be aware that you're not using your audio track to its full dynamic recording potential.

Impedance   The next thing to note when buying a mike is its **impedance**. There are either high impedance or low impedance mikes—also expressed as **hi-Z** or **low-Z**. Every mike, amplifier, tape recorder, and speaker has a certain impedance. Mikes have output impedances, amps and recorders have both input and output impedances, and speakers have input impedances. Impedances are a measurement of the resistance that the particular unit presents to the flow of the electrical signal voltage, and they are expressed in ohms of resistance. Low-impedance mikes have from 50 to 250 ohms. High-impedance mikes have from 20,000 to several million ohms (expressed as megaohms). Occasionally, you'll run into one other impedance rating: 600 ohms, or **line level impedance**. A 600-ohm impedance is a low impedance. So remember, 600 and below is low-Z, anything above is hi-Z. You *must* use a mike whose impedance range is the same as the input impedance range of the unit you're plugging it into. If you have a VTR or audio mixer that has only hi-Z inputs, and your mikes are all low-Z, you can buy what is called a **line matching transformer** (for about $10) to connect between the end of the mike cable and the VTR or mixer input to transform the low-Z impedance to hi-Z. There are also line matching transformers to change hi-Z to low-Z if your input is low impedance.

The difference between low- and high-impedance mikes has to do with the output level or signal strength coming out of the mike. Hi-Z mikes have a higher output level, but when you use more than twenty feet of cable between the mike and the input, the high frequency end of the signal gets lost, is **rolled off**. Low-impedance mikes produce a lower output signal, but they can be used with hundreds of feet of cable. For most applications, use low-Z mikes with line matching transformers on the ends if you need a hi-Z impedance at your point of input.

The final consideration when choosing a mike is its handling potential. Some mikes are especially made to be hand held; others are designed to be used only on mike stands. The hand-held mikes have been constructed so that the internal elements are not affected by the handling of the outer casing, so you can use the mike in a field situation without worrying about handling noise becoming part of the signal you're recording. The mike should have **wind screens** or **pop filters** (wire mesh caps over the element end of mikes) if you're planning a lot of voice recording or if you plan to use the mike outdoors.

Recording Audio

Once you sort out the various microphone options available you have to decide how you'll use the mike to record the audio portion of your tape. The most basic audio recording format is that in which the mike is built into the camera of the portable system. This mike is often the first introduction people have to the audio process in video, and its limitations as well as its potential color the attitudes the novice develops about recording audio. This mike is easy to use, it blends with the lens/vidicon tube assembly mounted below it to create one giant eye-ear, which is pointed at the subject to capture the audio-video dynamics taking place. This setup is probably the prototype for the personal video camera of the future, in which the recording of what is seen and what is heard will be combined into one process, simple both technically and philosophically. You won't think about the sound or the picture as independent of each other, each needing special treatment. Technology will provide your ability to record both at optimum quality in one unified gesture. Unfortunately, the conceptual promise that the mike/vidicon camera holds is not reflected in the quality it provides to the serious video producer. It's fine for recording bar mitzvahs and birthday parties, but that's it.

Built-in Mike

The farther away a mike is from the source of the sound it's recording, the more likely it is that other noises will combine with that sound as it travels to the mike. The result is a loss of clarity exemplified by the high-end hiss you often hear on tapes made with built-in mikes. Almost all built-in mikes have an **automatic gain control** feature (known as **AGC**, or **ALC** for **automatic level control**), which senses variations in the strength of the sound waves you're recording and attempts to maintain a certain signal level for those sounds, theoretically eliminating the need to manually ride the volume control as you record the sound. This is the germ of a good idea, but at the moment the mikes being used are not directional enough to ensure that the AGC will maintain a volume level on only the sound source. AGC circuits tend to pick the loudest noise coming into the mike—they are especially partial to mid- to high-frequency sounds—often producing poor audio recordings by amplifying things like typewriters, street noises, and other off-subject sounds and blending them together with the sound you want to record in a rather neat symphonic cacophony.

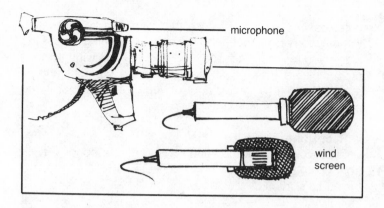

microphone

wind
screen

The built-in mikes also have a tendency to overload when the sound you're recording is too loud for the volume level at which you're recording it. The result is distortion, which sounds like someone being sucked into the third dimension, very compression-oriented science fiction noises, which can't be controlled except by moving the mike (and the camera) away from the source of the sound.

If you want to record sound that will be audible, use mikes that are independent of the video camera, placed where they will best capture the sound, just as the video camera is placed where it will get the best picture. Occasionally, you may want to use the built-in mike for what it is: a fairly good omni-directional mike. I once had the chance to tape the London Royal Philharmonic, and the built-in mike performed beautifully, capturing the string instruments to perfection. Very effective results are also produced with the built-in mike when recording background street noises, crowd scenes, and other high-noise-level occasions. But people's voices, live electric music, and other sounds that must be isolated and possibly combined at particular individual levels need a mike assigned to them that will provide the best possible recording of the sound they're producing, and each of those mikes must be placed in a position peculiar to the source of the sound.

Mike Connections   The next step, after choosing a mike, is to connect it to the VTR so that the signal it produces can be recorded on the tape. This can be done either directly or indirectly. A mike connected directly to the VTR is just that; the plug at the end of the mike cable is inserted into the audio-in or mike-in receptacle on the recorder. You are then equipped with two recording devices when making a tape, the camera and the mike. If you want to use more than one mike, you have to use an indirect method of connecting the mikes to the VTR by getting an **audio mixer**. The mikes plug into the mixer and the mixer plugs into the VTR.

Mixers vary in quality and price—from $10 to $1000. They all perform the same function, but as the price goes up so

THE VIDEO PRIMER

does their efficiency and flexibility. A basic audio mixer consists of input receptacles for two to six mikes (more expensive models have even more inputs). The signal coming into the mixer from *each* of the mikes is run through a separate level control circuit, which is adjusted by the operator using a volume control, or **pot**, on the mixer control panel. This is also called a **gain control**. After passing through these level controls, the signals from each of the inputs are *combined* into one total signal and sent through a **master volume control** circuit with its corresponding master pot. Once all the signals have been adjusted and regulated in this manner, the combined signal is fed out of the mixer through a signal output receptacle. By connecting a single audio cable between this output and the audio input of the VTR, you can combine signals from a number of mikes into one total signal for recording on the VTR.

The simplest mixers are **passive mixers**. They don't change the signals passing through the mixer except to **boost** (turn up) or **attenuate** (turn down) the strength of each signal relative to the other signals. Passive mixers require no electrical power of their own. They do reduce the strength of the signal somewhat as it passes through the internal circuits to be regulated and combined. More expensive mixers are **active mixers**; they compensate for any signal losses resulting from the signal being sent through the mixer circuits by slightly reamplifying the signal to maintain a certain signal strength. More complex active mixers have **equalizer** circuits, which allow you to "treat" the signal frequencies by selectively boosting or attenuating them.

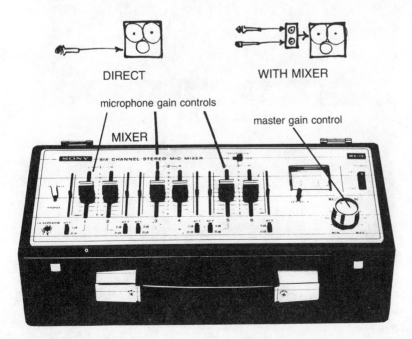

DIRECT

WITH MIXER

microphone gain controls

master gain control

MIXER

Equalizers are also available as independent units. They're connected between the mikes and mixer (to equalize one particular mike signal) or the mixer and VTR (to equalize the total sound). Equalizers are sensitive, with selective bass and treble controls allowing you to boost or attenuate low, midrange, and high frequencies. If you have a signal coming into the mixer with a lot of bass boom and not enough treble, you can **eq** the signal by attenuating the bass frequencies and boosting the treble frequencies to suit your ear.

You should have at least one single-channel equalizer if you want to package your tapes properly. With such a unit you can get rid of hiss, increase vocal clarity, and decrease high-frequency background noises and low-frequency hum. Shure makes an excellent single-channel equalizer, the Audio Master, which is quite inexpensive. The stereo graphic equalizers sold in hi-fi stores are also inexpensive and work very well. If you buy a graphic equalizer, be sure it has at least six to eight adjustable frequency bands per channel.

Expensive active mixers may also feature **compression** and **echo** and **reverb**. Compression functions much like an automatic gain control. It boosts weak portions of the signals running through its circuits while attenuating strong signals to achieve a pre-set level; the resulting sound has a

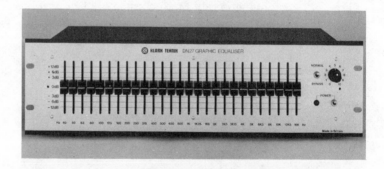

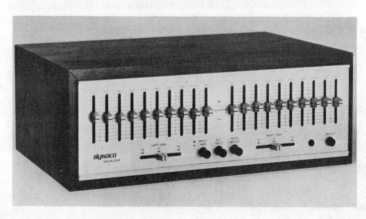

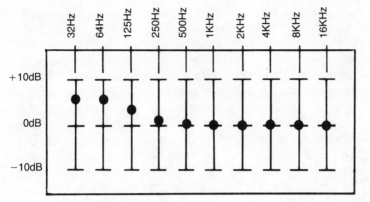

BASS BOOSTED

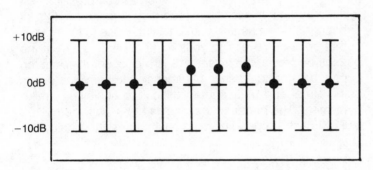

VOICE AREA BOOSTED

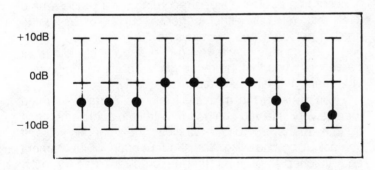

BASS AND TREBLE
REDUCED

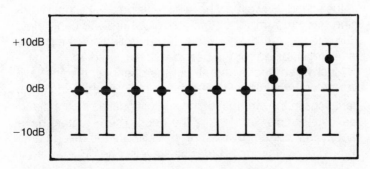

TREBLE BOOSTED

compressed, almost explosive feel to it. Echo is the repetition of a sound after it has been made. Real echo is accomplished with an echo chamber, a very resonant room in which a speaker is placed at one end to produce the original signal and a mike is placed at the other end to record it seconds later. Echo can also be produced with a tape recorder—the signal is bounced between two tracks so that the sound is repeated just after it is made. Reverb, on the other hand, is the repetition of an original sound over and over again, each repetition being of less strength and duration than the one before it. Reverb is usually accomplished by running the signal through a set of fine wire springs and then combining the result with the original signal. Compression, reverb, and echo units are available separately if your mixer is not equipped with them. Whereas compression isn't really important unless you're recording in an area with a lot of background noise and you're talking close to the mike—at that point compressing your voice separates it from the background—you may want a reverb or echo unit. A touch of reverb can add a sense of dimension to the voice that is both natural and pleasant but difficult to achieve by mike placement. The prices of these units are usually from $100 to $200 each, depending on manufacturer and quality.

Operation

Recording audio for video recordings is a process unto itself that you can either perform yourself or assign to another person. Once the mikes have been located in relation to the sounds they're going to pick up, the audio must be controlled so that the recording is free of distortion and other annoying lapses in technical quality.

If you're using the mike built into the portable camera there is very little you can do to control the audio other than try to keep the recording area clear of extraneous noise. If you are using one mike that is independent of the camera and attached to an input that has AGC (such as the mike input on the portable deck), again there is not much you can do but make sure the mike is precisely directed at the source of the sound. However, when you're using a mike that is running through a mixer or into a deck where you can **defeat** (turn off) the AGC, the audio can be controlled and the proper levels maintained for the best possible recording.

Most video decks and all but the cheapest mixers have **VU meters (volume unit meters)**, which indicate visually the variations in the strength of the audio signal as it enters and leaves the mixer and enters the deck. These meters have faceplates that are divided up or calibrated into segments representing various **dB (decibel)** levels. The dB is an arbitrary and subjective measure of the strength of a signal in terms of the volume of the sound it produces. Decibels are used in audio as a ratio between two sounds. The sound produced at 1000Hz is usually used as the standard reference frequency. A tone generator is used to produce the 1000Hz tone and the loudness of the tone is set so that it is comfortable to the human ear. The tone signal is run into the VTR or mixer and you monitor the sound it produces, adjusting the volume so that it remains at a comfortable listening level. The tone at that level is assigned the value of 0dB. Now, if you turn the volume of that sound up so it gets slightly louder, you have raised the gain 1 dB. As you turn up the volume (increase the gain), the value in dB's (relative to the original value you established) increases logarithmically. A 3dB increase from the strength of the input signal to the strength of the output signal (mike into mixer, signal out of mixer; signal into amp, signal out of amp, etc.) means you'll hear the sound at twice the volume. A 6dB increase means you'll hear the sound at four times the volume, 12dB produces sixteen times the volume, and so forth.

There are three major categories of audio levels in terms of dB. The signal produced by a mike is said to be between −70 and −10dB. The signal produced by a mixer or VTR recording amp is between −20 and +10dB. By the time the sound is amplified and reproduced by a speaker it is from +20dB to several hundred dB, depending on how loud you play it back.

Decibels are confusing to someone who's just trying to record the sound being picked up by a mike. Most manufacturers understand this, so besides calibrating their VU meters in dB's, they also divide the meter faceplate into two sections: a "safe" recording area and a "distortion" recording area, which is marked in red. When the needle, which represents the strength fluctuations of the signal level, goes into the red, you know that you're overloading the circuit and distorting the signal. The point at which the meter dial is divided from safe into red distortion is 0dB (don't forget, the mike signal is coming in at anywhere from −70 to about −10dB). As you record, the gain control, which regulates the strength of the signal coming into the recorder or mixer, should be adjusted so that the *loudest continuous sound* moves the needle up to 0dB. Occasionally, louder sounds can be allowed to go past this 0dB point, and the great majority of sounds will be below this 0dB point.

When you have a single mike connected to a VTR, you only have to worry about the recording level of that mike. The volume or gain control pot allows you to adjust the signal strength as it comes into the deck by watching the VU meter and boosting or attenuating the signal (turning the gain control up or down), depending upon the fluctuations

Decibels

Volume Controls

SOUND

155

of the needle across the dial of the meter. You should never ride the gain control in an effort to keep the meter steady just below the red distortion level (0dB). If possible, find out what the loudest persistent sound is going to be by testing prior to recording, and set the gain control so that the meter goes to 0dB when that sound occurs. Then leave the gain control alone, checking it occasionally during recording to make sure that there have been no drastic level changes. Any drastic change in sound level will require an on-the-spot readjustment of the volume control. There's no way of anticipating this, but if it does happen just adjust the control as soon as possible to prevent distortion. If you stand over the meter, watching the needle and turning the gain pot up and down every time the signal strength changes, you'll produce a flat recording, devoid of the dynamics that the sound variations of the event are producing; you'll also discover that by the time you've reacted to changes in sound level they'll be over and you'll be turning the gain up just when it should be turned down.

The same gain/VU meter relationship is to be followed when using a mixer. The least expensive passive mixers have no meters at all, so you have to use the meter on the VTR to gauge the amount of total signal being recorded. More expensive active mixers have at least one meter to register the total strength of the signals as they leave the mixer (and thus the strength of each signal as it enters the mixer when you turn down the gain on all the other inputs). The most expensive mixers have a meter for every signal input plus a master meter to measure the level of the total combined signal. It's probable that you'll be working with a $100 to $200 active mixer with only one meter. This master meter indicates the strength of the total signal going out of the mixer to the VTR. The level of this combined signal can be changed by boosting or attenuating any of the gain pots that control the individual mike signals coming into the mixer and can be boosted or attenuated by changing the position of the master volume control gain pot as well. Changing the individual gain pots will result in that particular signal getting louder (stronger) or softer (weaker) in relation to the total signal of which it's a part. Turning up the master volume control will result in the total signal (and all the signals of which it's composed) getting louder or softer as a whole. If one mike is recording a person singing and a second mike is recording a piano being played, you could make the person louder than the piano by turning up the person-mike gain pot more than the piano-mike gain pot until you have achieved the balance you wanted. Of course, you would watch the master meter to see that the total wasn't putting the meter in the red. Once a suitable balance is achieved between person and piano, you might adjust the combined

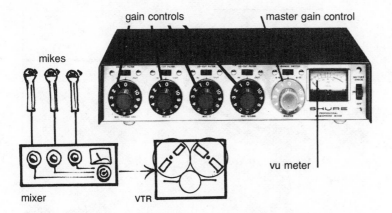

gain controls   master gain control

mikes

vu meter

mixer   VTR

signal with the master gain pot. If the person started to sing louder *and* the piano was played louder, you'd turn down the master gain. If the piano was played at the same level but the person started to sing louder, you'd have to go to the person-gain pot and readjust the level relationship between person and piano. The master gain control can also be used to fade out the total signal at certain points in the program such as at the end and during transition periods. The studio jargon for this is to **bring up** and to **fade** the signal although **fade in** and **fade out** are also used.

Audio Techniques

The techniques involved in creating good audio depend on your ability to properly choose and place mikes in relation to the sounds you're recording and your ability to mix the various signals as needed to balance the sounds. Simply, this means that you don't record a voice with a mike placed a hundred yards away.

The major concern of most audio is recording the human voice. Other sounds do not have the same crucial need for a proper level and clarity if they are to be intelligible. The roar of a jet plane or the squeal of a car rounding a fast corner project an audio impression; the voice presents information that must be decipherable to be understood. The first factor in recording voice is to choose a mike that will provide the best possible service in this situation. A number of mikes are available to do the job.

If the person is speaking in a high traffic area, you need a *Mike Options* directional mike, which you'll probably want to hold in your hand and point directly at the person's mouth. A hand-held mike of the cardioid variety, either dynamic or electret condenser, would be suitable. If you don't want the mike to show in the picture, pinpoint the sound with a shot-gun mike, a hand-held model with a super-cardioid pattern and probably an electret condenser element.

Mike options expand if the person speaking is in a quiet

SOUND

157

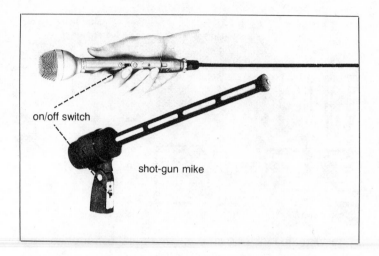

on/off switch

shot-gun mike

area with no background noises. Use either of the mike types described above, or a less directional mike—any uni-directional or omni-directional mike such as the built-in camera mike. If the mike is omni-directional you'll have to place it very close to the subject. Even if it's a simple directional mike, the closer it is to the subject the better. You could also use a mounted mike in this situation. Mike stands are inexpensive, running from $6 to $10, and a unit called a baby boom is available for the same price. The baby boom mounts on the top of the mike stand and has a metal shaft that extends horizontally about three feet (.91m). By raising the mike stand to its full height and extending the boom arm down toward your subject's mouth, it's possible to place the mike out of camera range but just above the head of the subject. This is an effective setup, but you have to watch your camera movements so that you don't suddenly widen the picture to include the boom and mike. For practical purposes it is easiest to mount the mikes on small desk stands in the manner of TV news and talk shows. They become part of the picture and create the effect of a formal talk show setting, just as a hand-held mike in a location taping has the effect of an informal news interview.

When people are talking in a group another alternative is to use **lavalier mikes**. These mikes are hung on strings around the talkers' necks or clipped to jackets or ties. They are usually very small, and they rest against the speaker's chest, getting a certain amount of resonance from the chest cavity as well as sound waves from the mouth. Lavaliers can also be hung under clothing so that they aren't visible. The only drawback to lavaliers is that they limit the movement of individuals wearing them since they're on a leash of sorts connected to the mixer. This can be solved by using FM wireless mikes, lavaliers hung around the neck and leading to a small FM radio transmitter mounted on the person's belt or in a pocket out of view. The unit transmits the signal to a receiver near the recorder; from the receiver the signal is then fed into the audio input of the recorder or mixer. There are problems with FM mikes, starting with their cost, their reduced dependability with rough usage, and the possibility of interference within the system and from the system to the VTR. All the same, if you can afford a good FM mike system, it is by far the most convenient method for recording a single person talking. Just load it on your subject, and you get perfect sound in the most difficult-to-mike places.

LAVALIER MIKES

Regular lavalier mikes run from $30 up, depending on whether they're dynamic or electret condenser (more expensive) and on their size. FM wireless mikes start at about $200. The cheaper dynamic lavaliers are of good quality, although they look neo-fifties and are a couple of inches long and an inch in diameter. The high quality electret condensers can be less than three-quarters of an inch (2cm) long and less than half an inch (1.3cm) in diameter (the battery is located at the end of the cable in the plug assembly). Even the cheap lavaliers have surprisingly good quality and are very dependable.

One possible application of lavaliers in an interview situation is to use two of them will a small passive two-channel mixer such as those made by Switchcraft and Lafayette for $5 to $10. Each of the lavaliers is plugged into the mixer and the mixer is plugged into the audio input on the portable deck. (The mixer is smaller than a pack of cigarettes and weighs only a few ounces.) If you're using the camera and interviewing simultaneously, you wear one lavalier and the person you're talking to wears the other. The audio balance can be preset (just lower the levels of each lavalier to slightly under its highest point) to make conversations sound balanced and to eliminate the distortion that usually occurs when the built-in mike is used—the camera oper-

SOUND

ator's voice booming and the subject's voice at a lower volume. Two people in front of the camera can both wear lavaliers using this setup. You'll find that when you use lavaliers your video will look much more natural than if a mike were stuck in someone's face, especially if you hide the lavaliers in the subject's clothing.

Handling Techniques   Certain mike handling techniques should be observed when a hand-held mike is used to record a person speaking. First, maintain a reasonable distance between the mouth and mike. If you hold it any closer than six inches (15.2cm), the voice will distort. A good distance is eight to twelve inches (20–30cm) from mouth to mike. Speak past the top of the mike rather than directly into it to eliminate much of the popping and sibilance that those unused to using mikes produce as they speak. As you work with the mike try not to handle it too much. Hold it firmly in your hand and don't slip your grip around, for this can produce mechanical noise. If you're taping outside and it's windy, the wind will blow across the head of the mike and just about blank out all other sound. This can be alleviated by using a wind screen—usually made of foam rubber—over the head of the mike. If you haven't got time to buy one (they cost only a couple of dollars) you can make one out of a small block of porous foam rubber. A thick, woolly sock will do in a pinch.

Recording group conversations in which more than one person is talking at a time requires a number of mikes and a mixer to isolate each voice and ride the gain on each mike as the conversation shifts from person to person. Someone must concentrate on monitoring and adjusting the sound in this kind of recording. If you're using deck stands, number each stand and number the corresponding channel on the mixer. If you're using lavaliers, assign a number to each chair the people will be sitting in and number the corresponding channels on the mixer. Then you won't get lost when one person stops talking and another starts. You'll know instantly which mikes to turn up or down. If you haven't got enough mikes to assign one to each person, use one shot-gun mike and, rather than mixing the sound, shift the aim of the mike from speaker to speaker as the conversation takes place. Don't try to use one ordinary cardioid or omni-directional mike for group-talk recordings. The results will be general babble with occasional moments of audio clarity.

Recording Music   The recording technique used for live music is the exact opposite of that for speech. Speech is limited to a narrow frequency range, which allows the mike operator to concentrate on eliminating as much extraneous noise as possible in an effort to concentrate on the principal voice sounds. Music, on the other hand, covers almost the entire audible frequency range. There are two methods of recording live music: with just one mike or with a number of mikes, each

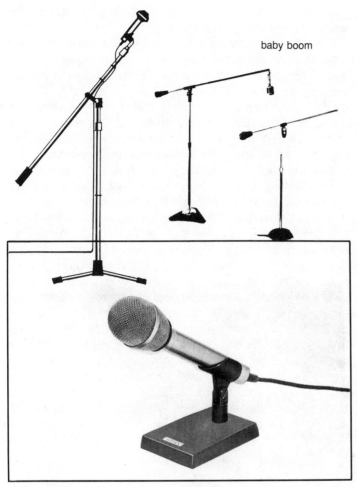

baby boom

desk stand

MIKE STANDS

assigned to a particular instrument or set of instruments.

If you're working with a portable unit and only one mike, recording live music is a matter of getting the mike far enough away from the source of the sound so that the mike isn't overloaded. This is especially important when recording rock or other loud electronic music. It's almost impossible to use the built-in mike in these situations. An independent mike plugged directly into the VTR is suited to such events. As a general practice, you should place yourself midway or farther back in the concert hall/audience area and then point the mike in the *opposite* direction from the music source. The result will sound somewhat removed, but it will not be distorted and the AGC of the portable VTR will produce a rough mix of the various elements of the total sound. Cheap mikes with limited frequency responses are especially useful for one-mike recording of live music. They effectively cut off the bottom and top of the incoming signal,

SOUND

those frequency ranges that are most likely to overload the audio circuits and create the most distortion.

In a more controlled situation you should use a mixer and a set of mikes to record live music. This can be done in two ways, depending on the number of instruments you're recording and the number of channels available to you through your mixer. If you have half a dozen instruments and only four channels in the mixer, you'll have to assign a mike to each of the general sound areas—near the drums, near the guitar amps, near the keyboards, and so forth—plus a mike for the vocalist. It will then be a question of experimenting with the exact placement of each mike to capture the sound coming from the instruments near it. An alternative is to take a portion of the signal out of the vocal p.a. amp's monitor output and combine it with the signals from the other mikes. If you have enough mixer channels and enough mikes available, the ideal miking procedure is: mike in front of the bass drum, mike over the snare, mike over the tom-toms, mike in front of each amp, mike for each of the vocalists and background vocalists. To obtain the best possible recording in this situation, isolate the mixer from the area where the music is being made by running the cables to a mixer located in another room so that the mixer signal can be monitored and the balances adjusted properly. If you try to mix live music through a mixer located near the source of the sound, you won't be able to tell which sound you're monitoring and which is coming from the stage.

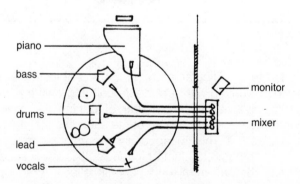

Once you master voice recording and live music recording, you'll have little trouble with other types of sound, since other sound sources are unlikely to need the exacting attention of these two recording situations.

Audio Monitoring

The word *monitoring* has been used throughout this section to denote listening to the sound signals while they're being mixed and recorded. Monitoring is accomplished in a number of ways. If you're located away from the source of the sound, you can monitor on speakers. This is done by taking a monitor signal out of the mixer (if you're using a stereo mixer and you're mixing mono only, you'll have a second output of the signal; some mixers have a monitor output as well) and running it into an amplifier and then to a speaker. The amp and speaker don't have to be of very good quality, although the better they are the more critical an evaluation you'll be able to make of the sound, the miking, and the mix. If you're running one mike straight into the VTR, you can monitor by taking the signal out of the audio line-out on the back of the deck and running it into an amp and speaker.

In some situations you can't use a speaker to monitor the sound. For instance, if you're close to the source of the sound, feedback from the speaker to the mike and then back to the speaker again will produce a howling noise. In such a case you'll have to use headphones. Most mixers have a headphone output jack. Some decks have an earphone-out jack which can be used. Larger VTRs require that you take the signal out of the audio line-out in the back of the deck, possibly amplify it slightly, and then run it to a set of headphones. The small earphone that plugs into one ear that comes with the Sony and Panasonic portables is just the opposite of what you need to monitor by headphones. Get a large pair of heavy-duty phones with well-insulated earcups, which will keep noise out and let you hear the signal. You may have to rewire the jack on the end of the headphone cable from a stero phone plug to a mono mini plug to use it with the portable deck, but that is easily done.

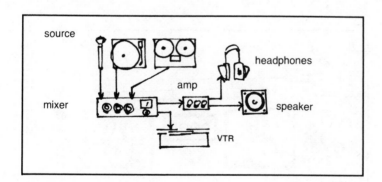

Audio Post-Production

It is possible to alter the audio portion of the tape once you've completed your total video recording. The most extreme way of doing this is to erase the audio present on the tape and rerecord from scratch. The **audio-dub** function on half-inch VTRs erases the audio track of the tape and allows you to record on it again without disturbing the video and control track portions of the tape. The uses of this function are limited since you loose the total original audio. But if you decide that your visuals would go better with music than with the original track, you can replace the track with a music track. Or you might want to remove the original wild track and replace it with narration of some sort. It's possible to redub the original effects and voices as is done in film production, but getting your participants back to speak their lines again in sync with the movement of their lips is more trouble than it's worth.

A more practical method of post-production audio is to retain the original sound track and to add additional signals to it. This is done by using two VTRs and making a copy of the original tape on the second deck. Connect the video signal from VTR A to VTR B using the video line connection. Run the audio signal out of VTR A from the audio-out jack into one channel of your mixer. Run the output signal from the mixer into the audio-in of the second VTR. Now when you play the tape on VTR A and record on VTR B, the video signal will go directly, while the audio signal will go through the mixer before being recorded by VTR B. It will then be possible to add additional signals to the audio track at the mixer, balancing them with the original track or even eliminating the original track at times. The result will be a new combined sound signal going into VTR B to be recorded. This technique can be used to add music, voice-overs, and sound effects to the audio track while maintaining the original audio. Connect a phonograph, tape recorder, or mike to other inputs of the mixer and combine the signals with the original audio.

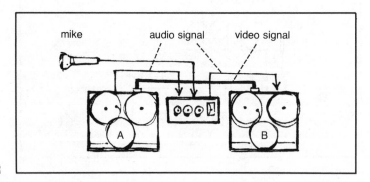

VOICE-OVER WITH MIXER

This can also be used during editing in the same manner—video goes from playback VTR A to editing VTR B, audio goes from VTR A to mixer to VTR B.

Using this system it's possible to introduce an equalizer or compressor into the audio signal path and reshape the tones of the original audio signal. For instance, if you have a lot of treble hiss on the original signal you can equalize it. You can also add tape echo or reverb.

Once you get into this type of post-production audio you'll want to collect a music and sound effects library to use in conjunction with your tapes. A cassette machine is ideal for this kind of work, since the limited frequency response of cassettes is not a drawback in terms of the system's capabilities. You can get a friend to play a few simple guitar chords into a cassette machine and use that as your musical logo or as a background. Sound effects records such as the series produced by Elektra Records, "Authentic Sound Effects," can be added to the audio track. You might even want to steal a laugh or applause track off TV to create your own comment on such programming aids. One thing you must be careful about is the use of copyrighted music and words in your tapes if you intend to use them for broadcast or other theatrical display. You can't use copyrighted material from a record or a radio broadcast without securing a license from the copyright holder (publisher of the music) and the record company that recorded the work. Even if you get someone to perform a published song for you, you're up against the copyright laws and must get permission to use it from the publisher.* Copyright-free musical soundtracks are available—you buy the right to use them when you buy the recording—but they cost about $100 per record. If you want to add music to your tapes, get local amateur musicians to play while you make a recording and make sure that the work they're playing is not copyrighted.

Audio post-production can be a lot of fun and result in greatly improved tapes. If you're planning to edit your tapes there's no reason why you shouldn't play with the audio just as you play with the video segments in assembling them.

Mike Notes

Mikes can get out of phase, causing an imbalance in the various volume levels. To check the phase of your mikes, follow this procedure: Pick one mike as the reference mike, call it mike 1, and plug it into the mixer; now, plug a second mike into the mixer. Turn up the level on mike 1 and have somebody talk into it. Bring up the level on mike 2—if the total sound gets louder, the mikes are in phase. To correct the phase of the second mike, reverse the ground and conductor leads at the plug. Continue this procedure with your other mikes, checking them against mike 1.

Sponge-rubber caps or lengths of cardboard tubing

Checking Mike Phasing

*This is most important if your tapes will be broadcast or displayed in any way that will be profitable. Private showings for family and friends are a different matter. But it's a good idea to anticipate the display potential of your tape before you use copyrighted material.

placed over the end of the built-in portable camera mike will change the frequency response of that mike. In windy or noisy areas these trappings may help to maintain proper audio levels. If you're not satisfied with the response of the built-in mike, but still like it because it's handy and out of the way, get a separate, more directional mike and tape it to the top of the camera so that it extends off the top front end just above the built-in mike. Then run its cable along with the regular camera cable (securing them together with tape) into the mike input on the portable deck.

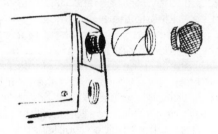

Mike Care   You should always handle your mikes with care. Rough handling, bumping, and other shocks can ruin the element. Keep them out of high humidity and heat. Also, don't leave the batteries in an electret condenser mike if you're not planning to use the mike for a long period of time. Label the batteries with the date you first inserted them, take them out when the mike isn't in use, and change them every few months whether they're run-down or not.

Checking Acoustics   A rough way of checking out room acoustics before you start to record is by ear evaluation. Get an audio cassette machine and record a treble signal such as a telephone dial tone. Play the signal on the machine as you walk around the area in which you're going to tape. You'll notice that in some places the sound is muffled and flat, in other places it's alive and full of echo. This is a measure of the acoustics in the various areas. If you're working in a very live area with the portable camera mike, you'll get a lot of high-end noise and hiss. If it's a dead area, the built-in mike will work fine. In a dead area you may want to add a little echo or reverb to the signal as you record it. In a live area you should work as close to the source of the sound as possible, using a hand-held directional mike or lavalier mikes.

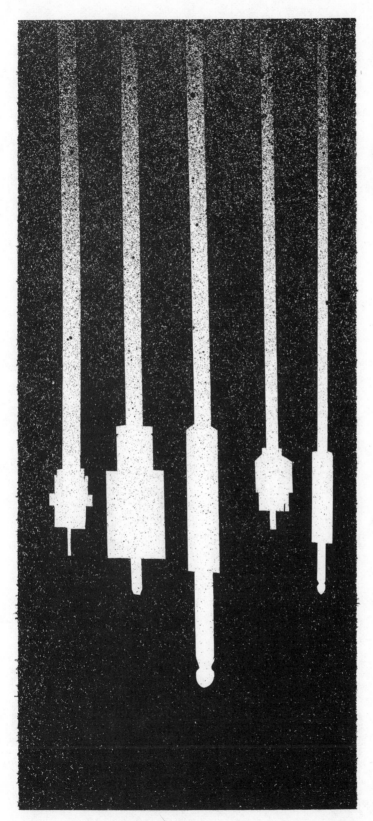

MAKING CONNECTIONS

CABLES

If electronic devices have any inherent failing, it's that they must be connected up in order to interact. They're not self-sufficient; they require interfacing with other components to perform their designated tasks. Which leads to the need for cables, those little wires that seem to run like a maze, snaking in and out between pieces of video equipment, stretched to their limit here, coiled up in huge quantities there. Most video people who work with their equipment frequently begin to see cables as part of a plot. There seems to be no way of keeping the cables from getting tangled up, intertwined, or broken. Cables are needed to connect the camera to the VTR, to connect the VTR to the electrical outlet, to connect the VTR to the monitor, to connect the monitor to the electrical outlet. And if you're using a microphone to record the sound, a whole other set of cables is needed for that. To keep you from wreaking vengeance on your equipment and from committing media hari-kari by plugging the camera-to-VTR cable into the electrical outlet, each cable varies in design according to the connection it makes.

There is one more cable problem that the video maker must contend with: cables are indecently expensive for what they are (two strands of wire and some rubber insulation) and what they do. You might spend $1,500 on a VTR and monitor only to find that it's going to cost you another $30 to connect them to each other. The impoverished video maker soon discovers that he can make a $30 cable in the privacy of his own home for about $5. Besides the savings, homemade cables can be exactly the length needed for the function they'll perform, making for a few less feet of wire to trip over when you're trying to use your equipment.

Some of the more expensive cables—usually a number of cable connections all bundled together inside one rubber skin or jacket—are quite complex and it is cheaper to buy them than make them yourself. But there are any number of cables you can easily make yourself, some of which will perform interconnections that commercially available cables have not been designed to do.

Components

The two basic components of every cable are the wire cable and the connecting plugs that are attached at each end. For every signal you want to run between two points, there are two cable connections to be made. These connections correspond to the two separate lengths of metal wire you'll find inside any particular cable when you cut it open. These wires are the conductors along which the signals flow, and they're described as the signal path. In most cases, one of the two wires inside the cable is considered the ground wire and the other is considered the hot conduc-

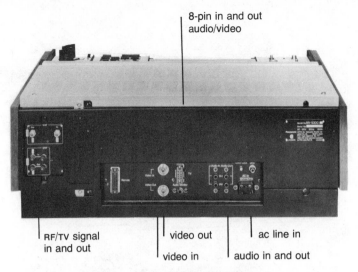

8-pin in and out
audio/video

RF/TV signal
in and out

video out

video in

ac line in

audio in and out

remote control connection

tor. The ground wire is usually wrapped around the hot wire, with a rubber or cloth insulation between to keep them from touching. For a circuit to be completed so that the signal can flow from point A to point B, both the ground and hot wires must be touching the points provided for them where the signal originates and where it is being delivered.

Audio cables are less sophisticated than video cables in terms of their ground-hot wire configuration and the care you have to take in making your own. To make audio connection cables you need a spool of audio cable. Usually described as "microphone cable," it can be either single- or two-conductor cable of the shielded variety. What this means is that there are one or two wires enclosed in a shield of wire or metalized foil. Single conductor shielded cable has one wire in the center and another wrapped around it to form the shield—there are actually two electrical paths inside the cable. A two-conductor shielded cable also has two wires inside the shield. One of the wires in the center of the two-conductor cable must be designated as the ground wire and the other as hot. Two-conductor shielded cable provides a third electrical path for the shield instead of using the same wire for shield and ground.

Examine the jack panel located on your VTR, monitor, and other equipment. The input-output holes marked for audio and, possibly, for microphone will be of a certain diameter and size depending on the model and the manufacturer. Before you start assembling cables, or buying them, you must determine what kind of plugs the equipment will take.

The configurations commonly used for audio plugs are mini plugs, phono plugs, phone plugs, Cannon connectors, and Amphenol connectors. Each type of plug varies in size

Audio Cables

Plugs

hot conductor

jacket   ground   insulation

CABLES

169

and design although they all perform the same connection function—they mount on the end of the cable and attach the internal components of the units to each other through the two conductors of the cable. The total plug unit includes a male jack (the plug itself) and a female receptacle (the mounting hole that goes on the equipment). The female receptacle is usually installed on the jack panel or patch

PHONE PLUG

MINI PLUG

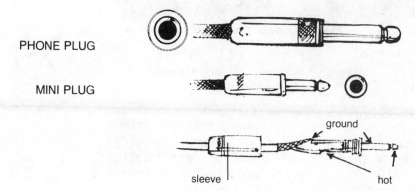

ground

sleeve   hot

panel on the equipment. The act of plugging a jack into its receptacles is known as **patching** and the two jacks plus the cable in between them are known as a **patch cord**. When you want to connect a VTR with a monitor you "patch" them together.

The mini plug and the phone plug are the two most familiar forms of audio plugs. They resemble each other in design; the mini plug is really just a much smaller version of the phone plug. When buying mini and phone plugs, make sure that they are the two-conductor type—especially the phone plug, since it also comes in a three-conductor configuration for use in making stereo audio connections.

If you examine the phone or mini plug you'll see that the metal shaft of the plug is divided into two segments. The tip of the shaft is sectioned off from the rest of the plug by a rubber washer or plastic disc. This tip is one of the conductor terminals; the rear portion of the shaft is the ground. The shaft is connected at the far end from the tip to a sort of base, and around that base is screwed a sleeve, which must be unscrewed to attach the cable to the plug. When you unscrew the sleeve you'll find two posts, one extending farther from the base than the other, to which the wires of the cable will be connected to attach the plug to the cable. When that connection has been made, the cable wire ends will be attached to the plug and the plug will become the extension of the hot wire and the ground wire. The hot wire should be connected to the terminal that connects to the tip; the ground wire should be connected to the terminal that connects to the upper portion of the shaft.

THE VIDEO PRIMER

The phono plug is also a familiar unit, sometimes called an RCA phono plug. It is most often used to make audio connections between record players and hi-fi amplifiers. The phono plug is a much shorter, smaller unit than the phone plug (phono standing for phonograph, phone plug having been originally designed for use in telephone switchboard connections). The hot conductor unit is the small shaft that sticks out of the middle of the phono plug and is inserted into the hole in the receptacle; the ground segment of the phono plug is the metal cup that encircles the hot conductor shaft and rests around the outside portion of the receptacle.

You'll also find Cannon plug receptacles on some equipment—usually monitors and one-inch VTRs. These plugs, also called Switchcraft or XLR plugs, are heavy-duty plugs originally designed to connect microphones to control consoles in recording studios. The plugs and their receptacles can be either male or female, depending on the whim of the manufacturer, so you must check carefully to see how they're set up on your equipment. You may find two different Cannon receptacles next to each other on a patch panel, one for audio-in and one for audio-out. It is most likely that the audio-in will be a female receptacle and audio-out a male receptacle—so even if you can't read what they're labeled, you still won't make the mistake of patching audio-

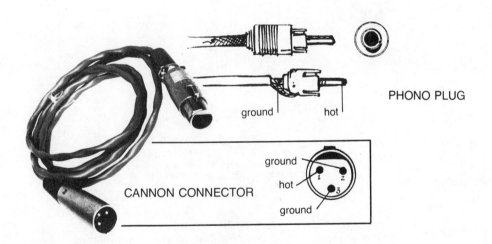

PHONO PLUG

ground   hot

CANNON CONNECTOR

ground

hot

ground

in into an audio-out jack. All Cannon plugs used in audio are three-conductor plugs, with three pins sticking out of the male jack or receptacle, three holes in the female jack or receptacle. Pin number one of the three-conductor terminals is designed to be used as an overall shield connection, pin number two is used for the signal conductor, and pin number three for the ground. With single-conductor

CABLES

shielded cable, connect the outside wire, which serves as shield and ground, to pins one and three.

Occasionally you'll run into an Amphenol connector as the cable-to-mike connection. Amphenol connectors screw onto their receptacles, making a very secure connection. But they are almost never used on the other end of the mike cable, which plugs into the audio mixer, amplifier, or VTR.

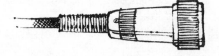

AMPHENOL CONNECTOR

Making Audio Cables   You're ready to make your own audio cables once you've determined which audio plugs are being used on your equipment and once you've bought the necessary plugs and a spool of mike cable. You may find that one piece of equipment takes a mini plug for audio-out, but the other end of the connection you want to make takes a phone plug for audio-in. This is common and a good reason for custom-wiring your cables. It is possible to buy audio cables with different connectors on each end; it is also possible to get adaptor plugs—such as mini-to-phone, mini-to-phono, phono-to-phone, and so forth—which fit over the existing plug and provide a second plug connection. But these adaptors are to be used only in emergencies since they

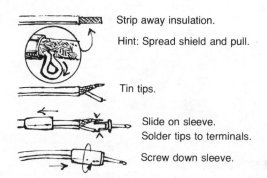

Strip away insulation.

Hint: Spread shield and pull.

Tin tips.

Slide on sleeve.
Solder tips to terminals.

Screw down sleeve.

MAKING AUDIO CABLES

have a habit of slipping off. If you are planning to buy rather than make cables, be sure that the plugs on each end of the cable have metal sleeves. Most of the cheaper cables made today have molded rubber for sleeves, which will smash when stepped on. If the connections inside are broken (which happens through normal use), there is no way to take them apart to repair them.

Plugs run from $.50 to $2.00 each, depending on quality. Since you can reuse them forever it is best to buy the most expensively made heavy-duty plugs you can find.

THE VIDEO PRIMER

To connect a plug to a cable do the following:

1. Cut off about an inch (2.5cm) of the rubber insulation at each end of the cable, using a pair of diagonal pliers or a wire stripper. Then strip the insulation away so the two conductor wires are exposed. If there is a conductor with a ground wire encircling it, pull the ground wires to one side and twist them together, making sure they are not touching at the point where the ground and conductor enter the cable.

2. Using a 25-watt (or less) soldering iron, **tin** the two exposed wires by touching the soldering iron to each wire and applying a bit of solder so that solder flows along and coats the exposed length of wire.

3. Unscrew the sleeve from the plug and slide the sleeve onto the cable.

4. On the exposed upper part of the plug assembly, you will see the two terminals, each with a hole in it. Some of these terminals have screw-down posts. If so, wrap the wires around each screw shank and screw them down. If there are just holes on the terminals, attach the inner conductor wire to the shorter of these two terminals and the ground wire to the longer, by pushing the tip of the tinned wire through the appropriate hole.

5. Solder each wire to its terminal.

6. Slide the sleeve back down the cable and screw the sleeve into the plug assembly.

Repeat this procedure at the other end to attach the other plug.

Video Cables

As with audio cables, there are several types of video cables and a number of different plugs. Unlike audio, however, video cables are not so easily adapted from plug to plug because the cables are more standardized to the functions they perform and cannot be interchanged with adaptors or by changing the plugs.

The most common video cable is called **coaxial cable**. Coaxial cable, usually referred to as **coax** (co·acks), is a one-hot-conductor, one-ground cable. The hot conductor is a solid copper wire in the center of the cable, and the ground a mesh network of wires woven together to encircle the copper wire. There is an insulating material between the hot wire and the ground known as the **dielectric**, and the ground itself acts as a shield against extraneous signals, which might otherwise reach the conductor. The word *coaxial* means common axis, and it refers to the position of the conductor in relation to the ground.

Impedance

Every coaxial cable has a certain impedance—it will resist the flow of a signal along it to a certain degree and thus, in reverse logic, will maintain a certain signal strength as that signal is conducted along it. The impedance is determined by the construction of the cable, the thickness of the hot conductor and ground, and the distance between the inner core conductor and the outer ground.

Since a 75-ohm signal is what is produced by both the composite video output and the RF generator output, a 75-ohm coaxial cable is needed to carry this signal. There are a number of 75-ohm cables available, the most common being the RG-59/U coax. The other 75-ohm coaxial cables include RG-59B/U, RG-35B/U, RG-11/U, and RG-13A/U. They vary in their construction according to the use they were designed for: indoor, outdoor, burial underground, and so forth, but their signal-carrying capabilities are the same as the more universal RG-59/U coaxial cable.

Most VTRs and monitors have video-out and video-in or line-out and lin-in plugs that use a UHF connector and coaxial cable. The UHF connector has a central shaft and a sleeve that screws down over the receptacle. It has nothing to do with UHF frequencies except that it was originally used to make connections for cables carrying those frequencies.

Making Video Cables

To make a UHF coaxial cable:

1. Get a couple of male UHF connector plugs and two UG-176 cable adaptors.

2. Unscrew the plug assembly and slip the sleeve over the end of the cable.

3. Slip the cable adapter over the cable.

4. Strip the end of the cable. You'll find it a bit more difficult to strip the ends of coax cables because of their thickness.

5. Fold the braid back over the cable adaptor.

6. Insert the conductor wire into the bottom half of the plug so that it goes right into the shaft and comes out the hole in the shaft's tip.

7. Screw the plug onto the cable adaptor and tighten with pliers.

8. Solder the conductor to the center post (this takes only three hands).

9. Cut off the excess wire from the center post.

10. Screw the sleeve back over the assembly, and the cable will be ready.

Care must be taken to keep coaxial cable as short as possible. Coax will pick up stray electronic signals more readily than other cables and therefore should be kept away from RF generators, power lines, and any other equipment that might be generating an electrical field. Don't bend coax at right angles or step on it too often or you'll break the wires inside and have to wire up a new cable. Because of the fragility of coax and the advisability of using only the lengths necessary to get from one unit to another, you should definitely learn how to put together your own UHF coaxial patch cords.

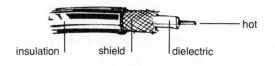

insulation | shield | dielectric | hot

COAXIAL CABLE

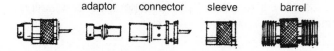

adaptor   connector   sleeve   barrel

UHF CONNECTOR

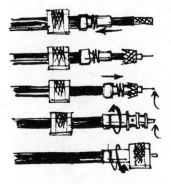

Slide on sleeve and adaptor.
Trim cable.

Fold back braid.
Trim center wire.
Slide down adaptor.
Tin center wire.
Screw on connector.
Solder tip.
Screw down sleeve.

**MAKING A UHF
CONNECTOR**

Most RF generators come supplied with an RF receptacle-to-TV-antenna cord, but you can improve the quality of the picture and extend the length of the connection if you replace it with coaxial cable. When making an RF-to-TV cable, be sure to install a plug at the VTR/RF-out end of the coax that fits that output receptacle, be it a mini plug or other jack. Since the signal coming out of the RF is 75 ohms and the input terminals of most TVs are made to take a 300-ohm signal, you need a 75- to 300-ohm transformer between the TV end of the coax and TV antenna terminals. These transformers, known as baluns, cost a couple of dollars and can be purchased at the hi-fi or TV store.

At the far end of the transformer are two spade lugs that fit under the screws on the antenna terminal. At the near end is a sleeve with threads on it and a hole in the center of a plastic dielectric, which is encased by the sleeve. To match the coaxial cable to this, you must get what is called a **coaxial connector** (also known as an **F connector**). It is about one-third the size of a UHF connector (they have a similar screw-on sleeve, which binds them to their female receptacles) and is used in cable TV and other 75-ohm cable applications. This type of connector is not soldered to the end of the cable, but is **crimped** on. To make one:

1. Strip the end of the cable to expose the center conductor.
2. Slip the crimping ring over the cable.
3. Insert the inner conductor into the F connector shaft;

CABLES

the outer ground wire is contacted by the thin rear shaft that slides under it.

4. Shove the end of the cable into the connector as far as it will go.

5. Slide the crimping ring over the entire assembly and squash it with a pair of pliers to hold the connections in place.

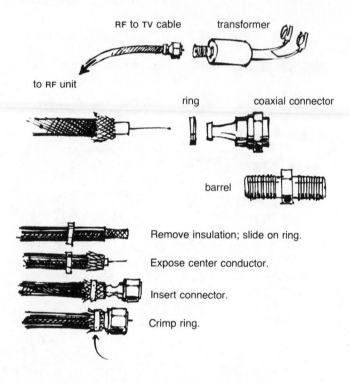

RF to TV cable   transformer

to RF unit

ring   coaxial connector

barrel

Remove insulation; slide on ring.

Expose center conductor.

Insert connector.

Crimp ring.

Specially made crimping tools are available to strip the wire and then crimp the connector in place—you may have seen cable-TV servicemen wearing them in holsters on their hips—but they can cost about $50 and aren't recommended unless you're planning to start your own cable company. A good pair of pliers will do the job, not as neatly, but just as well.

Up to 1000 feet of coaxial cable can be run between units without any serious signal loss. But when planning to run greater lengths of coax, you should use amplifiers along the way to keep the signal up to strength. The various types of amplifiers available are described on pages 278–80.

Besides audio and video cables, there are various cables that the Japanese have developed to combine a number of connections into one cable. These include ten-pin cables, eight-pin cables, six-pin cables, and four-pin cables.

The ten-pin cable is used in many portable units to connect the portable VTR to the portable camera. Each of the "pins" is the end of a conductor wire, so the ten pins can carry a total of five signals. Ten-pin cables are used to carry the following signals between the portable camera and VTR:

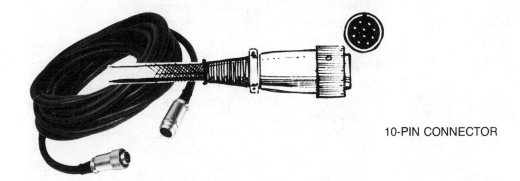

10-PIN CONNECTOR

DIN PLUG

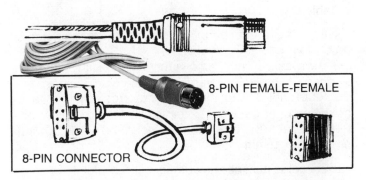

8-PIN FEMALE-FEMALE

8-PIN CONNECTOR

used for half-inch equipment. It has been standardized by the EIAJ and carries composite video-in, composite video-out, audio-in, and audio-out (thus eight conductors) between VTR and monitor.

The six-pin cable is a camera-to-VTR or switcher cable video-in and video-out run on one conductor/ground pair (the direction of signal flow reverses on playback), sound-in to deck, horizontal sync pulses to camera from deck, vertical sync pulses to camera from deck, trigger switch signal from camera to deck, and power to camera from deck. This ten-pin connector has become the standard cable for most manufacturers' portable systems, but that does not mean the cameras they're attached to are interchangeable. For instance, you should never attach an AKAI camera to a Sony portable deck, even though the camera has a ten-pin cable coming out of it and the Sony has a ten-pin receptacle.

The eight-pin cable is the standard VTR-to-monitor cable used by Sony. It carries video, sync, and other control signals between the camera and VTR or switcher. The six-pin connector cable has a **DIN** six-pin plug on each end (**DIN** is a European industrial standard). **DIN** plugs vary in number of

Standard Plugs

CABLES

pins, so if you want to replace a six-pin DIN, make sure that you get a replacement with the correct pin configuration.

Four-pin DIN plugs are used as power supply cables on many portable systems, including Sony and Panasonic units. They run from the ac adaptor to the portable deck or from the external battery packs designed for video to the portable deck.

There are other camera-to-VTR or switcher connecting cables around—ten-pin cables and larger ones—that are used in more expensive studio cameras. You will need a wiring diagram to understand what they do and how they carry the signals. Occasionally you'll also see square four-pin cables, which are used for playback of a signal from a VTR to a monitor that has a receptacle for it. Four-pins are becoming less popular with the acceptance of the eight-pin as the standard VTR-to-monitor connector.

Cable Care

1. Try to keep from bending, knotting, twisting, or stomping on your cables.

2. Always insert and remove them from their receptacles by gripping the jack; never tug on the cable itself.

3. Store cables by looping them into large coils and hanging them out of the way. Exceptionally long cables whose full length isn't always needed can be stored neatly by winding them on 16mm movie reels.

Adaptors   Every cable plug has an adaptor available for it. It may be what is called an **octopus cable**, which has an eight-pin plug on one end and separate UHF-in, UHF-out, audio-in, audio-out connectors on the other end. Or it may just be a mating plug that converts a mini audio plug to a phone plug or a phone plug to a phono plug. It'll cost you about $100 to buy all the cables, plugs, and adaptors you'll need for video and audio, but you should spend the money. Get every conceivable adaptor you can find—eight-pin to eight-pin, six-pin to six-pin, video and audio converter plugs, and so on. Those you can't buy, wire up yourself (see pp. 172–76). This will save a lot of hair-tearing when you start working with your equipment.

The majority of video equipment comes with a three-pin ac plug—two conductors as in the normal ac plug in this country and a ground lead. Often you'll find that the wall socket you want to use has only a two-pin receptacle. You need a three-pin to two-pin adaptor or an extension cord with three-pins in and two-pins out. You can get both at any hardware store. Some adaptors have a ground wire (often green) with a lug on the end, which is intended for securing the ground wire to one of the screws on the faceplate of the wall socket. But your equipment won't be grounded unless

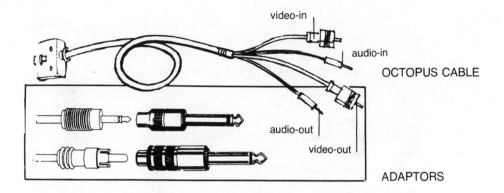

video-in

audio-in

OCTOPUS CABLE

audio-out

video-out

ADAPTORS

the socket casing is grounded—you'll have to have an electrician check that for you.

Extension cords for ac power are among the cable connections most people tend to ignore until they get caught needing a longer extension cord than they've got. Heavy-duty flat extension cords are available in lengths up to about twenty-five feet (7.6m), some of them on self-contained reels for storage. Get the best extension cords you can find with more length than you think you'll need. Store long cords on the reels sold in hardware stores for holding garden hoses.

If you find yourself constantly running out of outlets to plug equipment into, you can buy fused outlet boxes, 15-amp outlet boxes with from three to six outputs. All have three-pin receptacles and a heavy-duty power cable with a three-pin plug on the end connecting the box to the ac outlet. These boxes are both switched and equipped with fuses or circuit breakers—meaning that you can turn off the equipment by cutting off the power at the box, and that any power surges on the ac line will blow the fuse or open the circuit in the box and thus doubly protect your equipment. Don't confuse these units with the cheap plastic unfused outlet boxes selling for about half the price. A good fused or circuit-breaker outlet box made of metal will cost from $8 to $12 and is well worth the price.

By the way, when you wire your own cables be sure that no solder drips between the lugs and that the connections are made correctly. If you mess up, you'll blow fuses on your equipment. Be especially careful when working with the six-pin camera-to-deck cables used with the portable systems since these cables have a dc power current running through them.

Check cable plugs regularly for any signs of loosening from the jack or fraying of the cable at the end of the sleeve. Repair immediately.

A set of cables, each connecting a different piece of equipment and each a different length, will sometimes cause delay problems. Some of the signals will get from

CABLES

point A to point B sooner than others and result in improper switching, sync, and other time problems. Cure this by inserting extra cable lengths on the shorter lines to create equal delays with all signals. Occasionally you'll get hum on a cable line. That can be eliminated by using two cables to carry one signal: Run the hot conductor signal through the conductor of cable A and run the ground signal through the conductor of cable B, allowing the original ground shields of each cable to shield the signals. To accomplish this, connect the jack terminals (conductor and ground) to the conductors of two cables and ignore the shielded ground wires.

Noise in a picture can be caused by the length of the cable you're using, especially when going from SEG to VTR. Keep the cables as short as possible.

Cables and the various ways they're connected can be a continual source of confusion if you don't take the time to become familiar with the types currently in use. A recent Sony catalog is helpful since cables, cable adaptors, and the accessories available are listed along with illustrations so you can see what they look like. For audio cables and jacks, a catalog from Lafayette or any other electronics firm will show you what's being used and what the various units look like. For most of your video work you'll probably be concerned with only two connecting cables: the camera-to-VTR cable and the VTR-to-monitor cable. But if you understand the different types of cable and what signals they're carrying where, you'll find that you can make all sorts of connections that aren't mentioned in the instruction book.

10

WELCOME TO THE VAST WASTELAND

PRODUCTION VALUES

That's all television is, my dear. Nothing but auditions.
—Addison De Witt in Joseph
Mankiewicz's *All About Eve*

It is so easy to make video that those having the equipment often fall into one of two traps: They realize that it's possible for them to tape any event, any time, anywhere . . . so they don't bother to tape anything at all; or they realize they can tape anything and they do, using hours of tape as though they were making jars of electric soup to put on the shelf against the possibility of a media famine. These two extreme reactions to video are not uncommon. Often the same person will experience each at different times, eventually reaching the more moderate view that there are things to be taped and that taping should be done as efficiently and effectively as possible.

Since video tape equipment is relatively simple to operate, those making tapes often overlook the obvious planning and preparation that should be part of the taping procedure. Preproduction efforts can't help but create a superior tape. You'll discover that taping is much more fun, and the results more satisfying, when you get in the habit of starting to make your tapes before you turn on the machine. I'm talking about preproduction on all levels—whether you're planning to tape a birthday party or your own version of *Ben Hur*. Of course, the more complex your production, the more preproduction work is involved. But you should think of your tapes as video productions and proceed with an eye toward professionalism even if you're concerned only with recording personal events.

Preproduction begins with the determination of a topic or theme. That means deciding just what to tape, allowing for whatever taping possibilities may arise during the actual recording time-space, and providing yourself with as much information as possible about what you're going to be doing. Choose a subject and then investigate its potential. Even simple programs like *A Family Gathering* require consideration on these levels. Will there be adequate ambient lighting? How will the sound be treated? What equipment can be used without imposing its presence on the situation? And so forth, right down to where you'll display the tape for the family, who may demand to see it immediately.

To avoid repetition I'm going to assume that you have some sort of "professional" production in mind, an event which will be assembled on tape for group display and possible cable broadcast. All the work involved in getting such a taping ready can be applied in lesser degrees to more minor efforts such as personal tapes. Pick and choose what you feel will aid your tapes.

Two avenues of production immediately present themselves when you start making a tape; the choice between them depends on how much post-production you're planning. You may want to tape out of sequence, segment by segment, then assemble coherently after taping if editing equipment is available. If you can't edit, you have a choice between taping as live as possible with a portable unit and just allowing any blips from starting and stopping the camera to appear on the screen, or using an SEG multicamera setup. No matter which of these procedures is followed, the preproduction work is pretty much the same. It's just that an SEG or live situation demands that you make post-production decisions at the time of recording, and the flexibility provided by the editing option is lost.

Video Form

Video is a radical form. It can present information that is unexpected or unusual compared to what people are used to seeing on TV. Communication of information through video requires some sense of form as it relates to the audio/visual presentation of concepts. Here we must look at the two present examples: film and television. Both have found, through years of commercial development, that there must be a certain amount of form in the presentation in order for the audience to appreciate the information communicated through the medium. There is a neatness, the establishment of an entity, marking off of beginnings and endings, on which people have always depended. The ornate frame that holds the canvas, the cover of a book, the credits of a movie, the catchy title of a bestseller—all are concessions to form, to the acknowledgment of a format through which ideas can be presented.

In making video the urge is to launch into new directions, new ways of communicating thought. If this innovation includes the loss of any sense of format, it can be detrimental to the ultimate communication. You may know exactly what you're talking about, but most likely your audience won't.

And so we come to the main consideration here: the admission of an audience. If you're making video for yourself, home tapes and records of your daily activities, that's one thing. But if you're making video as a method of communicating, you must recognize the platform from which you're speaking. With that recognition comes the audience, the stage and backstage, and the edge of the stage, which is the dividing line between the presentation and the people out there watching.

You don't have to sit around and try to figure out how to make tapes that people will like, but you do have to keep constantly in mind that people will be watching your tapes. Whether they like what you're saying or not, they must understand it before they can react. Communication is a matter of form and format and of the refinement of form to produce content.

A Word from the Sponsor

The first consideration when making a tape is the story line. Everything has a story, a plot—even if you're just planning to set up a camera in a field and let the tape run on the world. If you have no plot, there is no reason for making a tape. A beginning, a middle, an end—you need those even to say, "I haven't got anything to say."

Some video people have a strong reaction to the preconditioning we've all had with television. We know it as a medium, as an event, as something that brings the stars into our homes. Each and every one of us has, at some time, been entertained and informed by television. We are familiar with its form, so familiar that we ignore it as we plunge into the tube to evaluate the content of what's being communicated to us.

The introduction of inexpensive video equipment came at a time when the media values of television were being questioned. The use of video equipment to create an alternative TV was part of the rejection of broadcast television. So we got our own television and we lashed back, consciously making nontelevision. And, when cable-casting became available, we tried to get our nontelevision on television. Now we're watching our own television on television. We're back to television; back to the admission that there is a television audience and that it is reacting not only to what we're trying to say, but to how we're saying it.

The video experience is part of the overall television experience. The two cannot be separated, although one has come to qualify the other as media. Television is an ongoing process, advancing a continuous stream of perceptions keyed to our reactions into a total life experience. Video is part of this process, but without television there would be no video.

We must learn as the young white guitarist learns from the venerable blues musician. We must go back to television to learn about television. And the first thing we find is that television is plot, story, communication of a series of events.

Video is like the early days of television, when it had energy, when it was just as much an alternative as video is today. There was a period when TV was part of the art deco experience, an experimental curiosity being examined by a disbelieving thirties. Then, after the Second World War, the concept of television was revised. The perceptions of the medium had a shell-shocked sense of excitement; the first era of electronic communications had arrived. Transistors, bulky equipment, crew cuts, and *From Here to Eternity* shirts . . . H-bombs and John Cameron Swayze . . . *Ameri-*

can Bandstand ... black and white ... The Lone Ranger and Tonto ... Cisco Kid and Poncho ... Amos and Andy ... the crazy fifties of Captain Video and Edward R. Murrow.

Everything was done live in those days, but if it had been done on tape it would have looked the same. It was the puberty of man's structural relationships to himself and his world. Everybody could know everything at once, share a common hypnotic experience whose impact really was mind-boggling.

This is where we must go with our video equipment. This is where we have to start our video experience if we are to fully appreciate what our equipment can do. It's like studying the development of any art form, like learning a trade, having a sense of a craftsman working with his tools. Tyrone Guthrie expressed it when he said that a person has to spend all his time learning how to be a skilled technician so that in moments of inspiration he has the ability to express his dreams.

At the same time, video is the introduction of a new element into the television process: the dissemination of media hardware to everyone who thinks he has something to say. The result is that more people will have the chance to be heard. Some will even catch on, and there will be a demand for what they have to say. Video is the ultimate platform, a way to speak out that can be more effective than newspapers, magazines, books, radio, movies, or street corners, because it is television.

Artists coming from other media often misunderstand TV as a medium. Their discomfort with it makes them critical of the TV medium. If you want to get into TV, you have to use video to develop as an artist within the medium in which you want to work. If you want to make TV, then you have to grow and work within video. Most important of all, you have to accept television.

Video is a form of television and to put down TV is like saying that books or records or movies are bad because they don't please you. I think it's unfortunate that so many of those using video equipment to make media have to work under the burden of a political and philosophical rejection of television. If you're not into TV and you want to get into video, the first thing you *have* to do is to get into TV. Watch it, participate in it as a mass medium, spend six months becoming familiar with the methods that have been developed to communicate with the mass audience. TV is the technical and theoretical textbook of the video maker. If you're going to be intolerant of the most successful manifestation of mass media ever invented, you're never going to make good video.

Your video tape must start with your having something to say—events and reactions to events that you're going to put on tape. Then you must be able to use your equipment to express your thoughts. This book tells you *how* to use your equipment; you have the responsibility of turning it to your best advantage.

Use the vision of early television as your frame of reference. Video can duplicate it because video, at this point, has many of the same qualities: It's live, electric, rough around the edges, and much more concerned with what you've got to say than exactly how you're going to say it. This, tempered with a sense of TV form and format, is where you must begin.

Getting Started

Plot and Scripts

It's kind of hard to escape your own realities and sit down and start writing out script ideas. Start with the story idea, the plot of what you're going to say. Then **block out** the script, arranging the reality you're going to set down on tape from the viewpoint of the camera, which will be your eyes. TV **story boards**, available at art supply stores, can be helpful. They come on a pad and each sheet has a number of blank TV screens printed on it with space below each screen for comments. You can block out your action, plot how your action will advance from scene to scene, mark down directorial thoughts, and note audio cues. The translation of an idea in your head into a series of steps that must be taken to effect this idea on the screen makes up the preproduction process.

Once you have a rough idea of what you want to do, write down a formal script. This can be tiresome and seem like overpreparation, but it is worth the effort. Video shooting

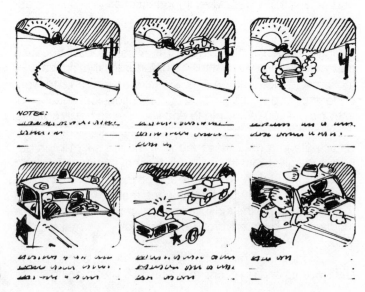

STORY BOARD

scripts are divided vertically with the video-camera directions in a column on the left and the audio dialogue and action directions on the right. The script includes camera cues and movements, audio cues, and directions on the technical execution of the show.

As you develop plans to make a tape, remember that getting across the message means making the medium inconspicuous. Consider the elements you have to work with: *The Setting* equipment, people, and physical location. Each plays a crucial part in how your tape will come out. The equipment must be properly placed in relation to the people and the location. Block out where the equipment will go. Make a diagram of lights, microphones, cameras, mixing equipment, actors, and props. Then, as you set up your scene, or at least bring your equipment to it, you'll have a reference from which to adapt to the circumstances. You'll learn this as you work with your equipment. Using a portable system, you'll get accustomed to walking into a room a certain way, holding the camera in anticipation of taping, setting the con-

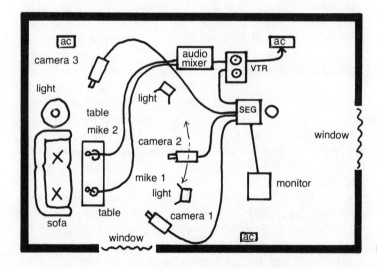

BLOCK OUT ELEMENTS

trols with an almost unconscious gesture, and placing yourself to take full advantage of lighting, sound, and other conditions. The same sense of procedure must be developed if you're setting up a full-scale production using an SEG, audio mixer, portable lights, and the like. Learn how to place your equipment in relation to what you're going to tape. Don't let setting up become a drag. Stay awake while you're doing it and attempt to understand how it has affected your past tapes and what possible changes could be made to improve your tapings.

During all this, and when you're taping, you should keep a rough log of the production. It's a handy tool to have when

PRODUCTION VALUES

you're editing and provides a set of production notes to which you can refer for future tapings.

There are two sets of dynamics at work in every camera scene. The first is the overall dynamics of the particular scene as it relates to the total flow of the video production. These external dynamics indicate how the scene will relate to the scenes that precede and follow it. From the viewpoint of the camera (you, your audience) you have the angle at which the camera views the subject and the potential movement of the camera either during that scene or between that scene and the one to follow. This camera viewpoint is important, although it is not an immediate consideration in the actual internal composition of the scene.

Once you understand the external dynamics that have led to the scene in question and will then move on to the next scene, you must consider the internal dynamics of that scene. Thinking of the scene prior to the scene you're about to record, work into the scene (external dynamics) and establish its components (internal dynamics), always, of course, keeping in mind the continuing flow of the production (external dynamics again). Working with the physical properties of the scene—and this includes you and your camera—figure out what you're trying to create visually. A scene is the space involved and what's filling it; the movement within that space; the sound accompanying the space and movement; the texture of the space. The components of the space are distributed both horizontally and vertically within it. You must determine the natural flow of the space as well as what you can add to it through the camera's view, all the while remaining conscious of how the visual was introduced from the scene before it and how it will introduce the scene to follow.

Action and Display   When you combine the space with the camera view there are several planes on which action can take place. Subjects can move left to right, right to left, up or down, backward or forward. By the same token, your camera can move in all these directions. The permutations of the two are what you have to work with as you tape.

You must remember that the eventual display of your tapes will take place on a screen whose area won't exceed 23 inches (58.4cm) diagonally. This makes it essential to simplify the scene-space-camera elements involved to the point that they are intelligible within the small area of the screen.

The proportions of the TV display screen are 3:4; the screen is three units high for every four units wide. Within these proportions is a **safe area**. The safe area is the part of the screen that will be visible on every monitor or receiver no matter how it's adjusted. About eighty percent of the height and eighty percent of the width are the **safe title area**—titles

SAFE AREA

TITLE CARD

11"/35.6 cm

14"/39 cm

viewfinder

safe area

should not extend to the edges of the screen. For action, ninety percent of the area from the center of the screen out toward the edges is considered a safe area. You've got to keep this area in mind when you're composing the scene. If you have some information too close to the edge of the viewfinder, it is possible that it won't show up on the monitor during replay. Even if your particular camera-VTR-monitor setup gives you a perfect match between area scanned and area reproduced, don't push your luck. The first time you use your RF adaptor and a normal TV receiver you'll find that the picture is not exactly the same (it may have shifted slightly to the right or left and slightly up or down) as it was on your monitor.

Russian director and film theorist Sergei Eisenstein suggested that film, being a combination of segments or elements, could either combine into a whole or collide. The same is true for video tape. There must be no collision between the elements of the scene, no disruption of the flow; a continuity must be established and preserved. Within a scene-space the camera must allow for the action that is taking place. For instance, a person is going to jump off a bridge. To communicate that action, you must allow a space into which the person can jump. A close-up on the face is no good if the audience hasn't been shown the person, the bridge, and the precarious relationship between the two and the ground.

A person walks across an area. There must be space for walking. Framing the scene with no area for the walking to take place in bars communication. The visual message of walking includes a sense of the area in front of the person into which he or she is walking. The person shouldn't be walking into the edge of the screen as the camera pans to follow the action. Allow the scene-space-subject to breathe naturally. Avoid the collision of the form and content, and avoid the collision of the flow of form and content from one segment to the next.

Video Construction

PRODUCTION VALUES

I don't mean to overuse the word *natural* here. Your final tape will be anything but natural, despite the fact that it should *seem* natural. There is a popular video concept that natural equals real time; that natural is when the amount of time devoted to an event on the tape is equal to the amount of time the event actually spans. This is a false perception of what natural is and an equally false perception of the video process. Sure, you can use up a half hour of tape to record a half hour of time and space. But you're still imposing artificial limitations on the time and space by recording only a half-hour segment of it. Realistically, the natural time flow is interrupted by starting and stopping the camera, be it while recording or during editing. Each start and stop of the camera is a "scene," a segment of recorded time.

.The internal time-space of each segment must take into account the whole of the composition. The length of each shot should be geared to what the shot is saying. It doesn't take ten minutes to say, "The door was opened"; it may take ten minutes to say, "The two people discussed the plans to rob the bank." Avoid using up time and audience attention by recording events at excessive length. Limit your recording to the establishment of the point of the subject-scene-space and the segments that will be used to construct the whole.

Media preconditioning comes into play here. A car trip is today expressed, and understood, by a person getting into a car, driving off, coming to a stop in a second location, and getting out of the car. There is no need to show the car in motion unless a part of the plot is linked to that segment. Time has become compressed by TV and movies; we expect certain events and actions to be only suggested on the screen as a montage of portions of events rather than as the entire event. You must take that into account as you tape. Your audience will demand it by not sitting still for long-winded explanations of what they've come to consider as obvious. Avoid collisions with your audience.

Real time is an overstatement that must be replaced by your own time statement.

The recording of a series of segments is still not a complete program. Each segment is part of the progression from the first to the last scene-space of your story. Each segment has an idea to it, a point that you are trying to make by recording it. In the same way, the joining of two segments presents a further idea, the production of a video concept in which the parts generate a greater whole.

A close-up of a male with a look of pleasure on his face followed by a close-up of a female with a look of pleasure on her face suggests that the male and female are sharing a pleasure. These close-ups repeated one after another, over and over again, with the length of each segment shortened

The Total Program

each time, suggest the culmination of a shared pleasure. But the same segments taken apart and added to other segments create something different: switching between close-ups of two females with looks of pleasure on their faces, any one of the above and a dog panting happily. . . . Each of these composites has a sense to it that none of the individual segments can possibly suggest.

With the various scene-segments of your program on tape, post-production begins—unless you've been working in a live studio facility. Allow a little breathing space between recording and the rest of the work that has to be done. Take a day off, put the recording out of your mind, so when you see it again you'll have a certain detachment.

Post-Production

The initial step in post-producing a tape is to review the footage. Watch everything you've taped, first for a general sense of what has been recorded and what you may have left out. Then watch it again for a more specific determination of just how the segments must be altered and ordered to build a program. If there is anything missing, tape additional scenes before you continue. Begin editing only when you've got everything you need.

Editing requires the ability to anticipate boredom, which can be produced by the overstatement of an idea, action, or scene. The predetermined real time of your recording must be changed to exclude information that is not central to the flow of the program. Introduce a new sense of time in order to compress information and to cut away anything that might detract or collide.

Editing

Assemble a **rough edit** by placing the segments in their program order. View this edited tape with an eye toward what you can remove to increase the impact of what's left. Keep a list of what should stay and what has to go. Now return to the original master tapes and assemble a second edit, this time shortening and eliminating until you have your program.

Once the edit is completed, it's a good idea to leave the tape for a while. You might want to preview it to a selected audience, watching their reactions to the show. As you sense what's right and wrong with the tape through the perceptions of others who are concerned only with the end product on the screen, you'll gain an insight into video as a communications process and see whether you've succeeded or not. At that point, you might go back and re-edit the tape to tighten it further. It often happens that something you find terribly interesting puts your audience to sleep. Since the audience is who you're making the tape for, correct this.

Don't be concerned with the total length of your tape. If it's only eleven minutes or if it's two hours, the content is what's important. There is no need to get locked into the television time formats. Make your show exactly the length it needs to be. Your audience will forgive you if you tell them everything you've got to say and no more, even if it takes only seventeen minutes.

Cutting   Cutting between segments can be accomplished in any one of four ways: **cutting on the action**, **cutting on the reaction**, **intercutting**, and **cutting to tighten**. The first three can be used during recording in a live studio-SEG situation or during editing. The last is pretty much an editing technique.

Cutting on the action is breaking away from whatever is happening to another event. A man is sitting at a desk, writing. A woman is speeding along in a sportscar.

Cutting on the reaction is joining segments so that they explain each other. The man looks up and stares past the camera. The door to the room opens and the woman walks in.

Intercutting is used to explain by detail, to break up the action. The man is writing at his desk, long shot. The paper on the desk and the man's handwriting, close-up. The man finishes writing and stands up, long shot.

Cutting to tighten can be used to shorten the duration of any of the above cuts. The man is writing at his desk for ten minutes. When the scene is cut to tighten, the man writes for thirty seconds, enough time to suggest that the man is writing. Visuals give ideas—there should be no laboring of the point.

Perhaps you've seen books on film or video production that have long-winded explanations of how the author would produce a program, complete with a detailed shooting script, the costs involved, and the expectation that the reader must somehow follow directions. This isn't really the

Video Experience   case when you actually get down to work. Producing a program is a combination of a good media imagination and a sense of equipment technique. You can learn by sitting and watching television, movies, and the video productions of others. Or you can learn by getting a camera, VTR, and monitor and going out and doing it yourself. The latter is the only practical method of approaching video as a creative process. If you've got something to say, you'll find a way to say it, whether you know the ground rules or not. Just be sure to keep in mind that video is a communications event, and speak clearly and distinctly through the medium so you can be heard.

Production Notes
Staging is a crucial part of every production. Every object in a scene will contribute a certain visual effect to the entire scene. You must be aware of the texture of objects and the

confusion they may present to the viewer if they clutter the scene. Try for simple, clean lines. If you have to, create your own backgrounds. They can be made from huge sheets of cardboard, sections of styrofoam, cloth, or even egg carton dividers. Just tack them up to a plywood backing or wooden frame. In b&w recording especially, the background and staging must be as carefully set as possible. The color of objects in b&w will show up as a certain shade on the gray scale. Stay away from very dark or very bright backdrops because they'll mess up the sense of the scene and distract the viewer from the subject.

Backgrounds

If you're working on a long program, try using a Polaroid camera during production to document each scene—on the back of each picture write down when it was taken and what it represents in terms of video footage by noting the reel and the position of the scene on the reel. You can review the material with the photos prior to editing and shuffle them around to help with your editing decisions.

Once you've finished your edited program, you may want to take stills of the production for promotional purposes. To make photos off a TV or monitor, use a 35mm still camera loaded with Tri-X (400 ASA) film, shooting at 1/30 of a second at f 8 or f 11. Set the monitor screen as brightly and with as good a contrast as possible. Use a lens on the camera that will take in the full angle of view of the screen without distortion, but make sure that the camera is placed far enough away from the screen with a long focal length lens—try 100mm—to get the screen and nothing else. Set the camera on a tripod and turn off the rest of the lights in the room.

Stills

If you're preparing a video tape for cable broadcast, a notice at the beginning of the tape to the following effect is helpful: "Dear viewer, if there is any flutter in the picture during this program, please adjust the horizontal hold control on your TV set. It will probably eliminate the problem."

Graphics

Titles are an important part of every production. Make title cards and logo cards on light blue cardboard or blue graph paper. Light blue isn't picked up by the vidicon, so it appears as white. The cards should be about 11" × 14" (38 × 35.6cm). All lettering and other graphics on the title cards must be big enough so that the system can display them with adequate resolution, keeping in mind the safe area limits of TV screens. Never use lettering smaller than half an inch (1.3cm) in height—larger if possible—to be displayed at least half an inch high on the screen.

Mount title cards on easels. The cheap wooden easels available at art supply stores are ideal and cost under $10. Any number of lettering techniques can be used to create

PRODUCTION VALUES

your title cards: black felt marker, stencils, the 3-D movie letters sold for home movie titles, the press-on letters sold for restaurant menu displays, the rub-on type letters that come on a sheet of wax paper and are burnished onto the title card.

The graphics you use must be lit properly so that there are no hot spots or glare. This can be done with small high-intensity lights, incandescent bulbs, or home movie lights if they are moved around so their hot spots don't show up on the display.

Production Equipment

The capabilities of a video system using only the basic equipment of a camera, recorder, and TV set are limited. The addition of a second and third video camera is desirable in some circumstances to add variety to the scene being recorded. When more than one camera is used, each of the cameras may be positioned at a different angle and distance from the subject, allowing a variety of views to choose from while the event is taking place. It would be possible to use three or more cameras and to connect each camera to its own video tape recorder. What each camera saw would then be recorded separately, and later the best segments from each could be edited together to form a program. There is, however, a less complicated and less expensive method used to create a video program when more than one camera is used. This is done by connecting all of the cameras to one video switcher. The signal from each camera goes to the video switcher, where it is monitored; the director of the program chooses one of the camera views of the scene at a time and through the electronics of the switcher holds the other camera signals and sends the selected signal to the video tape recorder. You can use three or more video cameras with the switcher interposed between them and the video tape recorder so that the VTR records only one signal at a time. When the event being video taped is over, there is a finished program on tape ready to be rewound and displayed. All the "editing" has been done as the event was taking place.

Other production equipment needed to create higher quality programming includes lighting units, which will illuminate the scene to its best advantage; additional audio equipment for sound recording; and video processing equipment to maintain the electronic signal at its peak condition as it travels from the cameras to switcher to video tape recorder to TV set.

11

TAPE-O-RAMA

MAKING
VIDEO

The how-to of video is only a prelude to actually making tapes. That is where it all happens, when you put your VTR in gear and point your camera. Most alternative TV is made on "location," and these locations are usually nothing more than the place where the event you want to tape happens to be at the time. I've made tapes in dimly lit bedrooms, sitting in my seat in the middle of a rock concert, taking the lift up the Eiffel Tower, in basement dance halls, in the back seats of cars, and anywhere else where what I wanted to tape was located. That's the thing about video: You often have to go where the action is. The technical problems presented by that fact are different from the technical considerations of a studio environment; it also allows a *cinema vérité* freedom. The combination of technical requirements and the absence of aesthetic boundaries is responsible for good and bad video, depending on how you approach the situation. I think the first time I ever really got into my equipment was one afternoon when I was standing in the middle of a street in London's Soho district, taping the goings on. Suddenly it started to rain. Now there's nothing in the Sony manual about taping under water, but one presumes that the tape gets soggy. On the other hand, what I saw through the viewfinder was a lovely shower and I wanted it on tape. I compromised the technical with the aesthetic by pulling my jacket over the portapak deck and keeping one hand over the camera's built-in mike. It worked fine, although the next time I go out in the rain I intend to put together a clear plastic tube to slide over the camera and extend out past the lens like a little awning.

That may sound like a ridiculous example of how to put equipment and environment together, but it serves to indicate the sort of difficulties you have to deal with on any particular taping. You have to prepare yourself for whatever might possibly happen, and what couldn't conceivably happen but does anyway. There are four basic taping situations: taping with a portable system in a live, who-knows-what's-going-down-next situation; taping with a multicamera unit in such a live situation; taping with a portable deck in a studio situation; and taping with a multicamera unit in a studio situation.

Taking the last situation first, working in a studio presents only a minimum number of requirements. Once you're in the position of having the equipment you need—cameras, SEG, VTR, lights, mikes, mixers, and so forth—you're home. All you have to do is set it up and then move it about until everything looks fine on your monitor screen. You can move the subjects, the equipment, and the setting, eventually creating an environment that looks "real" on the monitor screen.

Taping on Location

Location work is more difficult. You can't set about re-arranging things to establish what looks like reality. You've got to use reality to create reality, and that is often difficult.

The most frequent taping experience occurs when you show up with your gear over your shoulder to tape an event over which you have no control. You're simply the electric voyeur trying to capture what's happening onto your tape. This is the most difficult tape to make effectively, but it's also the most rewarding in terms of the training, the practice, and the sense of technique you'll get as you work your way

I'm often in such situations. I try to lug my gear along to press conferences and similar events to get them on tape. I remember once going to a press conference held by the Beach Boys before they left for Europe to make their "Holland" album. The circumstances created a typical setting for the wild taping you have to learn to do. I walked into the room to find the group spread across a long couch, surrounded by two dozen members of the press, all talking to a different Beach Boy at once. The lighting was poor at best, the noise level was unfortunate.

A Press Conference

The considerations I had were sound and picture. The noise level in the room made it necessary to work as close as possible to the source of the sound I wanted to record. A shot-gun mike held by an assistant would have been ideal, but I had only my portable camera mike. I knew the camera mike was going to make the entire event sound like rush hour at Grand Central Station unless I could place myself as close to the source of the sound as possible and use my body to shield noises coming from behind me. At the same time, the poor quality of the available light meant that a portable lighting unit would have been nice to have. But it was meant to be a "casual" press conference at which everybody sat around and rapped—glaring lights would not have been, shall we say, "groovy." I scanned the room with my camera, watching the light values through the viewfinder. At the far end of the room, just above Carl Wilson, a ceiling light was on, and next to his seat at the end of the couch was a table lamp. I moved down the room and got in as close as possible without sticking the camera in his face. There were two reasons for choosing that camera position: first, I had to get my mike in close; and second, the less dark area I had in my camera angle of view, the more light area and picture I was going to get. I zoomed in and framed Carl's face, which was reflecting enough light from the lamp and ceiling light to get the picture signal out of the gray noise area.

The resulting tape was satisfactory. I had twenty minutes of a Beach Boy talking about the group and its music, with picture and sound that were, at the best and least, satisfactory. It wasn't the tape I'd have liked to have made when I was planning to go to the press conference. But it is the tape that resulted from my understanding of the limitations of my equipment combined with my own aesthetic decisions as to what was of interest within the limited sight and sound range that I could capture.

Taping at a location in which you can't intrude your video presence is never easy. You've got to be totally self-contained and disturb the event taking place as little as possible. There are ways of obtaining the best possible tape under these conditions. The most important is, simply, experience. The more you use your equipment, the more sensitive you'll become to its capabilities. It's like doing anything over and over again, practice makes perfect, or at least it gives you a more professional eye toward the tape that will result.

Lighting   The most difficult factor in wild taping is lighting. Sound can be improved by the quality of the mikes you use and the addition of an assistant to your crew to monitor and direct the sound recording by using a shot-gun mike or well-placed dynamic or electret condenser. But you either use lights or you don't. Most video *vérité* situations are very *vérité*. Setting up even one light can destroy the event, which is taking place whether you're there or not. One way around this is to try to tape as often as possible during the daytime. You can't often ask someone to schedule an event for your convenience, but be aware that light coming in through windows can give you good fill-lighting and allow you to have fine signal-to-noise ratios on your tapes combined with setting up lights. If you have to work at night or in other situations without adequate light, impress upon those who are holding the event when you ask if you can tape that you do need some light—perhaps they'll let you come early and set up a couple of extra lights as long as you promise to conceal them from view (behind drapes or from another room flooding into the room where the event is happening). At the very least you might be able to change light bulbs in the lamps in the room to upgrade their wattage—75- or 100-watt bulbs instead of 60 watters. Every bit helps. Be careful you don't melt any plastic sockets or burn lamp shades with higher-wattage bulbs. You might find that a couple of lamps could be added to the room without making it look set up.

You may be wondering what kind of situations exist in which you couldn't do some extra lighting. Well, there are many, most of them having to do not so much with the event as with the people involved. I've found I have to be especially concerned with lighting (and the lack of it) when I'm taping "at home" events. If people get uptight in front of a camera, they get even more uncomfortable with glaring lights. If you want to tape a birthday party or other family gathering and capture the individuals as close as possible to who they really are, then try staging the event so that the video equipment is as inconspicuous as possible. This means no extra lights that look like "movie" lights, no mikes in evidence, and little video intrusion on what's happening. There is no real solution to the problems involved in these kinds of tapings, and the skill you develop in improving your video of them will be acquired only through practice.

Most tapings are of situations that are set up, one way or another. The least complex of these situations is exemplified by going somewhere to tape someone talking—a one-to-one video tape. I do this quite often and find that it is effective and makes interesting tapes. Once I did an interview *A Static Interview* with John Cale, former member of the Velvet Underground who is heavily into classical music. I wanted to talk to John about his most recent album. I planned to get additional footage of him recording, performing, and practicing, at a later date. For this particular taping I just wanted John to talk about his music. Some of the tape would be used as is, other parts would be used only as soundtrack with other visuals.

I didn't have a crew available, so I had to accomplish three things by myself: record the event, ask the questions, and run the sound. The first was the simplest. It's not really difficult to converse with someone while holding the camera or by setting it on a tripod just behind you and over your shoulder. Try sitting at a table so you can rest your elbows on it to gain more support for the hand-held camera. When you ask questions, move the camera away slightly (still keeping it pointed at the subject) so the subject can see your face when you talk; then concentrate on the viewfinder and your picture when you get an answer.

Have the subject sit across the table from you and set up whatever lighting is necessary to provide good illumination. The subject knows he or she is going to be taped, has agreed to do it, and understands that picture quality can be improved with artificial lighting. You don't have to blind him for this kind of work since there will be little camera motion (thus no **low-light lag** effect). A couple of lights, both overhead and off to the side in the two-light setup described on page 138, will do. Seating your subject next to a window so that the light from outside fills one side of the face and artificial light fills the other is effective in daytime work.

The big problem in taping this kind of setup is the sound. I *Sound* wanted to talk, to get answers, and yet not have to worry about the sound. Since the built-in camera mike on my portable camera was useless for achieving decent quality, I decided on two lavaliers and a small passive two-channel mixer. In that way both Cale and I could wear lavaliers, I could preset the volume levels, and I'd wind up with a soundtrack of excellent voice quality. The subject understood I was taping an interview, so wearing a lavalier was not objectionable. I find that most people prefer lavaliers to having mikes stuck in their faces. They don't think of the lavalier as a mike in that sense and often forget about it after a few minutes. In addition, my wearing a lavalier allowed my voice to be heard on the tape while I still had both hands free to work the camera.

The resulting tape with Cale provided footage that was as good as I've seen done with a full crew. I use this setup whenever the situation permits.

It is possible to work on location with more elaborate setups if you're working in an area you can commandeer for taping and if you have enough assistance to stage a full-blown production.

Sound and lighting are the two major improvements that you can make immediately, *if* you really want to make good video. The extent of improvements depends on the money available for extra sound and lighting equipment and the personnel you can muster to run it for you while you work the camera. It'll cost a few hundred dollars to get the stuff you need, but it's as crucial to your video making as tape or spray to clean the heads. Don't think of a lighting unit as an extra that it'd be nice to have some day. It's standard equipment that's absolutely necessary. The same goes for at least one directional mike.

No matter what the subject or event of your production, consider the action that is taking place as a performance, be it someone sitting in his living room talking or singing songs on a stage or roaming around outdoors. The actual event and action is what you must record on tape, but the preparation for the event and the setting up of your equipment should reflect a sense of theater that may not be implicit in the event you're recording.

A Performance

Two examples might serve to make this kind of production clear. The first is an actual performance, in the sense that the subject is doing something that is directed past your camera lens at an audience—real or imaginary. I once made an hour tape of Lou Reed singing his songs. The setting was my living room and the action centered around Lou and his guitar. Since Lou had agreed to play and sing in a posed setting, I set up my lights prior to his arrival—three light stands in the classic fill-, spot-, backlight setup. I asked a friend to sit in for Lou while I checked the lighting in terms of the area of camera movement I anticipated: moving in and out, from side to side, and so forth. Once I was sure that the camera angle of view and the lighting were in agreement, I set up my mikes and sound system.

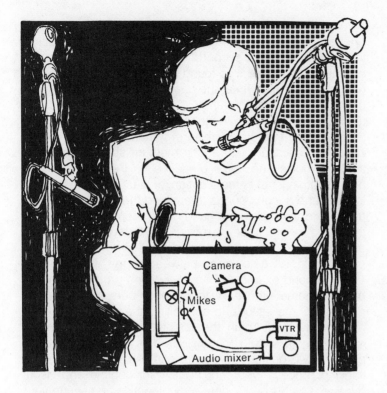

In this case I had to record both a guitar and singing. I didn't have shot-gun mikes, so I chose to use two dynamics on baby booms, one for the guitar, one for the voice. Because this setup made it difficult to hide the mikes, I decided to include them in the picture.

Mike Setup

Facing Lou as he sat on the edge of the couch, I set one mike (for the guitar) to the left, with the boom arm extending in and pointed at the sound hole in the guitar (about three feet—.91m—between guitar and mike). I raised the voice mike as high as possible on the stand and angled the boom arm in and over Lou's head, pointing down toward his mouth. This whole unit was set to his right (about four feet—1.2m—between mouth and mike). A long shot would show the mikes as well as Lou seated and playing; close-ups of his face or guitar would put the mikes out of the picture. So, though the mikes were there, they didn't obscure the performance.

I used a small Sony mixer for the sound, feeding the two mikes into the mixer and then the composite signal out of the mixer into the portable deck. I also added extra cable between the deck and the camera so I could leave the deck on the floor near the mixer and move around with the camera. Since the taping was going to last at least an hour and since the point of it was to capture a musical performance, I also used a tripod. I set up the tripod and mounted the camera on it. My crew consisted of two besides myself: one

friend running the sound mixer and equipped with earphones so he could monitor the two-mike mix as it came
through the mixer; a second friend as an auxiliary camera
operator to spell me. I've found that letting others work the
camera—especially in a one-camera situation—results in a
better variety of camera angles and action than if you try to
run it all yourself. I began the taping with the camera on the

Camera Movements tripod for the first song, limiting my camera moves to one
zoom-in, one zoom-out, and a medium shot. I presumed
that the music and Lou's presence on the tape wouldn't
necessitate a lot of camera action, at least initially. Between
the first and second songs, I removed the camera from the
tripod and worked with it hand-held. Then I put it back on
the tripod for other songs, moving the tripod position occasionally. Of course all this camera resetting was possible
since the tape was stopped between songs.

At-Home Interviews The second example of an on-location production might
be an at-home interview with a famous personality who will
give you some latitude in controlling the audio and video
aspects involved. Again, the personality and what he or she
is going to say should be considered as performance. If the
personality is going to give you a glimpse of his lifestyle and
thinking process, you should allow for audible and visual
dynamics on his part. Choose your location and set up your
lights. If the person is going to move around, set the lights
so that they don't intrude and so you can move the camera
and action around without catching a spotlight in the lens.
This can be done by setting the lights up as high as possible; use little Lowell lights and tape them to the tops of doors
or bookcases with gaffer's tape.

Next, decide how you're going to tape the sound. If the
person is going to move around a great deal, a lavalier or
dynamic just won't work as it did with John Cale. You'll have
to invest in a shot-gun mike. Another solution would be to
conceal a number of mikes around the room, which, as your
subject moved about, could be turned up or down through a
mixer to capture the sound, but this would be tricky at best.
A shot-gun is ideal. The sound operator can be stationed at
one position out of range and can simply follow the personality's movements with the mike, monitoring the sound
through a pair of headphones from the audio signal coming
out of the portable deck.

You should put the portable deck down and use an extension camera-to-deck cable so that you have total freedom to follow the personality around his or her environment.
Since you might be moving a great deal in this kind of situation, I'd recommend that you use a body brace to keep the
camera from wandering or jittering.

In all of these taping situations, overshooting is a necessity. I mentioned it in Chapter 10, but I'd just like to em-

phasize that you *must* overshoot to come up with a final tape that is ready for editing. Always start the recording at least six seconds before the action begins and let it run six seconds after the action has ended. This is overshooting for tape stability. Always record everything you think will be of interest, no matter how marginal you feel the footage might be, since it's easier to erase the tape if you have too much than to try to go back to do more because you have too little.

Taping in a Studio

It was a quite different experience when Bob Gruen and I made a tape of Blondie, which was used on NBC's "Rock Concert" and by TV networks in Australia and Japan.

Our budget allowed us to rent a production studio for eight hours. We wanted to get three songs on tape in that amount of time—eight hours to wind up with ten minutes of finished tape. Because it is virtually impossible with TV studio audio equipment to record a rock band playing live (don't believe it if they tell you different), we chose to lip sync the three songs from the band's album.

TAPING BLONDIE
(PHOTO: KAREN LESSER)

Video wiz John Brumage readjusted the color levels of the three Philips cameras we used so that lead singer Deborah Harry's skin was really pink, so reds were bright red, and blacks very deep black. Gruen and I spent a good deal of time arranging the set—a giant black-and-white blow-up of Deborah's face hanging behind the band; TV monitors showing clips from other "Rock Concert" shows; and two-foot-high Japanese robots stationed as props around the band's drums and guitar amps. For lighting, we worked with the lighting effects available in the studio.

Since the band hadn't made a tape before, we taped one

MAKING VIDEO

203

song, let them watch it, then sent them out to do it again—using the same technique as is used in recording a rock album. The band developed camera moves on the spot. Gruen worked one camera "video-style" on a shoulder brace; the two other camera operators used a conventional TV approach, working with their cameras on tripods with dollies.

A large hi-fi speaker in the studio—so the band could hear the lip-sync sound really loud—completed the arrangements.

Taping, whether you're just pointing the camera at yourself in a mirror or shooting the high school play, is a trial-and-error process. You learn as you work with your equipment. Just remember the variables you have to work with and try to compensate with additional equipment for any deficiencies in the setting, event, or subjects. It gets down to good lighting, good sound, and good camera work—not complex but requiring an appreciation of the video process, its advantages and disadvantages as a recording medium, and what you can do to upgrade the audio and video as you record.

With a certain amount of work and time you'll absorb the impact of video and begin to think of it as another facet of your daily lifestyle, as available as you want to make it. You'll become one of those video pros, an experienced media mechanic who can "talk" video. Mainly, it's a step forward, not that much more exciting than everybody getting an audio tape recorder for their home in the early sixties. Except that now the people are different; they have a little more of a sense of what media's essential nature is and what it can do.

This book is stuffed with information relating to one's first step into video. Everything in the world now has an instruction manual, which, if followed correctly, will provide new experiences until the batteries run down. So, since you've discovered video, you must wade through the coaxial connections, develop a casual expertise at head cleaning, practice until you can thread tape without looking at the threading diagram, and eventually get familiar enough to run your video like you run your telephone, TV set, or tape recorder. And with as much acceptance of the video machine as just another one of those entertaining, labor-saving devices.

This may seem to be at odds with the underlying drive some of us have to create visual experiences through video. It really isn't. Video isn't the movies or even the big TV production lots. Ultimately, video will be a miniaturized version of audio/visual communication. You'll make a video tape the way people write books today, assembling your own personal message to be added to the universal information bank. Those who want to see it will have access to it. Maybe some of the people who make the tapes will even earn a living doing it. The outlet will be there: a media form that provides an effective mode of expression with a minimum of trouble. (The book on how to make real television is thicker than this one.)

I can spend pages describing countless true-to-life, believe-it-or-not video stories, but that's just video shop-talk, nothing that you won't be able to figure out on your own as you begin to use the equipment. The important thing is to use this book so you can, literally, learn everything there is to know about video and then forget about it. Do the basic work it requires for something to become second nature. Video techniques have to be learned until they're unconscious habits. At that point—when you've done all the interviews that got messed up because of poor sound, poor lighting, poor direction; when you've learned what you need in order to make the video you want to make and you've figured out how to pay for it—you should relax, take a breath, and really get into video. Pretend you're interested in something else for a while. Watch some television. Talk to a friend. You know, calm yourself and reevaluate video, what it is, and what it can do for you. If the answer is, "Well, I

CONFIDENTIAL PRODUCTION ESTIMATE

Video Tape Production Breakdown

Video tape production of three songs, using lip-sync, pre-recorded sound.

    Five hours color studio time
    Two broadcast-quality color cameras
    Make-up artist
    Robinson-Gruen production team
    Four-person crew
    Lighting
    Tape Stock
    Post-production editing
    Master tape
    Three U-Matic cassette copies

                                              $3,672.
    Each additional U-Matic Cassette copy   *50.*

Transfer of 2-inch Quad video tape, approximately $100 for 15 minutes, plus approximately $6 per minute for 2-inch tape stock.

Transfer to 16mm color optical print film, approximately $300, plus approximately $60 per ten-minute print.

Terms: $2,000 deposit three days in advance of taping date; balance due on delivery of master tape.

VIDEO TAPE PRODUCTION PROPOSAL

Important: The following is as accurate as possible, but adjustments may have to be made pending price changes.

Video in Audio Record Studio Location

Client supplies audio recording studio and all related costs.

8 hours on location, including 2 color cameras, transport of video equipment, technical director, lighting, lighting director, camera operator, special effects generator, recording on 3/4" U-Matic system, director.

$5,105, plus tax

Same as above using 3 color cameras and additional crew.

$5,905, plus tax

Video in Location on Street

To tape on street, at home, or whatever.

8 hours on location, including 1 color camera, transport, operator, director, U-Matic recording system, etc.

$2,500, plus tax

Video Recording in Video Studio

This would use prerecorded band track with voice-over or lip-sync to prerecorded audio track.

8 hours in studio with 3 color cameras, lights, crew, director, etc., recorded on 2-inch broadcast Quad tape.

$6,900, plus tax

8 hours same as above, with 2 cameras using U-Matic instead of 2-inch Quad recording, also broadcast quality.

$5,100, plus tax

8 hours same as above, with only 2 color cameras using 2-inch Quad recorder.

$5,150, plus tax

8 hours same as above, with 2 color cameras and U-Matic recording.

$4,400, plus tax

Editing

Editing of materials into program.

$150 per hour plus tax

Transfer to 2-inch Quad from U-Matic

If shot on U-Matic, must be transferred to 2-inch Quad for US TV broadcast.

$250 per 15 minutes

Transfer to 16mm Color Sound Film, Optical Print

First ten minutes.

$400, plus tax

Each addition ten-minute segment.

$100, plus tax

like to tape movies off TV," then fine, use it for that. There's nothing wrong with that kind of approach. Video is a home entertainment item, after all. But if you see that video is a way to create some new form of information and that the quality of that information can be directly determined (for better or for worse) by the people making the video, then you should pursue that vision. At that point you won't need this book, or the instruction manuals, or news about what the next Sony portapak is going to look like. You've decided to become a media person and are worried about other things. The only way to get to that point is to use your video equipment.

VIDEO TECHNIQUES 2

It's advisable to be familiar with the principles and terminology of video even though making a tape isn't terribly
dependent on understanding how video and television
work. There are a number of systems and standards that
govern the functioning of video equipment, and you may
find yourself with incompatible pieces of equipment if you're
not aware of what works with what.

Like most electronic and electro-mechanical equipment,
video hardware is described by its ability to perform certain
functions. You must know what the terms mean and what
they imply as to the quality of the image that will be produced and what you can do with that image.

12

WHAT
YOU SEE
IS HOW
YOU LOOK

HOW
TELEVISION
WORKS

You've probably spent about as little time wondering how your TV works as you have pondering the internals of tape recorders, radios, and the other electronics that surround you—just assume it works and turn it on. Once you get involved in making video, your TV set will take on extra dimensions, it will become incorporated into the rest of your video gear, and you suddenly will realize that you don't know very much about it except what it does. You can get along quite well without understanding how the picture gets on the screen, just as you can use a cassette deck by following the numbered instructions in the Sony manual. But if you really want to delve into the potentials of video you'll have to gain some familiarity with your television set as an electronic device capable of doing certain things. This knowledge becomes especially essential when you begin to work with monitors and monitor/receivers since they are hybrid TV sets which require more attention than your home set. What follows is an explanation of the technical mechanics of TV. You don't have to memorize it, but I'd suggest that you read it carefully so you can begin to understand the general principles involved. As you work with your video equipment you'll find that the basic ideas presented here will begin to make more sense; the practical applications of this TV theory will reveal themselves.

Most of us were introduced to the principle of **persistence of vision** as children. If it isn't one of your school memories, try it now. Take a small, blank notebook and on the same spot on each page draw a stick figure. The posture of each stick figure should change slightly from page to page—an arm can be drawn in a series of positions from being held up to resting at the figure's side. When you have completed the series of drawings, take the notebook and riffle the pages rapidly. What you see is motion. The arm of the stick figure appears to be moving as the pages flip by. The phenomenon that allows this apparent motion is persistence of vision. The eye perceives each drawing of the stick figure and retains the individual impression for a period of time that is long enough to overlap the next perception. Motion is created out of the rapid exposure to the eye of a continuous series of still pictures. Physically, what is happening is that the retina of the eye is holding each image for about a 1/30 of a second. If a continuous series of still drawings or photographs is exposed to our eyes at a rate of speed faster than that, we will not see the individual pictures but instead blend them into the perception of motion. The motion we perceive is basically an optical illusion. The illusion of this illusion is that we perceive it as motion.

In the stick-figure flip book, the frequency with which one image is replaced by the next may not be sufficient to allow for a smooth sense of motion. We may see a certain amount of flicker, which is similar to the perceptible flicker in old-time silent movies whose jerkiness can be seen on the screen because the camera was hand-cranked and did not achieve a stable frequency in the change.

Animated cartoons provide another example of the persistence of vision. A ten-minute Mickey Mouse cartoon is composed of 14,400 individual drawings. What you view as the finished product is the continuous flow of the drawings, one replacing the next at the rate of twenty-four frames per second. This 24fps, twenty-four pictures for each second of viewing time, is the standard for 16mm film; 24 times 60—or 1,440—pictures go into one minute of movie motion. In an animated cartoon, each successive drawing is photographed onto a frame of film. In human action on film, the same principle holds true—the movie camera photographs twenty-four still pictures each second, 1,440 still pictures each minute, in sequence on a long strip of film that is divided into individual frames. The movie projector projects twenty-four still pictures at the same 24fps rate, re-creating the sense of motion that was captured by the camera.*

If all this seems either thoroughly confusing or highly improbable to you, get a strip of movie film and examine it. You'll see the series of small individual photographs I've been talking about and the way they change slightly from one to the next.

The television picture that you see on your home TV screen also relies on persistence of vision to create the impression of motion. The TV screen is a device that has been designed to work in conjunction with the human eye and brain to achieve its reality as an electronic screen that displays information we can comprehend. What you see on the TV screen is a combination of what is displayed on the screen and the way your brain works in relation to what is being displayed. Tv is a bi-sensual phenomenon depending on the human sense perceptions as much as on electronic principles.

Not to keep you in suspense any longer, a TV picture consists of sixty pictures every second, one rapidly replacing the next on the screen. Since the theory of persistence

PERSISTENCE OF VISION

*Twenty-four frames per second is the standard for 16mm film. Other film widths have other fps standards. It should also be noted that in the projection of any film, a shutter is used to further reduce the possibility of any flicker in the projected image. This shutter cuts off the stream of light from the image twice during the projection of each frame so that forty-eight separate pictures are projected each second, each frame being projected/interrupted/projected/interrupted/replaced by the next frame.

HOW TELEVISION WORKS

of vision is at work, the more pictures produced each second, the less likely it is that the human eye will perceive any of the separate pictures. The more flickers that occur, the less flicker we see! Television is of better quality than movies as *motion* picture, although in present, practical terms, the TV picture cannot be enlarged to the size of a movie screen without a serious deterioration in quality. That is only a problem of the moment, and the principles needed to solve it are already understood by those involved with the development of television communications hardware.

The picture portion of TV is based on interrelationships between electricity and light. The audio portion deals with similar relationships between electricity and sound. The capture, transmission, and reproduction of light and sound by electrical means is the purpose and function of TV.

Audio is dealt with in Chapter 8. It comprises at least fifty percent of television as a medium, but the concepts of sound reproduction and broadcast are somewhat easier to grasp because of our familiarity with radio, records, and other sound carriers. You may have trouble understanding just how a television picture is formed, transmitted, and re-formed. The basic principles aren't terribly complex, but most of us have an extremely limited foundation for understanding the sophistication of electronics, since we don't know how electricity works. The established parameters within which TV functions are such that even those schooled in the basics of electricity and television electronics have trouble understanding the results of the quantum leaps that have taken place during the last twenty years in the drive toward the technical perfection of television.

I don't propose to start an explanation of television with an explanation of the flow of electrons, which results in electricity. First of all, such an explanation would take up about half this book. Second, there is no practical need for the video maker to understand the theory of the universe and atomic structure from which the more particular theories of electronics with which we must deal have been postulated. Unless you're willing to devote a couple of years to the serious study of electricity and electronics, you'll have to find a starting point and take on faith all the facts that precede that point. There is nothing wrong with that—the great majority of those dealing with electronics as media tools have done the same. By more or less starting at the finish and working backward, you will, over a period of time, begin to understand more and more about why electronic apparatus functions as it does. The explanations that follow are the definitions of *what* particular components do rather than exactly *how* they do it. In electronics, what a component *does* is pretty much what it *is*, a fact you'll discover if you explore the development and principles of the components that make up the equipment you're using.

Any explanation of the workings of television must begin with a definition of the word **component**. An electronic component is a piece of equipment that performs a certain electrical function. A video camera is made up of a number of components, which include a vidicon tube, a video amplifier, pulse generators, deflection coils, and other units, each designed to serve a specific function. In turn, each of these components is a combination of smaller components. The video amplifier, for instance, is made up of transistors, resistors, capacitors, and other basic components, assembled in a manner to create a particular effect—in this case, the amplification or strengthening of an electronic signal.

Unfortunately, it is impossible to see what components do. You can't watch the flow of the electronic signal as it travels out of the vidicon tube into the video amplifier. There are test instruments that can tell you it's there and happening, but you can't see the signal actually being amplified by the components of the video amplifier. You can know what each of the components does, but you can only prove that the component is doing what you think it is by removing that component and observing that the function is no longer being performed. It's sort of like saying a car runs on gas because if there's no gas in the tank the car won't run; or proving that a TV set works by electricity by unplugging it from the wall. Not being formally schooled in electronics, I've come to deal with all this by thinking of the flow of electricity as similar to the flow of water. Water flows in a stream the way electronic impulses flow along a wire. If the water is sent through a funnel into a more confined space, the quantity of water will not be altered but the force (speed) at which it travels will be; if the water encounters a dam in the middle of the stream it will stop. Similarly, an electronic signal can be channeled through various electronic components such as resistors and capacitors, which effectively alter its state by placing certain barriers in the signal's path.

The electronic components used in television have been chosen because they perform specific functions. Television equipment isn't made out of spare parts. There is a reason for each component being where it is, and that fact of conscious assembly of components to create a whole makes TV comprehensible no matter how little prior electronic knowledge you have.

In 1908, a British gentleman, Alan Archebald Campbell Swinton, was the first to propose the all-electronic system of communication that later became TV. And for more than fifty years before that, attempts had been made to create what we know as television simply because people could envision the end result and apply what they knew in an effort to achieve it.

HOW TELEVISION WORKS

The Camera/The Screen

Television transforms a scene from reality into a corresponding series of electronic signals, transmits those signals over a certain space, and then reproduces the signals in such a manner that the scene can be comprehensibly re-created. There are certainly easier ways of accomplishing this than what has been developed as television, but none of them fulfills all of the requirements that are part of TV as a concept. There's telepathy, which is not particularly dependable at the moment. There's sound film, but although it can store and reproduce a scene, it can't be transmitted in and of itself unless it's put on a plane or train and carried from its place of origin to its place of display—not an instant process. There's theater, which involves moving the scene from place to place and having the scene repeated for each audience wanting to see it, but that limits the potential audience to the availability and size of an auditorium. The concept of television is unique, and its development was dependent on the fact that it is an electronic process using electrical signals to carry the scene from the point of origin to the point of display. The future may see the replacement of certain aspects of the television process by principles other than electricity, such as laser projections, but at the moment electricity is the basis of the TV system.

Recent advertising of TV sets by American manufacturers has often centered around the claim that they produce totally **solid state** television sets. Solid state means that all of the active components inside the television set are transistors, or integrated circuits, which function when an electrical current is applied to them. They do not have to be heated up to function and contain no gas in which their internal elements have to live in order to work. They are made out of solid materials (not gaseous or liquid) that modify voltage signals applied to them. If the manufacturer is being totally honest there will be an asterisk after the "solid state" claim and in small print we will be told that the TV is all solid state except for the picture tube.

Vacuum Tube   The original mass production of radios and televisions depended on the creation of components known as vacuum tubes. A light bulb is one form of vacuum tube. When these tubes are made, all the air is pumped out of them and they are often filled with gases. At the bottom of the tube is a connection through which the electrical signal charge is applied to the internal portion of the tube, usually some form of metal element mounted in the middle of the tube. The tubes and their elements are heated by an electrical current, and they modify the electrical signal sent through them according to their design. The light bulb has a metal element in it that glows when enough electrical current is applied to it. In the open air the element would quickly blacken and burn out, but in a vacuum tube it remains in its stable, glowing state for an indefinite period of time. Vacuum tubes were used as the components of TV sets until the invention of the transistor in the late 1940s. Whereas a vacuum tube is large, not totally dependable, and gives off a great deal of heat (necessitating a certain amount of space around it), the transistor is a solid state device. It is also less than one hundredth the size of the vacuum tube whose function it replaces. Many transistors can be packed together in a small space, yet produce the same results as the larger, more fragile tubes. As a bonus, the transistor requires much less electrical power—most of the electricity that is used to make a tube work goes into heating up the tube's element. Integrated circuits and large-scale integrated circuits (*ic*'s and *lsic*'s) have recently begun to replace the transistor. These refinements of the transistor allow a miniaturizing of components while maintaining the transistor function.

Transistors

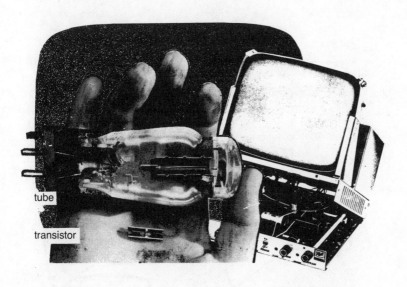

tube

transistor

The TV picture tube is the last hold-out in the solid state world of electronics. Eventually it will be replaced, possibly by liquid crystals or some other creation still in the laboratories, but right now it and its counterpart in the video camera are the two tubes to be initially dealt with if we are to understand how the TV and camera work.

Television, internally, is an electronic medium. But at each end of the television system there is a translation of light values either to or from electronic signal values. The video picture signal is an electronic reading of the light values of a scene. The lens at the front of the video camera collects the varying intensities of light reflected by the objects in the

HOW TELEVISION WORKS

scene and forms an image of that scene as light and dark and the variations of gray in between. Chapter 6 explains this action more fully, and Chapter 17 contains an explanation of the addition of color values to this reading. For the moment, we are concerned only with the black-to-white light value image that comes out of the far end of the lens and is read by the TV system and converted into a series of electronic signals. The two vacuum tubes that are essential to present-day TV are tubes capable of sensing light values in terms of electronic signals; one tube converts the light into the signal, the other reconverts the signal into light values.

CRT   Each of these tubes is called a **cathode ray tube** or CRT. The camera cathode ray tube translates variations in the intensity of light into variations in the intensity of an electronic signal. This CRT is placed directly behind the camera lens so that one end of it is exposed to the image formed by the lens. The surface of this end of the CRT is coated with a light-sensitive element. At the other end of the CRT is another

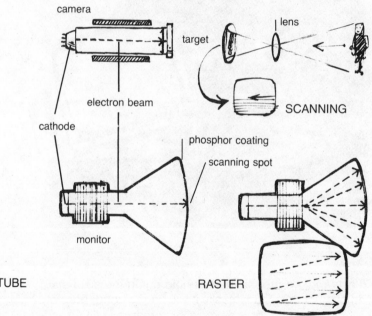

camera

target

lens

electron beam

SCANNING

cathode

phosphor coating

scanning spot

monitor

CATHODE RAY TUBE

RASTER

element, which is known as the cathode. The cathode is heated by the application of an electrical voltage and, as a result, emits a stream of electrons, which travel down the length of the tube and strike the light-sensitive element.

This is the basic action of the cathode ray tube: a flow of electrons from one end to the other. As these electrons strike the light-sensitive coating on the end of the CRT, the amount of light that has fallen on that end is measured as an electronic signal and the strength of this signal is used as an evaluation of the corresponding light values of the image formed by the lens.

The signal is then sent out of the camera to the TV set. In the TV set is another cathode ray tube, which also has a cathode at one end of it and a light-sensitive surface at the other. The electrons are emitted by this cathode and sent along the tube until they strike the light-sensitive surface, causing it to glow to various degrees, which correspond to the strength of the signal. The result, as viewed on the surface or "screen" of the CRT, is the image we know as the television picture.

The mechanics of the CRT and the manner in which the beam of electrons is directed at the face of the tube is in keeping with a predetermined system. Both the camera and the TV CRT use the same system so that the capturing of the light image and its reproduction coincide and the integrity of the image is maintained throughout its travels as a series of electronic signals.

As the electrons leave the cathode they are focused into a beam in much the same way a powerful light source like a spotlight can be focused into a beam of light. The beam of electrons is then directed in an ordered manner down the length of the tube toward the light-sensitive surface at the far end of the CRT. The end of the beam that hits this surface is deflected by magnetic fields to which it is sensitive so that it covers the face of the tube in a pattern. The act of tracing that pattern is called **scanning**, and the point of the beam hitting the face of the tube is called a **scanning spot**. The pattern itself is known as a **raster**. The way your eyes are reading this text can be used as an illustration of the scanning spot as it scans the raster. To read a page, you start at the upper lefthand corner and your eyes travel across each line, determining the information on that line. When your eyes reach the end of each line at the righthand side of the page, they return to the lefthand side of the page, moving down slightly as they return, and begin to read the next line of copy. If you drew an imaginary line from the center of each eyeball to the particular letter on the page on which the eyeballs were focused at any particular moment, you would duplicate the action of the scanning spot as it is emitted from the cathode at one end of the tube (the imaginary line is emitted from your eyeballs), as it travels to the other end of the tube (travels the distance between your eyeballs and the printed page), and then strikes a particular spot on the scanning surface or raster (comes to rest on a particular letter of a particular word on the page). The movement of the eyeballs and the imaginary line extending

Scanning

HOW TELEVISION WORKS

from them is the same as the movement, or scanning, of the scanning spot from left to right. The area of the printed page that contains the printing is the same as the area of the face of the CRT that contains the light image formed by the lens and called the raster.

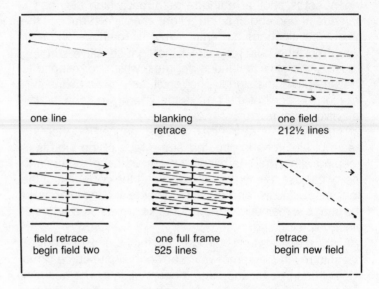

one line

blanking
retrace

one field
212½ lines

field retrace
begin field two

one full frame
525 lines

retrace
begin new field

If you leave the book open on a table and step back ten feet or so, you'll see that the letters on each line blend into each other, looking at a distance like solid black lines. If you move back an even greater distance, the solid black lines and the white spaces between them will blend together into one gray mass.

A video picture signal is made up of a series of lines, one below the next, in much the same fashion as the lines of print make up the page of a book. The scanning spot scans the light values of the image from the lens that is formed on the surface of the face of the CRT. Starting at the upper lefthand corner, it moves across the raster to the right, in a slightly downward direction, until it reaches the righthand edge of the raster. Then it returns to the left and scans another line across the raster to the right, just below the area scanned previously. A total of 525 lines are scanned in this manner to form a complete picture. At the end of the last line scanned, the beam's scanning spot is at the bottom of the raster area. It then returns to the top lefthand corner and begins to scan the picture again. It takes only 1/30 of a second for the entire raster area to be covered once by a series of 525 scanning lines.

The end of the CRT at which the image is formed by the lens is the crucial part of the CRT for the purposes of this explanation. The light-sensitive element is a photoconductive material, which coats the flat surface of that end of the CRT. This area is known as the **target area** in terms of the electronic function it performs, and as the *raster,* in terms of the pattern formed by the scanning spot. The target is where the light variations of the scene are focused by the lens and where the electron beam coming from the cathode is also focused by the various magnetic coils that surround the CRT. The target area has an electronic voltage applied to it, which varies with the amount of light striking it through the lens. The beam hits this target as it traces its raster pattern and completes a circuit—each spot that the beam hits has a particular voltage according to the amount of light hitting the target at that point. The voltage value becomes the electronic signal for that point and is then sent out of the camera through a series of amplifiers and processing equipment to the TV receiver.

The tube of the TV set works in the exact opposite direction of the camera tube. The voltage variations of the electronic signal fed into the CRT of the TV set are sent out of the cathode as an electronic beam toward the inside face of the TV screen. The inside of the screen is also a raster/target area. It is coated with a phosphor, which glows when it's hit by the electron beam. The amount of glow (seen as light on the other side of the screen, the viewing side) is determined by the strength of the beam (the voltage of the beam as it strikes any particular spot on the screen raster). Again the beam's scanning spot starts at the upper lefthand corner of the TV screen raster and traces its zigzag pattern across the screen until a total of 525 lines have been traced during a period of 1/30 of a second. The result is a set of glowing dots making up a set of glowing lines making up a glowing picture, which varies in intensity from place to place depending on the light values of the original scene as they were recorded by the camera CRT.

At this point I suggest that you have a look at the screen on your TV set. If you turn it on and look at it from only a couple of inches away, you'll be able to see that the screen is indeed divided into a series of parallel horizontal lines. You can count them and they'll add up to about 490—the other 35 lines are outside the viewing area or used during the time when the beam returns to start a new trace.

If all this makes sense, good. Unfortunately, there's one more step that takes place in the picture scanning process in both the camera and TV. Let's go back to our persistence of vision concept. Remember that our vision of any one instant is maintained for only about 1/30 of a second before it has completely faded away. Since the CRTs used in video scan their target areas only once every 1/30 of a second, it is more than likely that we'd see a flicker in the picture if the thirty-pictures-per-second standard were used as the fre-

HOW TELEVISION WORKS

quency for picture change to obtain motion on the TV screen or to record motion via the video camera.

To remove the possibility of flicker, the scanning system scans only half of the 525 lines in each pass of the scanning spot from upper left to lower right of the target area. It does that twice every 1/30 of a second so that 525 lines are scanned in total, but only 262.5 lines are scanned every 1/60 of a second. Given 525 lines as the total horizontal resolution of the video camera and TV CRT, the first sweep of the scanning beam is of lines 1, 3, 5, 7, 9, 11, and so on, to line 525. Then the scanning beam returns to the top of the raster and scans lines 2, 4, 6, 8, 10, 12, and so on, to line 524. At that point the scanning beam returns to the top of the raster and begins scanning again with line 1. Because 262.5 is not a whole number, the final scanning line of the odd-numbered lines is not a whole line; the scanning beam stops in the middle of the final line and travels back to the top of the raster, where it then starts scanning in the middle of the first of the even-numbered lines to get two sets of 262.5 lines, or 525. This is done to avoid creating confusion between the odd- and even-numbered scanning lines as the raster is scanned by the scanning spot.

The complete scanning of the entire raster pattern area by the scanning spot takes 1/30 of a second, but the raster is actually scanned twice, or every 1/60 of a second. Each scan covers half the target area and does deliver a picture, although it is not as detailed as the picture that results every 1/30 of a second from twice as many lines. The scanning of 262.5 lines every 1/60 of a second is known as a **field**. The scanning of the 262.5 odd-numbered lines plus the 262.5 even-numbered lines for a total of all 525 lines of the raster is known as a **frame**. Two fields make a frame, two fields make the complete picture. Each field is one half the total light variation information of the picture spread across the total scanning area. Since the two fields are fed to the eye at a rate of 1/60 of a second each, one alternating with the other, the eye cannot perceive the field-by-field sequence that is taking place. Instead, the eye sees the entire frame, or complete picture, for a moment (1/30 of a second) before it is replaced by the next two fields. A high rate of motion stability is assured because of persistence of vision and the fact that new information is being fed to the eye every 1/60 of a second, even though this information is only half the total information the camera tube and TV screen are capable of delivering.

The process of scanning alternate lines on each pass of the scanning spot across the target area is known as **interlace** scanning. There are two forms of interlace scanning: random and **2:1 interlace**, discussed on pages 240–41.

It is important to note that the interlace scanning process

of the 525 lines that make up the electronic video signal is exactly the same in both the CRT of the video camera and the CRT of the TV. This is a necessary equality, but it is also one that makes it much easier to understand all the processing and controls that are applied to the signal between the time the light image from the lens hits the CRT of the camera and the time that the light image is reproduced on the face of the TV screen.

The Video Signal

An examination of the structure of the video signal, how the video signal is controlled to maintain its parameters, and how the video signal is conducted out of the camera to the receiver are the next steps in understanding the technical video process. If you have had any trouble understanding the basic scanning process, review it now, since terms like *scanning, electron beam,* and *raster* will be used without being defined again. Those terms relate to the internal components of the equipment. We will now concern ourselves with the second level of components which the internal components constitute—the governors of the video system, which will be under your control and which you'll have to manipulate as you use your video equipment.

The frequency with which the scanning of the lines of the field and frame takes place and the frequency with which each complete field and frame takes place are our first foci in understanding the controls applied to the video signal by the ancillary components that surround the CRT in both camera and TV set.

Frequency

Standard household alternating current, called **ac**, reverses the direction in which it flows 60 times every second. This is expressed as 60Hz; **Hz** or **Hertz** is just a laboratory term meaning cycles per second.*

The 60Hz of the ac power line is the reference frequency to which video equipment is coordinated or locked. We already know that the frequency with which each field of the video picture changes is every 1/60 of a second. Thus there are sixty fields of information scanned by the CRT in the camera every second and sixty fields of information scanned by the CRT in the TV set every second. The **field frequency** of the American TV system is 60Hz. Two fields of picture information make one complete frame of picture information, so there are thirty frames of information scanned every second. This makes the **frame frequency** 30Hz.

Besides the frequency with which the fields and frames occur, there is also the frequency with which the lines occur

---

*Hz, named after H. R. Hertz, the German physicist who discovered radio waves, is the term internationally agreed upon for frequency, the number of times something happens in one second.

during scanning. There are 525 lines that have to be scanned within a period of 1/30 of a second. Frequency is measured by the second. We must multiply 525 lines times 30 to obtain the total number of lines that will be scanned during one second. This makes the **line frequency** 15,750Hz. In other words, 15,750 separate lines are scanned each second.

Math majors and those owning pocket calculators will observe that, given the line standard of a particular TV system, such as the 525 lines in the American TV system, and either of the two frequencies, line or frame/field, it is possible to determine the other frequency.

In video, line and field frequency are described also in terms of the horizontal and vertical direction that they take. The **horizontal frequency** is the line frequency, since the scanning action moves in a horizontal direction from one side of the raster to the other. The **vertical frequency** is the field frequency (or frame frequency), since the scanning spot must move from the bottom of the raster to the top again in a vertical direction as it marks the end of one field and the beginning of the next.

Two controls are introduced to maintain these frequencies so the scanning spot moves across the raster a certain number of times per second and moves from the bottom to the top of the raster a certain number of times per second (15,750Hz and 60Hz respectively). For the vertical movement, or field frequency, the oscillation of the alternating current coming from the ac power line is used as a reference. If the field frequency matches the ac frequency, 60Hz, we know that the vertical field frequency is occurring the proper number of times each second. Since every electrical outlet in America provides a 60Hz alternating current, any piece of video equipment plugged in anywhere in America can be locked to the same vertical frequency. A video camera in a New York studio and a TV set in Amarillo, Texas, will both be synchronized in terms of the number of fields per second that they originate.

The line frequency is a little trickier. Crystals are used which oscillate at 31,500Hz. These crystals are internal components of the video camera and the TV set. The 31,500Hz signal is electronically divided in two, which yields 15,750Hz* and this pulse is used to control the line, or horizontal frequency, of the scanning beam. To make sure that the crystal is vibrating at the proper rate, a check can be made by dividing the horizontal line frequency pulses by 525 (don't bother). If it equals 60 and so matches the power

*The reason for using a crystal that oscillates at twice the frequency needed to govern the line scanning process and then dividing it electronically is that the faster a crystal oscillates, the more dependable it is, since crystals are sensitive to atmospheric conditions.

line frequency (60Hz), everything is running properly. In fact, it is possible to do away with reliance on the ac power line 60Hz signal to determine the field frequency by using the line frequency control and electronically dividing the 15,750Hz pulse by 525 to get the field frequency pulse.

The output from these frequency oscillators is sent to **deflection coils**, electromagnets surrounding the cathode end of the CRT. There are two sets of coils, horizontal deflection coils and vertical deflection coils. Each coil governs the direction and speed of the electron beam. The horizontal coils create a magnetic field, which moves the beam from left to right at a scanning frequency of 15,750 lines per second. The vertical coils create a similar field, which moves the beam from the bottom to top of the raster area at a frequency of 60 fields per second.

The rate at which the line and field frequencies occur is known as the **time base** on which video operates. **Time-base stability** is a major factor in the functioning and compatibility of video equipment, and a possible failing of inexpensive video recorders. Time base is one of those typical video terms that can be confusing since it is really less complex than it seems. The time of one field of video information is 1 divided by 60, or 0.01666666 of a second. Expressed in milliseconds, the time of one field of video information is 16.7 milliseconds. So for each second, the field time base is 60 fields per second, or 60Hz. There would be a lack of time-base stability if we counted the frequency of the fields over a period of one second and found that it was not 60Hz. As with all other functions of the video scanning system, there is not only a field time base but also a line time base. The time base of the lines is the time it takes the electron beam to scan one line, or 16.7 milliseconds divided by 262.5 lines, since there are 262.5 lines scanned every field and one field takes 16.7 milliseconds to be scanned. The line time base is 0.063619 milliseconds. In practice, you'd probably go crazy trying to time the individual scanning lines to see if they started and finished within a period of 0.06 thousandths of a second. But that they do is the basis of line time-base stability.

Having established that the electron beam is focused into a scanning spot, which scans the target area of the cathode ray tube's faceplate in a pattern called a raster for a total of 525 horizontal lines every 1/30 of a second, and realizing that the time spent scanning each of these lines and each of the fields that they form is the basis on which the formation and the stability of the picture depends, there are two more major factors to be considered before we can allow all this technical nonsense to sink into our subconscious.

The scanning spot, as we have noted, moves from left to right along a horizontal line and then returns to the lefthand side of the target area to scan the next line from left to right.

The motion of the spot returning from its left-to-right scan and the time that return takes are known as the **retrace** or **flyback** period. Returning to the analogy of eyes scanning the page of a book, we can see that retrace occurs as the eye reaches the last letter of the last word on the right end of the line being read/scanned and then moves rapidly back to the left side of the page and down slightly to begin reading the first word of the next line. The scanning spot does the same thing. But, unlike the eyes, which remain open during the retrace period, the scanning spot is turned off during retrace so that no signal is sent out of the camera during that time. The turning off of the scanning spot between the end of one line and the beginning of the next is known as *Blanking* **blanking**. The scanning spot signal is blanked out between lines so that only the orderly reading of the target area, line by line from left to right, is used to create the video picture.

At the end of each field of information, the scanning spot must return from the bottom of the raster pattern to the top to begin scanning the pattern again, analogous to the action of the eye when reading the same page twice: The eyes stop at the last word on the page and then move back to the top of the page to begin reading the page again. In video it's called the **field retrace period**, and the signal from the scanning spot is again cut off during that interval. Blanking takes place between each field while the scanning spot is moving back up the target area.

These two retrace periods when the scanning spot is flying back to begin again are known respectively as **line blanking** and **field blanking**. Occasionally you'll find blanking referred to as **suppression** (**field suppression**, **line suppression**).

Since the blanking intervals between each line being scanned occur on a horizontal plane, line blanking is often called **horizontal blanking**. For the same reason, the vertical retrace of the scanning spot from the bottom of the raster to the top is known as **vertical blanking**.

These blanking periods last for only a fraction of a second, but that is enough time to use them electronically to control the video picture signal. The electron beam scanning spot starts at the left and moves across the target area scanning one line of information. During that time an electronic signal corresponding to the voltage variations of the light image being scanned is sent out of the camera. Next, the scanning spot is blanked out and retraces its path to begin scanning the next line. During this time, no electronic picture information is being sent out of the camera. The developers of the TV system took advantage of this free time to add synchronizing pulses to the video signal that allow the camera CRT and the TV CRT to begin their line scanning and their field scanning in step with each other. In short, they developed a system whereby the video signal remains

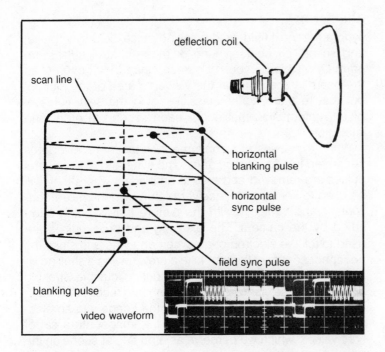

deflection coil

scan line

horizontal blanking pulse

horizontal sync pulse

field sync pulse

blanking pulse

video waveform

in **sync** from its point of origin to its point of reproduction.

There are two kinds of sync needed to maintain the integrity of the video signal. They are **horizontal** and **vertical sync**.

Sync Pulses

The horizontal sync pulses and horizontal blanking pulses are the electronic controls that keep the scanning spot on its proper horizontal course from left to right, line by line, across the target area of the CRT in a manner that can be reproduced on a TV screen. The blanking and drive pulses control the direction and speed of the scanning spot from left to right. The sync pulses tell the TV receiver CRT exactly *when* this scanning process beings and ends. Similarly, the vertical blanking and drive pulses are the governors of the change from one field to the next while the vertical sync pulses tell the TV receiver CRT when the change of fields is taking place so that it will remain in step with the source of the video signal, in this instance the camera.

The process going on in the camera is this: The scanning spot starts across a horizontal line, it stops at the end and there is a blanking pulse, which turns off the signal as the scanning spot returns to begin the next line. During the time when the spot is turned off and not sending a signal out along the signal path, the sync pulse replaces it on the signal path to inform the TV receiver what is happening. Thus, line by line the pattern is: scan line, blanking pulse, sync pulse; scan next line, blanking pulse, sync pulse; scan next line, blanking pulse, sync pulse. At the end of each field of information the same process takes place although the signals are slightly different so that they can be distinguished in the TV receiver. It goes: field, field blanking pulse, field sync pulse; next field, field blanking pulse, field

HOW TELEVISION WORKS

sync pulse; next field, field blanking pulse, field sync pulse; and so on. Of course, none of these control pulses are normally visible on your TV screen since they occur when the beam is scanning off the viewable area of the picture, but they're there nonetheless, keeping the sixty fields of picture information that occur each second in order and intelligible.

When the series of signals reaches the TV receiver, it is separated into its various parts. The sync and other control pulses are removed and sent to amplifiers, where they will be used to keep the TV receiver picture synchronized horizontally and vertically with the picture that is being originated by the camera. The blanking and drive pulses are used to tell the electron beam when and where to strike the phosphor screen of the CRT. Like a crew rowing a shell down a swift river with the coxswain shouting out the stroking orders, all these elements work together to record the image coming through the lens, translate the image into an electronic signal, and then retranslate the image into a set of light values, which we perceive as a particular scene on the TV screen.

Composite Video Signal

What the TV receiver is getting is called a **composite video signal**. That means that the signal contains all of the elements that we have described: composite blanking signals (vertical and horizontal blanking pulses), composite sync signals (vertical and horizontal sync pulses), horizontal and vertical drive signals, plus the actual video picture. A video signal can also be **non-composite**, which means it contains all the picture information except the sync signal. Non-composite signals are frequently used internally in video systems before the signal is sent to the TV receiver. For example, when more than one camera is being used it is often desirable that the video signal coming from each camera be non-composite; only later will the same synchronizing information be added to all the non-composite camera signals so they will be in sync with each other. We'll examine the reasons for doing this in Chapter 15.

This discussion started with the light coming in through the lens and forming an image on the face of the camera's cathode ray tube. From there we've covered how that light image is transformed into an electronic video signal. I have purposely avoided getting into particulars about the peculiarities of each of the components of the video signal and the place each occupies in what is known as the **video waveform**. This is highly technical information that you can't do too much about even if you take the time to understand it. What is important is that you have a general idea of how the video signal is formed and of the principles used to establish it in that form.

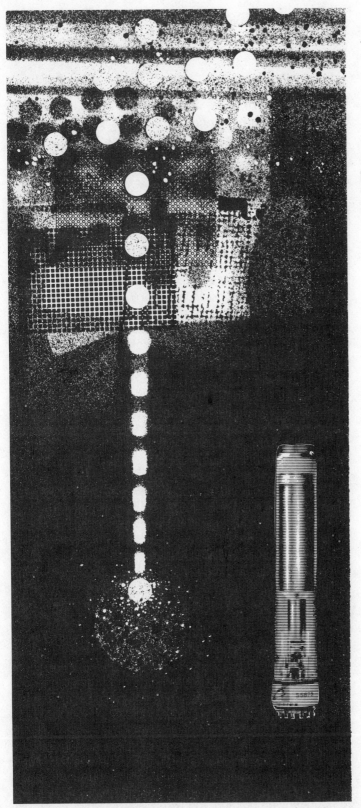

13

THE
PICKUP
TUBE

VIDEO
CAMERAS

We should have listened to Orwell.

—Roy Hollingworth
New York, 1972

The camera is the starting point of the video process, the thing that people most associate with media recording, which is reasonable since that's what's pointed at them. What the camera sees, the video screen displays. The intermediate steps don't really affect the underlying progression from camera to TV set. At this point then, let's see where it all begins.

This chapter is concerned with what is known as the **monochrome**, or black-and-white, video camera. Black and white are an aspect of color; each color in the spectrum has a light value, which corresponds to a different point on what is called the **gray scale**, the representation of every value of light brightness from black to white. Black is the absence of any of what we call color in an object, and white is the presence of all the colors of the spectrum at once. The shades of gray in between correspond to the addition of color to black or the subtraction of color from white. In other words, every color has a **brightness value**, which can be expressed as tending toward either black or white and thus being a particular shade of gray. The reason for the technical title *monochrome* for b&w television is because color is described as **chrominance** and mono means one. You'll also see b&w TV referred to as *achromatic*, which means without color.

As we have already established, the cathode ray tube of the video camera reads the light values of a scene in terms of their relative brightness or darkness, in terms of their tendency toward either black or white. That is the function of the video camera that will be available to most of us in the near future. The video color camera is similar, except that it reads not only the brightness values of a scene but also the chrominance values. This is accomplished through the use of more than one cathode ray tube as well as filters that divide the signal coming to each of the tubes into the primary colors involved. This is explained more fully in Chapter 17.

Components

The components of the video camera vary with the quality and purpose of the particular camera. Most people are surprised when they see their first portable video camera because it is so small and compact. Often, because of its size, they can't conceive of what it is technically. This is mainly due to our preconditioning to the huge color-capable cameras used in network television. Although they share components in common, the two types of cameras are designed for vastly different purposes and do have differences that are important to understand, especially since some of the forms that these differences have taken will have to be considered when buying video equipment.

The internal components of the video camera are centered around the cathode ray tube, which changes to an electronic signal the light values of the image cast on it by the lens. The components of the video camera obtain, control, shape, and amplify the video picture signal before they allow the signal to leave the camera for the video tape recorder, transmitter, or TV receiver. There are two major categories into which the component circuitry surrounding the CRT can be divided: first, the control circuits that govern the action of the CRT and its internal parts, such as the beam and scanning process; and second, those circuits that govern the validity of the video signal as it leaves the camera.

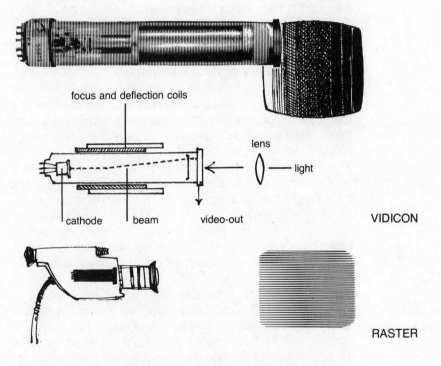

focus and deflection coils

lens

light

cathode beam video-out

VIDICON

RASTER

The major components of the internal control circuits of the camera are the cathode ray tube, **deflection and blanking circuits**, sync generator circuit, and focus circuits.

There are two deflection circuits, one for horizontal and one for vertical, which control the path of the scanning spot on the target area as it forms the raster. These circuits are metal coils mounted around the cathode end of the CRT. If the camera has an internal sync generator, it will be connected to the deflection circuits to provide sync control information to the scanning process. There is also an electron beam **focus coil** surrounding the CRT, which shapes the beam into a pinpoint spot. In some CRTs the focus coil has been replaced by an **electrostatic focus**, which does the same job—making the beam as coherent as possible so that when it reaches the target area it is one small spot.

Once the video signal corresponding to the light values hitting the target of the CRT has been obtained and shaped by the circuits, it is sent along a wire to the other set of camera components, those which control its passage out of the camera. The signal is processed first by a **preamplifier** circuit, which provides equalization, and then by the video amplifier, which strengthens the signal before it leaves the camera.

Most of us will probably never concern ourselves with the intricacies of these internal components except as they pertain to the particular camera we're using. But it is useful to have a general idea of the types of cathode ray tubes that are used in the various video cameras on the market, since the type of CRT determines the overall size of the camera, what the camera is good for, and how much it will cost.

A Short History of the TV Camera

What a video camera does, or should do, was understood well before a way of doing it was invented. The concept of a camera that could read light values and turn them into electronic impulses was understood in the early years of this century. Many of the earliest devices invented, such as those developed by John Logie Baird in England, used a light-sensitive cell and a mechanical scanning apparatus consisting of a rotating disc with a series of openings around the perimeter. The cell was placed behind the holes in the disc, and the disc was spun around to provide a frame-by-frame scanning of the scene. Baird's electro-mechanical TV system became the first commercially viable TV system in the world. In 1928 he used his apparatus to transmit a TV signal from London to New York City, and in 1929 the British Broadcasting Corporation adopted Baird's system to begin their TV service. The quality of Baird's invention was limited. The picture was made up of thirty lines, and the rate of picture change was about twelve per second. By the end of 1932 about 10,000 receivers for the Baird system had been sold in England, but the quality never improved substantially because of the limitations inherent in the system.*

*Baird's contribution as a media visionary should not be underrated. His TV system gave us the first public TV service, and by the mid-thirties he had successfully invented and demonstrated color TV, stereo TV, and large-screen TV projections in color. His electro-mechanical inventive genius was defeated by the advent of pure electronic communications.

While Baird's TV was being adopted by the BBC in the late twenties, two American-based research scientists were developing the first totally electronic TV systems. One was Vladimir Zworykin, a Russian immigrant who had been schooled in the concept of TV as a pupil of Boris Rosing in Russia. Rosing had come up with an electro-mechanical TV system similar to Baird's in 1907. As one of Rosing's pupils, Zworykin saw the possibilities and the limitations of the mechanical factors involved. By 1929, Zworykin had settled in America and was working for Westinghouse where he demonstrated his all-electric TV camera tube. Called the **Iconoscope**, the tube was capable of transforming brightness values from a scene into electronic signals. Zworykin's Iconoscope also provided blanking and sync signals during the retrace period of the scanning spot. A year earlier, a young man named Philo Taylor Farnsworth had also come up with an electronic vacuum tube for converting light to electricity. Called the **Image Dissector** tube, Farnsworth's invention paralleled Zworykin's in terms of the basic concepts involved and in its demonstration of the possibility of all-electric TV. As to who was actually first (it has been reported that Zworykin's Iconoscope was in existence as early as 1923), there is really no way of telling. Both men deserve credit for their creations and both can be considered among the founders of the modern TV system.

There were inherent problems with both the Image Dissector and Iconoscope camera tubes. Neither tube was terribly sensitive to the changes in light values of a scene and each needed an intensely illuminated scene before it could pick up and convert it into an electronic signal. But these two tubes did provide scientists working on the development of TV with a place to start from that was in the right ballpark. By the late 1930s an improvement on both tubes had been introduced. Called the **Orthicon** tube, this was succeeded by the **Image Orthicon** tube in 1945. Both came from the RCA labs and were pioneered by a gentleman named A. Rose. The Orthicon and the subsequent Image Orthicon eliminated many of the problems of the earlier tubes. They produced a bright, crisp video picture with a signal that was well above the level of electrical noise of the system's components. In 1934, two large British electronics firms, EMI and Marconi, created an all-electronic TV system using their version of the Orthicon, which was adopted by the BBC as a replacement for Baird's mechanical system. In early November 1936 commercial TV service as we know it began.

There were still problems with the Image Orthicon. First, the diameter of the tube's faceplate target area was three inches (7.6cm), which meant a large TV camera housing and large lens were required. Second, the internal components needed to operate the tube were complex and again added to the size of the camera body. Third, the life span of the tube was very short, in the range of a few hundred hours. And fourth, a lot of illumination was needed on the scene for the tube to work. There were also less critical problems. The color red, for instance, was not picked up by the tube—red lips would come out white, and so black or blue lipstick was required. Ladies who dyed their hair blond also ran into problems since the tube would transmit real blondes as blondes, but fake blondes as brunettes (hair dye being in the ultra-violet range, which the tube did not pick up).

In 1952, RCA introduced a new camera tube, that was, in many ways, an improvement on the Image Orthicon. Called a **vidicon** tube, it was one-third the size of the Image Orthicon with a target area only one inch (2.5cm) in diameter. The vidicon was extremely durable, it could be expected to perform from 3,000 to 5,000 hours without replacement, and it cost about one-tenth what an Image Orthicon cost. In operation the vidicon camera required less light on the subject to obtain a picture and it could be operated to adjust automatically its scanning of the light values of a scene from the brightest light value down to the darkest. The vidicon could be housed in a much smaller camera casing because much less ancillary equipment was needed to operate it, and a couple of the major electronic components needed to run the Image Orthicon were eliminated altogether. The only problem was that the quality of the picture image produced by the vidicon was not up to the standard of the Image Orthicon. There was less definition in the picture, and, under certain lighting conditions, moving objects appeared blurred around the edges. But the economic factors and the technical ease of use established the vidicon as a useful camera tube, especially in the area of non-broadcast television, the picture quality and stability requirements of which were lower than those established by the FCC and the networks.

By the 1960s, both the Image Orthicon and the vidicon had been improved. A larger Image Orthicon tube had been invented with a 4½-inch (11.4cm) target area that was more stable and had better sensitivity to contrast than the earlier tube. The N.V. Philips Company of Holland had come up with what they called a **Plumbicon** (their trademarked name)

PLUMBICON TUBE

tube, an improvement on the vidicon that eliminated picture lag, gave more uniform contrast, and generally brought the vidicon up to the broadcast standards established by the Image Orthicon. The Plumbicon maintains the vidicon's one-third size of the Image Orthicon and, although it is expensive, it has been adopted for use in many high-quality TV cameras, since it reduces the size of the camera. It is especially useful in color cameras that use three or four tubes.

The Vidicon

The inexpensive video cameras that have become available in the last few years all use vidicon tubes. Image Orthicons, Plumbicons, and a few other vidicon variations are used in broadcast TV, while the vidicon is the standard of the alternate TV system. There are two vidicon tube sizes presently in use, known as a **one-inch vidicon** and a **two-thirds-inch vidicon**. This refers to the diameter of the target area; naturally, the two-thirds-inch vidicon is the smaller tube. The more portable and compact video equipment becomes, the more likely it will be that the vidicon will be two-thirds of an inch (1.7cm) in diameter. In general, one-inch vidicons are used in the more expensive, larger studio cameras and the two-thirds-inch vidicons are used in portable cameras and less expensive studio cameras.

The most sophisticated small vidicon available is referred to as a "two-thirds-inch, separate mesh vidicon with electrostatic focus." The separate mesh refers to a mesh screen with approximately 1,000 tiny openings per square inch (.07 sq. cm), which improves the focus and resolution of the video picture by adding a further control to the electron beam as it travels from the cathode to the target area. The result is that the beam hits the target area at an exact right angle, which produces the most effective scanning of the target by the beam's scanning spot.*

There are important technical limitations to using the vidicon as the camera's CRT. The most visible is **image retention**, which is much like persistence of vision in the human eye. The vidicon will sometimes retain an image it has been scanning long after the image has either moved from its original position or been replaced by another image. Image retention occurs at times to such a degree that the image is burned into the target area and becomes a permanent sort of ghost on the picture area. The solution for minor image retention problems is to increase the light levels, which eliminates the smearing or streaking effect created by the image being retained. At the same time, if

---

*All vidicon tubes have a mesh assembly located between the cathode and the target area. A "separate mesh," an improvement on this assembly, delivers especially good focus and resolution.

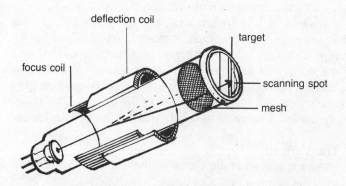

deflection coil

target

focus coil

scanning spot

mesh

VIDICON

too bright a light hits the vidicon target, there is a danger that the light will burn its image into the vidicon permanently. You may occasionally see this effect during a TV broadcast of a b&w movie—as bright white scenes replace dark scenes in the film, there seems to be a burst of shifting, hazy light. If a solid black figure moves quickly across a white area, the figure leaves a trail of ghostly light behind it. What this means is that a vidicon is being used in the TV camera that is transferring the film to video. This is done for two reasons: it costs considerably less to use a b&w vidicon camera rather than color camera to transfer b&w film to video for broadcast; and a certain amount of image retention in a CRT helps the film-to-video transfer by blurring any film frame flicker that might occur.

You will have some of the same problems in using a vidicon camera, but they are not so terrible that you can't get around them if you understand what they are and when they occur.

Camera Standards

Any video camera can be connected to any video tape recorder or transmitting and display equipment. There are differences between various types of cameras, however, even within groups of cameras that use the same type of CRT. These differences are essentially levels of complexity of the cameras. Any vidicon camera, for instance, will produce a video picture signal that can be displayed on a TV screen; the most expensive network quality color camera could be used to provide a picture signal to the lowest quality home video recorder. But for the purposes of this book, our area of interest is the vidicon tube video camera. These cameras offer a variety of options depending on the purpose and cost of the camera and produce good, acceptable pictures. The vidicon camera has become the standard for alternative TV because of its durability, light weight, and low cost.

The first thing you'll notice when confronted with the wide

range of vidicon cameras available is that some have viewfinders and some don't. Many industrial-use vidicon cameras have no viewfinder facility—there is no way of viewing, from the camera, exactly what the camera is seeing. Such cameras are used for warehouse surveillance, which doesn't require anything more than a camera sweeping a predetermined area. The viewfinder is superfluous since the picture being scanned by the camera will be monitored on a TV screen at a location remote from the camera.

Viewfinder

Other vidicon cameras do have viewfinders. They are more expensive than the non-viewfinder cameras, but they are much more practical when used to make television programs. The viewfinder is a small TV screen, which displays the image being scanned on the target area. This is one of the wonderful things about using the video camera—you are actually monitoring the view picture. No other type of camera allows an instant view of what is actually happening inside the camera. If you see a picture on the video camera's viewfinder monitor, you know that the camera is working and you can see exactly what the picture is going to look like on the TV screen when it is displayed. You can see if the picture is too dark or too light; if the subject is in or out of focus. Any other critical appraisals can be made at the camera, so you can instantly adjust the camera to correct the picture if there is anything wrong with it. Often the viewfinders of video cameras are called **electronic viewfinders**. This refers to the fact that you are watching what the camera is scanning on a small TV screen and your evaluation of what you're seeing is an evaluation of the electronic video picture signal being produced by the camera and being sent to the video tape recorder, monitor, or TV set.

Viewfinder sizes vary with the size of the video camera. Most small, portable video cameras have viewfinders that measure only about one inch (2.5cm) diagonally. This type of camera often has a magnifying eyepiece located between the eye and the monitor screen, which slightly enlarges the picture produced to make it easier to view. Less portable cameras have larger viewfinder monitors ranging between three to five inches (7.6 to 12.7cm) diagonally. The larger monitor allows you to stand behind the camera and watch the picture casually without having to press your eye to an opening. Several companies produce larger cameras with no viewfinder included but with the option of adding one when you can afford it.

The second difference between various vidicon video cameras lies in the lens. One camera may have a zoom lens, another a small wide angle lens. This should not really influence your evaluation of the camera since lenses are interchangeable. It is often advisable to buy a camera with

Lenses

VIDEO CAMERAS

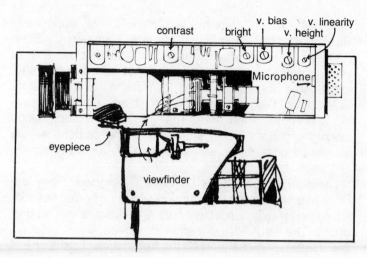

contrast   bright   v. bias   v. linearity
                               v. height
Microphone

eyepiece

viewfinder

no lens at all and then add whatever lens you need for the purpose to which you're going to put the camera.

Controls   The final major difference between vidicon video cameras is the actual size of the camera and the controls that are built into it. Most portable cameras have a full complement of the necessary controls included in the camera. More expensive, studio-size cameras have a separate control unit, which is usually mounted near the general control console of the video production center. The totally self-contained portable camera may weigh five to ten pounds (2.3 to 4.5kg) and be easily hand held; its power comes from the portable, rechargeable batteries located in the video tape recorder it is used with. The larger studio cameras weigh from ten to twenty pounds (4.5 to 9.1kg), require ac power, and are not at all portable in the hand-held sense, although they are much smaller than their network counterparts and can usually be broken down and transported in one or two large suitcases. Most of the really portable video cameras are available only in conjunction with a portable video recording deck. Although the power for the camera comes from the deck's battery power supply, they can be used independently of the deck with an external power converter-camera adaptor, which plugs in between the camera and a nonportable VTR or switcher and supplies the camera with dc power plus sync signals.

When evaluating cameras you should consider whether portables will function with a variety of decks with the addition of an adaptor and if such an adaptor is, in fact, available. If you are considering a studio camera, you should determine whether or not it has or will take a viewfinder, where the controls are located and just how much you're going to have to adjust them, and what the sync provisions are. Many of the bottom-of-the-line video cameras have internal sync generators but do not have provisions for

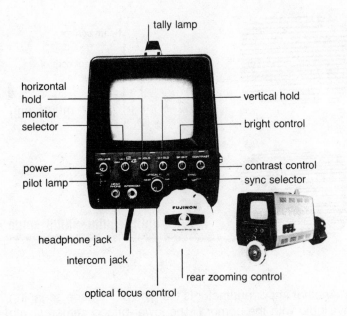

tally lamp

horizontal hold

vertical hold

monitor selector

bright control

power

contrast control

pilot lamp

sync selector

headphone jack

intercom jack

rear zooming control

optical focus control

STUDIO CAMERA

switching over to external sync generators. Such cameras cannot be used in conjunction with other video cameras to build a more than one-camera studio. (Multicamera setups are covered more fully in Chapter 15.) If you intend to use two or more cameras at the same time and mix their signals or switch between them, you must be sure that your cameras have provisions for accepting an external sync signal. This is usually not a problem with portable cameras since in their original use (with a portapak) they get their sync signals from the battery-powered deck, and when they are adapted to be used with other equipment their sync is external, coming from whatever piece of equipment they're being used with.

Another consideration is the video output facilities on the camera. Many of the least expensive cameras have only an RF-out jack, which means that the video picture signal coming out of the camera can be used to supply the signal to a normal TV receiver only, not to a video monitor, a video switcher, or a VTR. The only way to record a signal from a camera with an RF-out jack only is to feed the signal into the antennas of a monitor/receiver and then tape the signal out of the monitor/receiver as if it were a normal TV broadcast. This procedure will lower the quality of the final recorded signal. You'll need a camera with a video-out jack to provide a pure video picture signal, either composite or non-composite, depending on the sync source (internal or external), which can be mixed, switched, and/or fed directly into a video tape recorder or video monitor. Most medium- to higher-priced video cameras have both RF and video outputs.

VIDEO CAMERAS

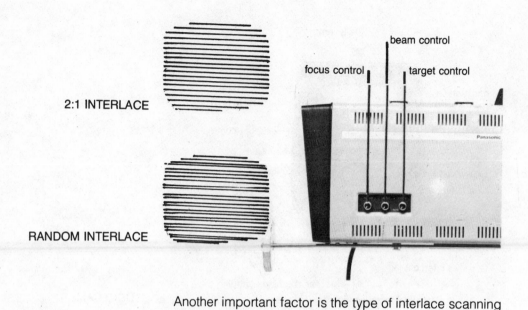

2:1 INTERLACE

RANDOM INTERLACE

beam control

focus control

target control

Panasonic

Interlace Scanning

Another important factor is the type of interlace scanning available with the camera. The lower-priced cameras with internal sync provide less scanning stability than the higher-priced cameras. This stability is measured in terms of two successive fields of video information and how they combine to make a frame. If the 262.5 lines of field 1 and the 262.5 lines of field 2 do not mesh together *exactly* to form a 525-line frame, then the scanning is considered unstable and the resulting picture will be jittery. The cameras on the market with internal sync often use random interlace scanning, sometimes known as **industrial sync**. This is a cheap sync control that does not guarantee an absolutely stable picture; but if only one camera is being used in a system, this is not critical to the recording being made. Problems may eventually occur with random interlace scanning if you intend to mix a prerecorded random interlace tape with a live signal through what is called a **genlock**; nor will you get as much stability if you freeze one video field (still frame) or run the signal in slow motion. The actual mechanical factors involved with random interlace are the circuits that generate the horizontal and vertical sync pulses—these are not right on the 15,750Hz and 60Hz needed for stable scanning.

More expensive cameras and those with provisions for external sync have 2:1 interlace scanning. 2:1 interlace means that there are exactly two fields for each frame and that each field has a 15,750Hz line frequency and a 60Hz field frequency. These are generated by circuits that can be locked to the ac line frequency of 60Hz, and by the use of binary mathematics and integrated circuitry they can constantly check and adjust line and field frequencies. When more than one camera is being used in a system, 2:1 interlace must be used so that switching between camera signals can take place. Both cameras are controlled by the same 2:1 sync pulses (usually from an external sync generator), ensuring that their horizontal and vertical frequencies are the same and their scanning is running at the same place at the same time. Without this it would be impossible to switch between cameras.

Putting the extra money into a camera with internal 2:1 interlace scanning is not terribly important, but getting cameras that accept external sync is. With a camera that takes external sync you can use it either with a video deck that supplies 2:1 to any camera it's hooked to (most of them do), with the internal sync of an SEG, or with an external sync generator.

External Control Units

Top-of-the-line studio cameras often have separate control units not built into the camera. These control units allow for precise adjustment of the camera component operations, so that the video picture signal coming from the camera can be adjusted to very close tolerances. You can adjust the electronic beam and focus of the vidicon, the **pedestal** (beginning of black level where the picture signal leaves off and the sync and other control signals start), and the video level (strength of video and control signals). In addition, a control unit will usually have provisions for a **tally light** (a light on top of the camera used in multicamera switching situations; it goes on when that camera's picture signal is the one being chosen as the program signal) and intercom so the camera operators can get directions from the director through headphones that plug into the rear of video cameras equipped with **tally** facilities. Except for tally light and intercom, *all* video cameras have the controls listed above, but lower priced and portable models have them hidden away inside the camera where only a service technician can get to them. In these cases, the controls are preset and locked at the factory with fairly good assurances that they will require only occasional readjustment. Most video people, especially those not overly concerned with the technical aspects of video and the absolute quality-capability of their equipment, don't want to constantly adjust controls to get the best possible picture, and certainly don't want to spend time learning what the controls do and mean. Manufacturers seldom expose the controls to view in cameras designed for this particular market.

Resolution

The last consideration when buying a camera is its **resolution**. This is expressed as the number of scanning lines that the camera is capable of recording. Video tape recorders and monitors, as well as cameras, have a certain resolution, which varies with the quality and purpose of the unit. It's important to remember that the lowest resolution of a component in a system will govern the resolution of all the other components in that system. If your camera has 400-line resolution, your video tape recorder 300-line resolution, and your video monitor or TV set 400-line resolution, you'll get only 300 lines of resolution on the screen when you play back a tape made with that video recorder and camera—the lowest resolving power is the one that sets the resolving power of the entire system.

There are two kinds of resolution: horizontal and vertical. Both have to do with the amount of detail that can be observed in the picture on a monitor or TV screen. Because the number of horizontal lines in a US standard TV picture is 525, the vertical resolution cannot be more than 525 lines at any point on the screen from the top to the bottom. It is possible for a video camera or monitor to have more than 525-line resolution, but you will get 525 lines (usually less) of vertical resolution when you observe a picture on such a high resolution screen if that picture has been recorded on tape or transmitted as a broadcast. If you were to attach a 600-line resolution camera directly into a 600-line resolution monitor, you would get 600 lines of resolution. But it is a waste to spend the money on such a camera or monitor since the intermediate equipment needed to create video programming would probably cut the vertical resolution to half that number of lines. It should be pointed out that although the standard number of horizontal lines (vertical resolution) of the American TV system is 525, the actual resolution is only about 250 to 300 lines, even on a home TV receiver that is tuned to a network-quality broadcast. A video camera with 300-line resolution connected to a video tape recorder with 300-line resolution is sufficient for most purposes. With color video, the resolution is even less, usually 240–260 lines.

Horizontal resolution is more of a subjective value than is vertical resolution. Horizontal resolution refers to the amount of detail you can see in the horizontal direction of the picture. This shouldn't be confused with the total number of scanning lines on the screen's target area, which is a fixed constant of 525 in this country. What we're talking about here is the amount of visible detail, expressed in lines, on a 525-line screen. Using a test pattern with a group of lines running vertically, you can test the horizontal resolution of your camera and monitor. If the vertical lines are sharp and you can see each one in detail until they get very close together (the chart has lines spaced from inches apart to almost touching), then your horizontal resolution is good. If the vertical lines, even those spaced apart on the test pattern, look blurred and fuzzy, your horizontal resolution is poor. You may have trouble deciding whether it is the resolution of the camera, monitor, or video recorder that is at fault unless one of them (say the camera or the monitor) is of laboratory standard and has such excellent resolution that you know the other components are not up to that standard.

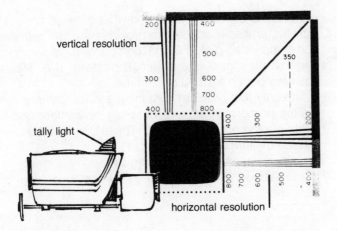

vertical resolution

tally light

horizontal resolution

There is no real need to worry about resolution when using inexpensive video equipment. Most manufacturers have established approximately 300 lines as the vertical resolution of their equipment with the horizontal resolution being a subjective determination based on the quality of the vertical resolution. A general rule is the more lines of vertical resolution the better, but if it's going to cost you twice as much to double the resolution of a piece of equipment, remember that the rest of your equipment will limit the total resolution of the system as a whole. In portable systems, for example, it's common for the camera to have better resolution than the deck can record.

Operation

Once you have a video camera, you must learn how to operate it. This means, essentially, making sure it's connected to a video recorder or monitor TV, and then checking to see you've taken the cap off the lens. After that it's a matter of opening or closing the f-stop of the lens to admit an acceptable amount of light and focusing the lens so that the subject appears sharp in the viewfinder of the monitor. That's it.

More sophisticated cameras may require a check of the function switches to see that the signal going out is correct (video or RF), that the sync pulse is correct (internal or external), and so forth. But even those cameras are operated in exactly the same manner as ones without external controls. The operation of a video camera is not a complicated process until you get into units that require pedestal, aperture beam, video level, and other fine tuning adjustment. It's unlikely that your video experience will ever include that amount of technical camera operation since you'd have to be building a major studio before you'd consider acquiring one of these big cameras; and if you did get one, you'd find that the majority of the controls could be preset with the visual aid of a good monitor and what is called a **waveform monitor**, an oscilloscope that provides a visual representation of the video picture signal components, allowing you to see that the signal is maintained properly. If all else fails, you also get an instruction manual.

The only real problem in running a video camera is making the proper connections between the camera and other video equipment you're using. These connections are discussed in Chapter 9.

Once the camera is connected properly and turned on, the technique with which you use it should be your major concern.

14

BRIDGING
THE GAP

THE VTR

Functions

The video tape recorder is the heart of the video production system. At first glance it looks like a rather bizarre version of a normal audio tape recorder. In fact, most half-inch open reel decks are just as easy to use as their audio counterparts although more attention is required when threading the tape and adjusting the deck to record.

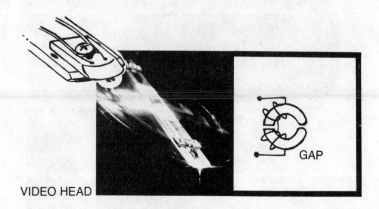

VIDEO HEAD

The key to the video recording process, in terms of the VTR, is the video heads. If you understand what they do and how you must take care of them you'll have few problems getting satisfactory record and playback out of your VTR.

The video head, like the audio head, is a ring of metal with a slit cut in it so that it does not form a complete circle. This slit is called the **gap**. The head is wrapped with wire; when an electrical current is sent through the wire, the head forms a small electromagnet as the current pulses through the coil to complete its circuit path.* The head gap is very small—invisible to the eye—but it is at this point that the tape and the head come into contact and that the transfer of the magnetic pulse from one to the other takes place. (From head to tape during record, from tape to head during playback.) The electronic pulse going through the head coil creates a magnetic pattern at the head gap; that, in turn, magnetizes the tape at the point it comes into contact with the head gap. Since the tape is constantly moving past the head, each magnetic pulse sent across the gap is transferred to a subsequent portion of the tape. After recording, the entire length of the tape has one pulse after another on it, each pulse corresponding to a segment of the electronic

*An electromagnet is a magnet that has magnetic properties only when an electrical current is applied to it. The simplest form of electromagnet is a core of iron with a wire coiled around it. When an electrical current is sent through the wire the iron bar becomes a magnet; when the current stops, the bar loses its magnetism.

signal that has been fed into the recorder and to the heads. This principle is the same in both audio and video; the major difference being that in video there is much more electronic signal information, covering a wider range of frequencies, to be put on the tape through the heads.

Audio heads, one of which is on every video tape recorder, need pick up only a limited **frequency range**, usually from 30 to 20,000 cycles per second (30Hz to 20KHz), between which all sound audible to the human ear occurs. A person whistling might produce a sound that is about 1,800Hz, for example. The frequency of the whistling is the frequency of the electronic signal into which the whistling sound waves are transformed by the microphone element.

Certain conditions must be met before the frequency signal can be pulsed onto the tape. First, the tape head gap must be small enough so that the frequency doesn't get "lost" in the gap. This may sound silly, and it's not the precise technical explanation, but does describe the overall effect when the gap is wider than the frequency being sent across it. The higher the frequency, the shorter its wavelength—thus the necessity for very "narrow gap" heads. The audio gap is capable of sensing frequencies to about 20KHz; the video gap can sense higher frequencies since the tolerances of the video head are closer and the gap smaller than in audio.

The second condition necessary to get this frequency on the tape is that the tape must be moving past the head. If the tape were standing still each signal in the series being transmitted would be impressed on top of the one before it and there would be no way of reproducing the recording. The third condition is that the tape move past the head fast enough for the higher video frequencies to be stored on the tape. A combination of electrical and mechanical factors control this last condition—physically, the head gap can be made only so small; economically, tape can be run only at certain speeds. Put the two together and a ceiling on the frequency response is set, following the principle that the faster the tape is running and the smaller the head gap, the wider the response range. In audio recording there is more treble and bass on a tape running past the heads at 15 inches per second (15 ips) than there is on a tape running at 1 7/8 ips. This is because the slower the tape is running, the more difficult it is to transmit a wide range of frequency pulses to it.

If you have trouble imagining the transfer of the signal from the head to the tape, a classic scientific demonstration might help you to visualize the process. You'll need a strip of paper, some iron filings (which you can make by getting a bit of iron and actually filing at it and collecting the resulting little metal bits of dust), and a magnet. Sprinkle the filings on the piece of paper and then take the magnet in hand. Pass the magnet under the paper and filings. As you do this, notice how the random arrangement of the filings changes to a coherent pattern. The pattern corresponds to the amount of magnetism (the magnetic field) created by the magnet as it passes the filings. The same thing happens when the tape and the head come into contact; each pattern created on the tape corresponds to a different magnetic field created by the electronic pulse that turns the recording head into a small electromagnet.

Video heads are the same as audio heads in their basic design, but there are major differences in the way they are used to get the signal onto the tape. As mentioned earlier, the video reaches higher frequencies than audio does. It starts at about zero and goes to about 4MHz (four megacycles, or 4 million cycles, per second). Since the video heads and their gaps can be made only so small, the speed at which the tape passes the heads must be as fast as is practically possible to get the total frequency range of the signals onto the tape without distortion or loss of **high end** (high frequency) information.

A Little History

The first try at video recording was made by Bing Crosby Productions in the early 1950s. They used a modified audio tape recorder and ran the tape from reel to reel past a stationary head. This is known as *longitudinal recording* since the tape travels in a straight course across the head. To record an acceptable video signal on the tape using this system, the tape had to be moving at about 100 ips, over eight feet (2.4m) of tape for every second of video information, so that a standard 10½-inch (26.6cm) reel of tape (2,400 feet or 731.5m) would hold only four minutes worth of video.* The tape had to move that fast past the head to ensure that the short wavelengths of the higher frequencies would be transmitted and stored on the tape. Longitudinal recording was far from practical, but it did show two things: that video recording via magnetic tape was possible and that a new system had to be invented that would allow the tape speed to be considerably less so that a reel of tape of manageable size could be used to store a realistic amount of video.

The Ampex Corporation solved the problem in a most ingenious way. Since the problem involved head-to-tape

---

*The Crosby machine divided the video signal into eleven different sections and recorded each section on one of eleven parallel tracks on an inch-wide (2.5cm) tape. A 14-inch (35.6cm) diameter reel of tape at 100 ips ran fifteen minutes. RCA followed this up with a VTR that recorded four minutes of video on half-inch (1.3cm) tape on a seventeen-inch (43.2cm) reel running at 360 ips!

speed (the speed at which the tape passed the heads) and since they wanted to maintain a low reel-to-reel speed (to reduce the total length of tape required), they hit upon the idea of moving the tape heads as well as the tape. If the tape was moving from left to right at, say, 15 ips, and the heads were moving from right to left at, say, 7 ips, the resulting head-to-tape speed would be 105 ips (7 times 15) or approximately that of the earlier Crosby VTR, without requiring an impossible amount of tape on which to store a reasonable amount of information. To improve the quality of the information and to make it possible to store an hour of information on a manageable reel of tape, Ampex used four rotating heads contacting the tape surface one after the next at a speed of 240 revolutions per second. The tape passed by the heads at 15 ips and the heads passed the tape at 240 ips. The result was an *effective* tape-to-head speed of 1,800 ips. (The actual head-to-tape speed is 3,600 ips, which is the field rate of 240 times 15. Half of that is the frame rate, or 1,800 ips.) If the heads were *not* rotating and the tape moved at 15 ips, it would take almost half a million feet (152.4km) of tape to make a comparable one-hour recording.

This process is known as **quadraplex** and is the standard video recording format of RCA and Ampex's high-quality VTRs found in network TV studios. These VTRs use two-inch-wide tape. The four rotating video heads are located on a drum that is perpendicular to the tape path. They contact the tape at right angles as the tape passes by the heads. Each head puts a certain amount of video information on the tape before it is replaced by the next head/next tape segment. The video information goes on the tape in stripes at right angles to the direction in which the tape is moving and each head induces about sixteen lines of picture information onto the area of tape it contacts. There are some highly crucial factors involved in this form of tape transport and head design, including a vacuum assembly that sucks the tape into a curve as it approaches the rotating head wheel so that the concave curve of the video tape conforms to the convex curve of the heads.

The development of quadraplex VTRs resulted in technically refined video recording and broadcast standards that could maintain excellent picture quality. The sync and time-base stability of this form of video recording is very good, and the picture signal from these recorders is of uniformly high quality, provided the machine is kept in top operating condition. To give you an idea of how much time and labor goes into using one of these machines, each machine is about ten feet high, six feet wide, and three feet deep (3.05 × 1.8 × .91cm) and each requires its own operator. The electronic components used to keep the signal

Quadraplex Recording

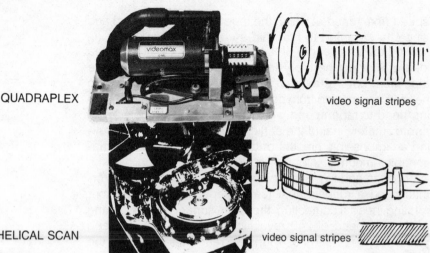

QUADRAPLEX

video signal stripes

HELICAL SCAN

video signal stripes

valid are truly mind-boggling. Cleaning the tape heads, which is done after each use, is usually a matter of drenching the heads with an entire sixteen-ounce (half liter) can of head cleaner spray.

Once quadraplex became the established VTR format for broadcast facilities, a number of companies began to research the possibility of a less expensive, more portable video recorder format. Keeping the principle of moving heads as well as moving tape, but varying the way the heads and tape came into contact, they developed a less expensive VTR. Originally called **slant track** recording, the method is commonly known as **helical scan video tape recording**.

Helical Scan Recording

Helical and slant track refer to the path of the heads in relation to the tape. The heads—usually two, although the original slant track VTRs used only one*—are mounted on a set of arms that revolve much like the propeller of a plane. These heads are contained in what is called a **head drum assembly**—a cylindrical metal drum with a slit cut around its girth from which the tips of the video heads protrude. The tape is wound in a helical pattern around this drum. In other words, the tape comes off the supply reel and is slanted down slightly as it is wrapped around the head drum assembly (usually around just half of the head drum), and then leaves contact of the head drum to go to the take-up reel. The screwlike movement of the tape from one level down to another is the helix configuration. The tape moves from left to right in the helix while the video heads revolve in the opposite direction on a horizontal plane that is halfway between the horizontal planes of the supply and take-up reels.

*The original helical scan VTRs, by Toshiba, had one head rotating at 3,600 rpm instead of two heads, each rotating at 1,800 rpm.

The video heads are mounted on the tips of an arm and are 180° apart. The tape wraps down around half the circumference of the head drum as the heads rotate so one head is in contact with the tape while the other is not. Because of the helical movement of the tape, each head describes a slanted or angled line from right to left across the tape surface as it comes in contact with the tape. The head comes into contact with the tape at the top of the tape and ends contact at the bottom of the tape. The result of this head-to-tape contact pattern is a stripe of magnetic pulses transmitted to the tape by each sweep of each head across the width of the tape. As each stripe is put on the tape, the tape advances just enough so that virgin area is exposed to the next head, which applies a new stripe of information.

The transmission of information is a process governed by the swing of the heads from right to left and the movement of the tape from left to right. The speed of the tape as it runs from supply reel around the head drum to take-up reel and the speed of the heads rotating in the opposite direction produce the effective head-to-tape speed: sixty signal stripes every second; that is, each head contacts the tape and moves across it thirty times each second.

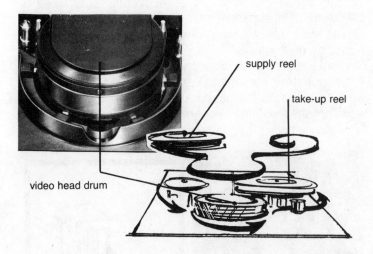

supply reel

take-up reel

video head drum

It should come as no surprise that each pass of the head in a helical scan VTR using half-inch tape corresponds to one field of video information. If you return to the equations explained for the video camera (pp. 240–42), the helical VTR's head speed makes sense: 1,800 rpm = 30 revolutions per second (1,800 divided by 60) times two heads = 60 fields/revolutions per second. There are sixty fields of information transmitted every second from the camera to the VTR, and each field must be stored on the tape. With helical scan, the tape head describes a path on the tape that is

THE VTR

long enough (lasts long enough) for a full field of information to be stored as a continuous, uninterrupted unit between the time the head contacts and ends contact with the tape surface during a pass. The head must rotate or pass across the tape at least thirty times a second; each of the two heads carries a field of information and the sum of two head passes equals one frame. This is the basis of helical scan video tape recorders as they have been standardized by the Japanese Electronic Association.

Once the tape has left the supply reel, wrapped around the head drum, and left the head drum, it follows a path that takes it past two other crucial components. The first is the **audio and control track head**, a stationary recording/playback head that has two functions and is actually two heads, one above the other. The top portion is an audio head. It transmits the electronic pulses of the audio signal onto the tape in the area between the end of the video signal stripe and the top edge of the tape. The bottom portion is the control track head. When recording takes place, this head transmits pulses onto a small area at the bottom edge of the tape. These pulses are read off the tape by the same head during playback to control the positioning of the video heads so that they can accurately trace the originally recorded signal. This ensures that the picture reproduced is the same as the picture that was recorded. The control pulses mark the beginning of every other field of video information on the tape.

Controls

Head-to-tape speed is controlled by leaving the heads to rotate at a fixed speed and varying the speed of the video tape as it passes the heads. Changing the tape speed is accomplished with the second major component along the

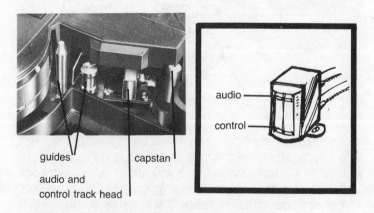

guides   capstan

audio and
control track head

audio
control

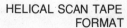

HELICAL SCAN TAPE
FORMAT

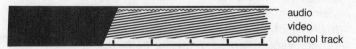

audio
video
control track

tape path, the **capstan**, a metal shaft and rubber roller (called a **pinch roller**) between which the tape travels on the way to the take-up reel. The capstan and pinch roller clamp the tape and pull it along. The metal shaft of the capstan is connected to the driving motor of the VTR and is regulated by the control track pulses, which are read off the tape by the control head during playback. If the heads get out of line with the tape path on the recorded tape during playback, the capstan senses this from the control track-servo information and pulls the tape through faster or slower to put the heads back in sync with the video tracks on the tape. If there were no control track and capstan servo control, the video picture coming off the tape would have a great deal of static in it and would be constantly rolling up and down on the display screen. This servo control is especially important during electronic editing (p. 82) when the capstan servo mechanism is a crucial factor in achieving stable edits.

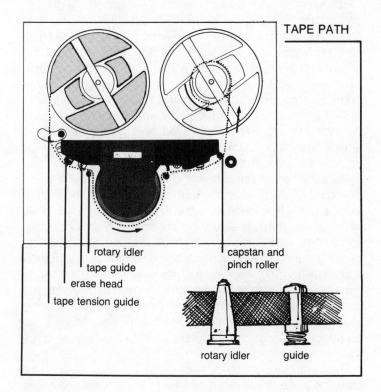

TAPE PATH

rotary idler
tape guide
erase head
tape tension guide

capstan and
pinch roller

rotary idler   guide

The supply reel, video heads, control-audio head, capstan assembly, and take-up reel are the major milestones in the journey of the tape through the VTR. There are a number of metal poles that stick up at various places along this **tape path** and they are more important than they might seem at first glance. These **tape guides** make sure that the tape is properly following the path set out for it. They're like the poles set in the snow for a slalom event through which the skiers wend their way down the course.

The first of these, the **tape tension guide**, is placed just after the tape comes off the supply reel. It is not stationary; it moves back and forth across a short space and controls the **skew** of the tape. This little arm keeps the tape taut as it comes off the supply reel and goes to the head drum. It is very important because the tension between the reels of tape, the head drum, and the capstan must be maintained or the picture will deteriorate into fuzzy b&w noise. When the upper portion of the picture appears bent on playback, that is a sign that the tension is not correct. This tension adjustment is known as skew on most VTRs, and there is a skew adjustment knob which you may have to use during playback if you see distortion in the picture. It is a critical control in helical scan video taping since the stability of the picture is dependent on the maintenance of proper tension.

Following the tension guide are two stationary guides, one on each side of the head drum assembly. Sometimes called **rotary idlers**, they are cone-shaped and have flanges, which hold the tape in position as they guide it across the head drum at the proper helix angle. They control the **tracking** of the tape around the head drum. Tracking is the second important control factor in the tape's movement. It is essentially the speed and angle at which the tape passes the video heads and is controlled during playback. If the tape is on the right track, the picture during playback will be up to standards. If the speed or angle at which the tape passes the heads varies from the speed or angle at which it passed the heads when the tape was recorded, the picture will have a great deal of noise in it, and it is said to be mistracking. There is a tracking control knob on most VTRs, sometimes with a meter, which allows you to adjust the tracking during playback if you see distortion in the picture.

Both skew and tracking adjustments have to be made most often when a tape has been recorded on one VTR and is being played back on another VTR that is of the same standard but isn't perfectly adjusted to the first. Minor skew and tracking variations are adjusted automatically by some decks.

The final component located along the tape path is the **erase head**. Usually placed between the skew tension guide and the first of the tape head guides, the erase head is an electromagnet which, when a signal is applied to it, disturbs or scatters the pattern recorded on the tape so that the picture disappears. The erase head has two functions—to erase the entire tape or to erase the audio portion only. It is two heads in one, one head gap covering the height of the video track area of the tape, the other covering the height of the audio track area. The erase head removes any signal on the tape during recording so there is no possibility of one signal being superimposed on another. While you're recording, the erase head "cleans" the tape before it reaches the video heads so that the newly recorded information will be the only coherent signal on the tape. Some VTRs incorporate **rotary erase heads** in their head drum assembly. These heads provide more exact picture erasing, even field by field erase, for the purposes of editing.

The other controls on the deck of the VTR are the start, stop, play, record, fast forward (ff), and rewind function knobs or buttons controlling the motion of the tape and the function that the heads perform (playback or record). More expensive VTRs also have pause, still, audio dub, edit, and other controls. The operation of these controls is covered on pages 43–7.

Components

The internal electronic components of the VTR perform three functions: the processing of the electronic signal into the deck and then to the heads for recording; the governing of the recording parameters to conform to a predetermined system; and the processing of the electronic signal from the playback heads out of the deck for display. I won't dwell on these internal circuits since there isn't much you can do about it if they malfunction except pack up the deck and take it in for servicing. It is possible to readjust the settings of some of the components if they get out of whack, but even this is a follow-the-numbers type of repair that must be performed with extreme care and with the service manual in front of you. More serious ailments, such as the deterioration of a component or the modification of a circuit board, are really out of the realm of anyone who hasn't been instructed by the manufacturer in the repair of the particular unit.

What is necessary with regard to the internals of the VTR is a general understanding of what happens to the signal as it enters and leaves the deck. The video signal, as it comes into the deck from the camera, is immediately amplified by a **video amplifier** to make the signal as strong as necessary. It is then passed through a filter that removes any signals that may have been generated or collected in its transmission into the deck. The filtering process is accomplished by a **low-pass filter**, which allows the lower frequencies to pass through it while removing any stray high frequencies, known as **high-end noise**, from the signal. Then the signal is **clamped**, which takes care of the other end of the video signal—the low-pass filter removes high-end noise, the clamping circuits remove any low-frequency noise such as hum, which may be caused by the 60Hz signal from the ac line.

Once the extraneous frequencies and causes of possible distortion are eliminated, **clipping** takes place. Clipping circuits cut off the video signal at both its bottom level (**black clipping**) and top level (**white clipping**), which insures that the black to white range of the video signal is within the limits that the particular VTR is capable of recording. In this way, the video picture goes from black to white with no disruptions or spurious signals—sync and blanking signals are kept below black level by the circuits and there are no excursions of the signal above the white level to produce distortions in the picture.

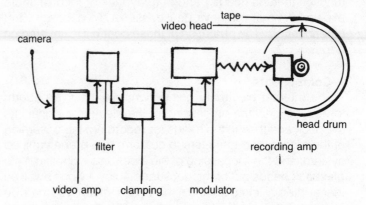

PROGRESS OF VIDEO SIGNAL FROM CAMERA TO VTR

Once the video signal is all tidied up, the next step is to make it presentable to the video heads so they can pulse it onto the tape. This is done by **frequency modulating** (FM) the signal. The frequency modulator is a circuit that produces a radio carrier wave on which the video signal is impressed. The frequency modulated signal is also referred to as **modulated RF** (for radio frequency) and **sideband FM** modulation. In essence, the modulator circuit produces a signal that is of constant strength but the frequency of which varies according to the video signal information supplied to it. It is the same process by which FM radio broadcasts are made—a carrier signal of an established strength is varied in frequency both above and below the carrier wave's original frequency in accordance with the signal it is being combined with. AM radio is also an RF signal, but the frequency of the wave remains the same while the strength, or amplitude, of the signal varies according to the information the signal is carrying.

The FM signal is fed to the **recording amplifiers** of the VTR, which are connected to the head assembly. From there the signal is run to the video heads and then to the tape.

During playback the process is reversed. The signals are induced from the tape through the video heads, and each head feeds its signal to a **playback amplifier**. Since each head brings in one field of video information, the signals from the playback amplifiers must be mixed together to re-form the total signal. Once the signal has been reconstituted, it is sent to **equalizing amplifiers**, which make sure that it has its proper strength values. After that, **limiters** are once again employed to remove any spurious noise that may have been picked up during the transfer to and from the tape, and then the signal is **demodulated**, meaning the carrier wave frequency is removed. We are left with the pure video signal, identical to the signal as it entered the deck and ready to leave the deck again for display after it has been brought up to full strength by a final video amplifier.

VTR Standards

Mating square pegs with round holes has been the traditional activity of the television industry when it comes to establishing standards to govern the operation of their hardware systems. This attitude has kept the people who build the square-peg and round-hole adaptors in business. In fact, it has made their dollar items crucial to the operation of millions of dollars worth of equipment. There are as many logical reasons for the proliferation of standards in video today as there are for there being but one standard. The loudest cry for standardization comes from equipment users. The most common nightmare for those buying video equipment is the experience of spending a fortune on a video tape recorder only to have it break down or the company that made it go out of business. The video tapes made on that machine become obsolete—there will be no VTRS available to retrieve the information stored on the tapes. On a more everyday level, it is desirable that a video tape be replayable at a location distant from its origin, even if that location is not equipped with a VTR. This, for the video user, necessitates standardization.

For the manufacturers, the very concept of standards is different. Traditionally, they have viewed standardization as a pain in the neck. They have labored to develop and produce a reliable, well-built piece of equipment that will perform certain functions. They have then entered the marketplace with that equipment and begun to merchandise it with an eye toward selling as many as possible. Their focus is on their equipment. That their competitors are producing similar equipment is to be taken as a challenge, not a draw. This is especially true at the time a category of equipment first comes into being. When Ampex introduced the first video tape recorder, their system was *the* system. There was no question of standards because if you wanted to make a video tape, you needed an Ampex machine. That RCA was also working on a VTR was not Ampex's concern.

The interchangeability of tapes recorded on one machine for playback on the other was not crucial to the product Ampex was selling. Fortunately, RCA realized that they were entering the marketplace second and that their equipment had to conform to Ampex's, at least regarding tapes' interchangeability, since all the major networks already had Ampex machines. But it was only the limited number of potential buyers—the networks and their affiliates—that kept RCA from going into total competition with Ampex by producing a machine that would replay only tapes recorded on it.

When helical scan came along a few years later, there was an automatic change in standards, since the very system by which helical scan functions was different from the RCA and Ampex quad VTRs. Helical scan became a second standard. The configurations of tape path, head formation, and tape size differed from the broadcast machines then available—the major reason being that helical scan was developed out of the need to build a less expensive, more portable alternative to quad.

The first helical scan machines used one-inch video tape, half the width of the broadcast machines. As they were developed, different helical scan machines came from different manufacturers. The broader-based market for helical scan, including schools, industry, and other self-contained or closed circuit applications, led many electronics firms to produce video equipment to sell to these markets. These companies went into competition with each other to produce the least expensive and most reliable units. This led to the use of several tape widths, including one-inch, three-quarter-inch, half-inch, and quarter-inch. Each successive reduction in tape size signaled a reduction in the size and complexity of the machine on which it was recorded and played back. With a number of manufacturers all creating their own versions of helical scan equipment, the initial question of standards could be considered only in the sub-categories of tape size and brand.

As the helical scan VTR business prospered, it became obvious that the less expensive the equipment, the greater the market. Once the market became all of us, ideas began to change. Originally, each helical scan manufacturer had entered the marketplace with equipment that would be used in isolated applications, in a factory or a college, for example. The concern was to make public the knowledge of the equipment's availability, not to convince potential buyers that the equipment was standardized to that of the competition. If a particular school or company introduced a close circuit video system, their needs and knowledge did not extend to tape interchangeability with other such installation. If a particular school or company introduced a closed tions. As the base broadened and more and more equipment came into use, however, the demand for video equipment began to point in the direction of the average consumer. At that juncture the need for standards became apparent. The manufacturers were no longer competing to establish helical scan; they were competing to sell more helical scan equipment than their rivals, and these potential sales could best be served by some form of standardization. If a Panasonic equipment owner was planning to acquire even more equipment, such as an editing deck to complement a portable system, then Sony was as eager as Panasonic to be in on that sale. Also, lesser manufacturers who hadn't been responsible for the development of helical scan, but wanted to get a piece of the action, were bound to imitate the technology of those who had first invented and marketed it. That most of these companies were located in Japan must be considered as well. The Oriental wisdom involved in consolidating the marketplace by standardizing the equipment to make it all interchangeable meant that the infighting between manufacturers would not include creating confusion about what they were selling. This is something that American manufacturers have yet to appreciate; their corporate greed is a winner-take-all kind of fight that has left the consumer with a lot of outdated equipment over the years. The Japanese have learned that aggressive coexistence makes everybody richer.

In August 1969, just a year after the introduction in the US of relatively inexpensive half-inch video equipment, a standard was established by the Electronic Industries Association of Japan (EIAJ) which made tape interchangeability between various half-inch machines a reality. The media impact of this standardization assured the consumer that he or she was not buying an experiment but a functioning piece of equipment that seemed dependable because it was standardized—as safe a bet as buying a color TV or an audio tape recorder. The first standard for b&w half-inch equipment was followed by a color standard for the same equipment.

This standard is known as the **EIAJ Type #1 Standard**. Both b&w and color come under this standard although the b&w is referred to as EIAJ Type #1 and color is referred to as **EIAJ Type #1 Recommended Color Standard**. Not all helical scan recording units comply with it; decks using one-inch tape, cassette and cartridge systems, and decks using quarter-inch tape have yet to be standardized within their tape format size. Only half-inch tapes made on EIAJ Type #1 VTRs are compatible with each other and can be played back on any other EIAJ Type #1 VTR. This, of course, includes b&w tapes being playable in b&w on EIAJ Type #1 color VTRs and color tapes being playable in b&w on EIAJ Type #1 b&w VTRs.

The EIAJ Type #1 standard describes a set of tolerances which video tape recorders conforming to the standard must maintain. The accompanying chart gives a rundown of the various factors that must be met by the manufacturer to produce a Type #1 machine. To explain this standard generally, the machine must record or play one hour of video with 2,400 feet (731.5m) of half-inch tape. The tape must run by the heads at 7½ ips and resolution must be better than 300 lines in b&w, 240 in color. The scanning system of the video heads must be two-head helical scan, each head revolving at 1,800 rpm. The positioning of the audio, video, and control track signals at the top edge, center, and lower edge of the tape respectively is also specified. Of course, the manufacturer can apply the company's own technology to create VTRs that conform to these standards. The internal works of any particular VTR from any particular manufacturer may vary drastically from those of the rest. This is one problem with any form of standardization: the assumption that all equipment complying to the standard is somehow going to function well. You should be cautious when buying equipment from other than established brand-name manufacturers of EIAJ standard equipment, since there are a number of lemons on the market, albeit EIAJ Type #1 lemons.

These standards, tape widths, and systems must not be confused with the ability to **interface** between any two or more pieces of video equipment. A signal recorded on a half-inch tape machine on half-inch tape can be transferred to tape recorded on a one-inch machine, a two-inch machine, or any other standard system, as long as the television system is the same (US standard signal, European

STANDARDS

Current video standards are: 3/4-inch (U-Matic) cassette; 1/2-inch (Beta/Betamax) cassette; 1/2-inch (VHS) cassette; 1/2-inch open reel (EIAJ Type #1); and 2-inch (quad) open reel.

U-Matic/U Type Standard
Tape: 3/4 inch wide
Tape housing: Cassette
Tape speed: 3.75 ips
Audio track: Stereo
Scanning configuration: Helical

Beta/Betamax Standard
Tape: 1/2 inch wide
Tape housing: Cassette
Tape speed:
    one hour mode—1.57 ips
    two hour mode—0.79 ips
Audio track: Mono
Scanning configuration: Helical

VHS Standard
Tape: 1/2 inch wide
Tape housing: Cassette
Tape speed:
    two hour mode—1.31 ips
    four hour mode—0.66 ips
Audio track: Mono
Scanning configuration: Helical

EIAJ Type #1 Standard
Tape: 1/2 inch wide
Tape housing: Open reel
Tape speed: 7.5 ips
Audio track: Mono
Scanning configuration: Helical

Quadraplex Standard
Tape: 2 inches wide
Housing: Open reel
Tape Speed: 7.5/15 ips (varies with manufacturer)
Audio track: One primary, one secondary of lower fidelity
Scanning configuration:
    Four video heads segment each field into four sections and trace them successively across the width of the tape.

NOTE: These standards are general since application of the quad principle varies between manufacturers.

standard signal, and so forth). The physical tape cannot be played on machines of different standards, but the video picture signal can be run from one machine to another without regard to the standards involved. Of course the quality of the signal will vary with the quality of the original recording, but it *can* be transferred and stored on other equipment. In this case, interfacing is the physical act of running a cable between two machines. It is the process of connecting equipment whereby the connection establishes the equipment in a functioning relationship. The Sony Betamax cassette and the RCA VHS cassette are not interchangeable, but the signals can be transferred from one machine to the other through a connecting cable.

15

VIDEO PROCESSING

SPECIAL EFFECTS

The dreams of your own studio become more frequent as you accumulate video equipment; you soon find yourself unable to resist the thought of all those knobs and dials, and the magnificent shows you'll produce, which will melt the cable wires with their splendor. You might as well surrender immediately; you have a case of terminal video. Video production equipment is the ultimate fantasy trip, the urge to have a television studio in your living room where you can mix, fade, wipe, and superimpose your vision of commercial television as it might be if you were running NBC.

In this chapter we will discuss the various pieces of equipment available for producing more involved visuals than are possible with the portable one-camera/one-deck system. These include video switching equipment for mixing and modifying camera signals, multiplexing and film chain equipment for adding film and slide visuals, genlocking units to combine prerecorded tape segments with live camera signals, and other special circuitry that will allow you, if you can afford the price, to make much more elaborate and complicated video productions.

Once you have a portable video unit and an editing deck you can produce neat tapes with proper edits. As you work with this basic equipment you'll become aware of the possibilities involved in making programs with two or more cameras simultaneously, switching between the cameras to choose the best view of the subject at any given moment.

The **video switcher** is the basic piece of video production equipment, although it is usually packaged as a **special effects generator** (SEG), a combination of a switcher with a **sync generator,** and keying, genlock, and wipe-insert circuits. Various switchers include some or all of these facilities. When you choose your SEG you can then determine which of the functions you're getting, which can be added on later, and which are not available with the unit.

Switching
Although the one-camera-to-one-VTR relationship seems natural for the initial involvement in video, it is not the way that video developed as a communications medium. Long before there were video tape recorders, cameras were in use, their signals being supplied to a transmitter for broadcast to home receiving units. At the time no way had been developed to record materials and then assemble them later through editing. Any change in camera view or position had to be accomplished during the actual time the program was being broadcast. The use of more than one camera developed out of this situation, the ideal number being three or four cameras, each trained on a particular view of the scene. The signals supplied by each camera were monitored in a control room and the best signal at a given moment was

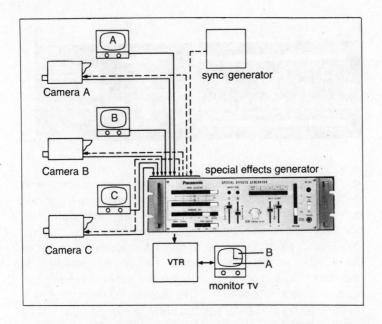

Camera A

Camera B

Camera C

sync generator

special effects generator

VTR

monitor TV

B
A

selected to be supplied to the transmitting equipment. The other camera signals were held in the control room, and when one camera signal was judged to present a better view of the scene than the one **on the line** to the transmitter, that camera signal would be switched on the line and the signal on the line previously would be switched off.

To understand switching, think of a camera plugged into a monitor: On the monitor is the view of the signal from that camera. To present the signal from a second camera on the monitor, you would unplug the first camera and plug in the second camera—the signal on the monitor has changed from camera 1 picture to camera 2 picture. Switching is not quite so easily accomplished since there is a disturbance in the picture signal during the time one camera is being disconnected and the second is being connected. To defeat this problem, switching equipment has been developed. The signals from the cameras come into the switcher and are there selected; the chosen signal is sent out of the switcher to the VTR, monitor, or transmitter without any visual disturbance of the picture information, since the switch to that signal takes place prior to its being fed out of the SEG.

Basically, a switcher is a unit that switches between camera signals. If you have three camera signals coming into the switcher, the switcher allows you to terminate two of those signals and lets one pass on out of the switcher to the VTR or monitor. Then, if you want to change the signal going out, you switch from one camera signal to another; the first signal is terminated within the switcher while the second signal is allowed to replace it as the signal leaving the switcher.

SPECIAL EFFECTS

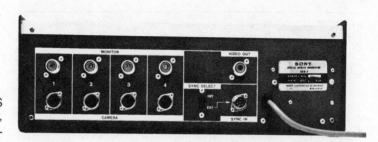

SONY SPECIAL EFFECTS
GENERATOR SEG-1,
REAR PANEL

Buses

There is a great deal of terminology associated with switching equipment. The first term you must understand is bus. Bus is both an audio and a video term that stands for one complete in-out circuit. When three camera signals go into a switcher and any one of them is selected to go out of the switcher as the chosen signal, *and* all are controlled by one master video gain control, the circuitry involved is described as being "one bus." This bus can be thought of as a channel along which the various camera signals flow. With a one-bus switcher you can regulate only the "volume" (gain) of the camera signals; you can't switch between them. A two-bus switcher has two channels, and the camera signals can be selected at the input point to be fed to either bus A or bus B. With two buses it is possible to mix together the signals from two cameras by adjusting the master gain control for each bus or by switching between bus A and bus B to determine which signal will be sent out of the switcher to the VTR. A three-bus switcher has three master channels, each with its own controls, and is usually found in switchers that allow you to take the final signals from bus A to bus B and feed them into bus C before feeding them to line-out. This allows a variety of special effects.

The first thing you'll encounter when using a switcher is the signal input patch panel. This is where the cables leading from the cameras are connected to the switcher. These inputs are known as *line-in*. Different switchers have different types of input receptacles, ranging from UHF connectors to special pin input receptacles designed for use with cameras that are made by the manufacturer of the switcher. Nippontism! There will also be a signal output receptacle on the switcher—known as *line-out*—usually a UHF connection from which the output program signal is fed to the VTR.

Switchers vary in their monitor facilities, some being more elaborate than others. Generally there is an output receptacle to feed a master or program signal out to a monitor so that you can see the final signal as it is leaving the switcher for the VTR through the line-out output. There are also monitor output receptacles for each of the buses, so the picture from the camera on each bus can be monitored. These are known as preview monitor outputs since the sig-

THE VIDEO PRIMER

ONE-BUS SWITCHER

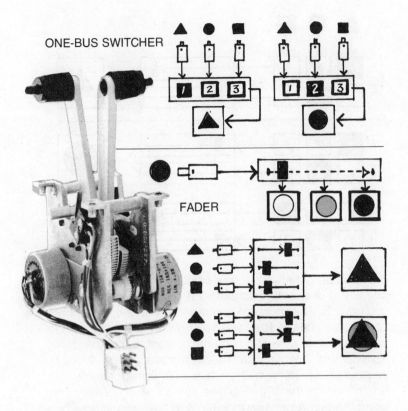

FADER

nals supplied from them are the camera signals prior to their being processed through the switcher. On more complex switchers that allow for special effects, there will also be other preview monitor signal outputs for the effects so they can be set up on the preview circuitry prior to being used as part of the program line-out signal.

The uses to which you can put a one-bus switcher are limited. It will have gain controls, known as *faders*, for each of the video cameras coming into it. A fader will increase or decrease the intensity of the video signal from a full picture to black (no signal). On a one-bus switcher it is possible to mix camera signals, or "switch" between them, only by fading out one camera signal and fading up another. It works in much the way an audio mixer does.

A two-bus switcher allows both fading and switching. Two-bus switchers have a set of buttons that lets you determine which camera signal goes on which bus. If you have three camera signals coming into a switcher and you have 2 buses, two of the signals are assigned to their own bus and the third is held for future use. Each of the buses has a series of buttons numbered from one to five (or the total number of camera inputs available on that bus). You **punch up** one camera onto each bus. With three cameras you might put camera 1 on bus A, camera 2 on bus B, and camera 3 on hold. You can switch between the two buses,

Controls

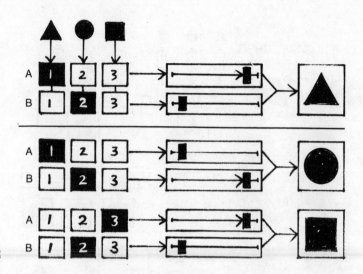

but you use only two camera signals, one per bus, at any one time. For example, camera 1 is on bus A and camera 2 is on bus B. The program might begin with the signal from camera 1 on bus A. Then you might switch to the signal from camera 2 on bus B. Then you might want to use the signal from camera 3. While your line-out signal is camera 2/bus B, you punch up camera 3 on bus A (moving camera 1 to hold), monitor the signal to see that it's what you want by using the bus A preview monitor, and then switch from bus B to bus A, thus making the line-out signal camera 3. What you are actually doing is switching between bus A and bus B, not between camera signals—the camera signals are simply being assigned to either of the buses and the switching is done between buses rather than between cameras. In a switcher with two or more buses, there are preview facilities for each bus, which allows you to preview the signal on one bus while the other bus is feeding the other signal to the program line-out. Sometimes there are also monitor outputs for each camera into the switcher.

A three-bus switcher has three channels. Two are for fading and switching the signal, one is for applying any of the signals from bus A or bus B to bus C, which is an **effects channel** where the signals are varied to produce special effects.

Two- and three-bus switchers are the types most often encountered in video production work. A one-bus switcher has such limited capabilities it isn't often used, and because it is inexpensive and designed for budget applications, its internal electronics are usually of inferior quality. The simplest form of switcher for most uses is the two-bus switcher; it allows the user to switch between two camera signals.

Switching has a problem inherent to it: As one signal is cut off and the other put in its place there is a loss of sync. If each of the two camera signals coming into the switcher is running on its own internal sync, the sync will vary between camera 1 and camera 2. This shows up when the two camera signals are combined or exchanged in the switcher as some picture instability, usually a vertical roll. The solution is to have a sync generator to supply the *same* sync signal to both cameras. Most lower- to medium-priced switchers have their own internal sync generators, which supply 2:1 interlace sync to each of the cameras connected to them. These sync generators are "external" in the sense that they are not in the cameras, but "internal" in that they're built into the switchers. The cameras must, therefore, be able to take external sync. It is possible to use a sync generator that is independent of both the cameras and the switcher. This is the most reliable, professional method of supplying sync, but no matter whether the sync signals come from the switcher or from an external generator, it's necessary for all of the cameras being fed into the switcher to be locked to one common sync source.

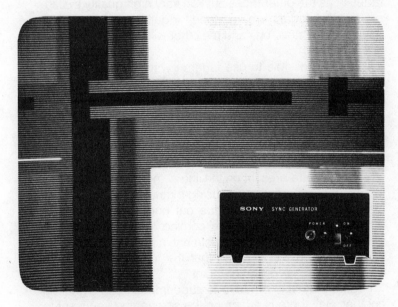

VERTICAL INTERVAL

With all the camera signals in sync, the problem of switching between signals is lessened. There will, however, still be some disturbance of the signal if the signal from camera 1 is replaced by the signal from camera 2 with no regard for field structure of the signal. If you were to switch from camera 1 to camera 2 in the middle of a field (even though the cameras were in sync and the point in field A was replaced by the exact same scanning point in field B), it is likely that there would still be some picture roll and momentary loss of sync. This problem has been solved by the introduction of the **vertical interval switcher**. This device ensures that when

SPECIAL EFFECTS

269

you switch between camera signal 1 and camera signal 2, the actual switching takes place during the change of fields, the vertical interval, so no disturbance to the actual fields-picture signal takes place. Since the fields are in sync, the change occurs at the same time in both signals.

Fading

There are various methods of changing from one camera signal to another besides the instantaneous direct switching via the buses. Using the fader controls is one such method. Each fader control is connected to a bus (in the two- and three-bus switchers) and allows the user to vary the intensity of the video signal from full to nil. When you see a video picture begin to bleach out and fade away while another picture begins to materialize in its place, fader controls are being used. Faders may appear in any number of configurations: dials, sliding levers, or the T-shaped levers used on professional equipment. Faders are commonly set next to each other; one operates in the reverse direction of the other, making it possible to control the two of them as one unit, pulling them down or up together so that one picture fades in as the other fades out. On very high quality switchers the two faders are "ganged" together so that all you have to do is move one and the other will move in the opposite direction.

It is also possible to use faders by starting with a blank screen and then fading in the first signal, going through the program, and finally fading out to a blank screen again after the last scene—a very professional look.

The faders on some equipment present a problem when used together. During the fade-in/fade-out of two pictures there is a point at which a considerable amount of signal 1 and signal 2 are present on the line at the same time, creating a signal overload and causing picture distortion. To defeat this it's necessary to fade out one picture a little in advance of the other picture coming in. Some switchers avoid this with an internal adjustment in the signal strengths at that point in the fade, so that the total signal strength doesn't reach the distortion point.

Superimposition

Faders may also be used to superimpose one signal on another. Say there is a face on the screen and you want to put a set of iron bars across the face. Camera 1 on bus A through fader A supplies the face signal. Camera 2 on bus B through fader B supplies the iron bars signal. Camera 1/fader A is on the line, then camera 2/fader B is faded up until the bars appear across the face. This effect will not be totally convincing since it will be necessary to reduce the signal on fader A (fade down slightly) as you bring up the signal on fader B to avoid overload distortion. The final combined signal, face behind bars, is a full-strength total signal, but it may still look a little washed out, since neither signal is at its full potential.

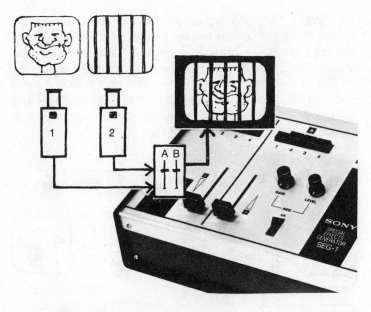

Special Effects Techniques

One of the standard special effects available on almost all switchers is the split screen—portions of one camera signal taking up part of the screen and portions of a second camera signal taking up the rest of the screen. Dividing the screen in this manner is referred to as an **insert** or **wipe**, and *Inserts* it is achieved by inserting one picture signal into the field of another picture signal. Inserts are very commonly used in television to display the faces of two people on the screen although the two people may not be actually next to each other.

Inserts can be set up in a variety of ways: circles, squares, or corners of the screen can be used as the area in which signal 2 displaces signal 1. Inserts are accomplished by an electronic replacement of one set of scanning lines with another set at predetermined points in the field. If you were going to insert a new picture into the bottom righthand corner of a picture, for instance, a specific area of picture 1 would be eliminated by halfing the scanning of the field of picture 1 at a certain point on the scanning field. This area would then be filled in by the scanning lines of picture 2. The two fields are laid on top of each other with a portion of field 1 visible and a portion of field 2 visible because a certain segment of field 1 is not being scanned.

Inserts are easier to comprehend when you realize that the picture information being inserted by camera 2 into the picture field of camera 1 is not camera 2's entire picture but just that portion which fits into the insert area. Let's say the full picture on camera 1 is of the side of a house and the full picture on camera 2 is of the side of a second house. An insert, say in the upper lefthand corner, of camera 2's signal on camera 1's signal would yield a picture of camera 1's house filling the screen except for the insert area in which the upper lefthand *portion* of camera 2's house would be visible. The entire house in camera 2's picture would not be visible in the insert unless, when filming, camera 2 were moved back and situated so that most of camera 2's picture was open space, with the house located in the upper lefthand corner.

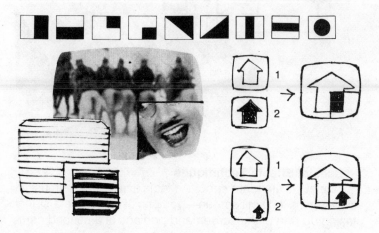

INSERT

It's possible to adjust the area of the insert signal with controls located on the front of the switcher. One control (usually a sliding control similar to the fader) allows you to set the horizontal edge of the insert (the scan line where the signals switch); the other control allows you to set the vertical edge of the insert (the points on the scan lines in the insert area where one scan line stops and the other signal scan line begins scanning). By moving the controls you can establish the size of the insert area. These controls are used along with the **effects buttons**; each has a design pictured on it, usually a black area on a white button, corresponding to the area and shape of the insert. For corner inserts (the most common insert effect on inexpensive SEGs) you would have four buttons—one with a white field and a black corner on the upper left; one with a white field and a black corner on the upper right; one with a white field and a black corner on the lower left; and one with a white field and a black corner on the lower right. Each button is connected to two buses—the black portion to one bus, the white portion to the other. To make an insert, you punch up the insert corner wanted and then adjust the size of the insert area with the horizontal and vertical area control pots. The size can be anywhere between a very small portion of the corner to full screen.

The black-on-white designation of the buttons is also

used to indicate which bus provides which signal to which portion of the insert picture. Bus A, for example, might be designated black and bus B white. Camera 1 is on bus A, and bus A corresponds to the black insert area; camera 2 is on bus B, and bus B corresponds to the original picture into which the insert is made. If you switch the camera signals to their opposite buses (1 to B, 2 to A) the insert will be interchanged with the picture it is being inserted into.

Keying is another special effect available on some SEGs. B&W keying is an electronic process that combines the black portions of one picture signal with the white portions of a second picture signal to create a total signal that looks very electric and spacy—a heavy, heady effect which can be overused if you get carried away while you're playing with the keying control. There are various types of keying, but they are all accomplished by the same process: The picture signal from one camera is replaced by the picture signal from another camera at any point that the signal from the second camera exceeds a preset intensity level—a predetermined value between black and white. The switching of signals is done automatically by the electronics of the keying circuit.

Keying

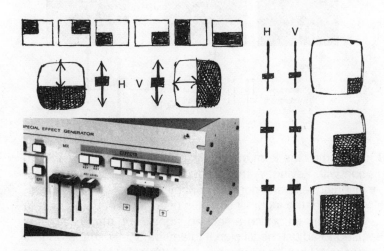

A simple keying effect is **internal keying**. Two cameras are used—one supplies the picture to be keyed into, the other supplies the keying signal. Internal keying permits you to display words or letters on a scene on the screen. This is often done in TV interviews or news shows to identify what's going on. The subject is viewed by camera 1. Camera 2 is focused on a black card with white letters mounted on it. The two signals are then combined through the keyer, and as camera 2 scans the black card, the white letters are of such a level and intensity that they automatically switch the

SPECIAL EFFECTS

scanning of the picture from camera 1's signal to camera 2's signal. The end effect is the insertion of the letters into the camera 1 scene, to replace the segment of that scene that the camera 1 signal is providing. A control knob or fader is used with the keying unit to adjust the signal level (from black to white) where this crossover takes place. This ensures that only the white letters—none of the black card—are inserted into the signal from camera 2.

External keying can provide the same effect. The difference is that the video **fill signal** (the white letters in this instance) is obtained from a camera being fed into a special keying signal input receptacle rather than using the signal from one of the cameras on bus A or bus B. Using a three-bus switcher with an external keying facility, it is possible to have one camera set up especially to provide a keying signal to switch the other camera signals. You'd have camera 1 on bus A, camera 2 on bus B, and the key camera coming in through the external keying input. You could then key between bus A or bus B based on the signal supplied by the keying input camera.

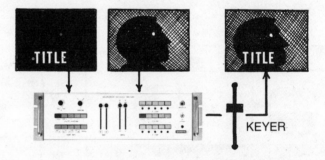

KEYER

Matte Effect

At times the term **matte** is used in place of or in conjunction with keying. A matte effect is yet another method for inserting letters or portions of another signal into a scene. In this case, the input fill signal is used to key out the picture areas and introduce a second signal. The matte control is used to adjust the fill signal to any particular shade of gray desired, from black to white, and thereby give the appearance of a solidly shaded keyed-in area, which contrasts with the original picture. If the original scene is all white, the matte could be adjusted to black so that the keyed-in material would stand out; if the original scene is dark gray, the matte could be adjusted to light gray or white so that the keyed-in material would stand out.

To avoid confusion, you should think of the matte control as that circuit which allows a b&w keyed insert to be tuned to any desired shade of gray. You may occasionally run into the use of the term *matte* to describe a keying effect; that is the result of the lack of standard terminology on the part of the equipment manufacturers. If you're confused about what it means on the piece of equipment you're using, check the manual. *Matte* is often used in video as the term to describe keying a person into a scene so that the person seems to become a part of the scene. This is accomplished by placing the person in front of a black or blue background and then keying the person sans background (which is below the key level) into the scene coming from the other camera.

USE OF MATTE

Inlay and Overlay Inserts

Two other insertion effects are defined by the type of subject you are keying into the picture. If the fill signal is of a predetermined shape and remains static at that shape so that the keying level can be adjusted to conform to that shape, this is known as an **inlay** insertion, also called a **keyed insertion** or a **static matte**. If the fill object being keyed in is moving and will determine its own fill parameters as it moves (such as a person walking across the scene for which you want to add a new surrounding background), an **overlay insertion** is used. It is also called a **self-keyed** or **moving matte insertion**. An overlay requires more attention to the keying level controls to keep the portions of unwanted original background below the level at which they would be included in the fill signal.

Keying is more fun and a lot simpler than it sounds here. Start playing with it and you'll develop any number of effects, from keying people into bizarre scenes to dropping your video logo into the picture.

The basic principle of keying is that the key control can be set so that anything darker than a particular level of gray (which level you determine) is not visible as part of the picture signal when it is combined with the second, non-keyed signal to form a composite picture.

Genlock   Genlock is not really a special effect, but it is one of the facilities available on some switchers. It's also available as an independent piece of equipment. The technology of half-inch video allows the mixing of two camera signals driven from the same sync generator. The switching unit or a separate sync generator drives the cameras. The signals coming out of the cameras are non-composite signals since the sync is added from an external source. If, however, you wanted to mix the signals of two prerecorded video tapes, you would have a problem since they're *already* composite video signals (having sync as an integral part of their total signal), but neither tape is in sync with the other. When you're editing the two tapes together, you can bypass the problem by locking the vertical sync of the editing deck to the sync of the deck supplying the next segment of picture information and making a simple cut from one segment to the next. If, however, you were to try to run the signals from two video tapes into the SEG, you would be asking the SEG to mix between the sync and picture signals of two unsynchronized video sources. External sync is not possible in half-inch playback. It is impossible to mix two prerecorded signals except by displaying each tape on a monitor, putting a camera in front of each monitor to scan the display, and then mixing the resulting live signals through an SEG. A certain amount of distortion would result from such a procedure since the picture screen is not perfectly flat. A genlock unit (sometimes called **conlock**) provides an alternative, allowing you to mix the signal from one prerecorded tape with one or more live camera signals. This is a very desirable effect and should be included on any SEG you buy. Genlock will add about $100 to the price. If you have to buy it later it will run from $300 to $600, depending on whether you buy a separate unit or have it wired into your SEG.

Genlock locks the cameras to the sync signal of the composite video tape being mixed with those live camera signals. The video output from the VTR is connected to the genlock circuit, which separates the sync signals from the rest of the tape signal. The sync signals are then fed into the sync generator from the genlock, where they are used as the basis for the sync signals sent to the various switcher inputs, including the cameras. You end up with the sync being provided from the sync generator as in normal operation, but this sync exactly matches the sync on the prerecorded tape. One input of the switcher is designated as the VTR input and you use the video tape signal as just another incoming signal, which you can mix, switch, and do effects with in combination with the live camera signals.

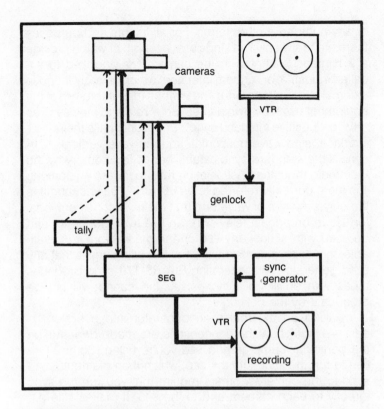

Sync Generators

As I mentioned earlier, sync generators are available as units separate from the SEG. Most SEGs and switchers have provisions for an external sync generator whether they have internal generators or not. An external sync generator is usually of better quality than an internal one, since it isn't a circuit thrown into an SEG but a separate piece of equipment made especially for the purpose. A good external sync generator (about $500) provides a more stable sync signal than the generators built into the under $1000 SEGs now on the market. This is especially true of the really cheap SEGs (under $700) whose internal sync generators provide extremely poor stability. It's interesting that the most expensive SEGs made for professional broadcast use ($3000 and up) don't include internal sync at all—the manufacturers assume that you'll be investing the money necessary to get a reliable external sync generator.

There are a number of sync generators on the market, and your choice will depend on the type of sync signal you want to supply to your SEG and cameras. All b&w sync generators provide four output signals to control the SEG and camera operation: horizontal and vertical drive pulses, blanking pulses, and composite (mixed horizontal and vertical) sync pulses. The type of overall sync provided by any sync generator can be described as "2:1 interlace," "EIAJ sync," "NTSC sync," "EIA sync," "positive interlace sync," or any number of other terms. For all intents and purposes there are really only two kinds of sync, both of which provide a 2:1 interlace control to the camera (as opposed to so-called industrial or random interlace, which is rarely found in external sync generators). The various manufacturers of the equipment use their own names. Sony refers to their sync as "HV" or "positive interlace sync"; Panasonic calls theirs "2:1 interlace, sine wave horizontal, square wave vertical." **EIA sync** is the standard sync established in this country by the Electronic Industries Association for broadcast equipment; it is more correctly referred to as EIA RS-170 sync, and it is the only sync signal waveform that the FCC considers legal for use in broadcast television. An EIA sync generator can be used with almost any SEG on the market, but it is available only as an external unit. What this all means is that any sync generator not generating EIA RS-170 sync produces tapes with an illegal sync pulse, and cannot/will not be broadcast by the networks.

Installing an external sync generator into a system is easy—exactly how the connections are made depends on the particular cameras and SEG you're using. Some allow you to run the sync into the SEG, which then distributes the sync to the cameras; others require that you run the sync directly to each camera and to the SEG. It makes little difference.

Some sync generators are available with a built-in genlock facility. If you don't have a genlock on your SEG and are planning to add external sync, it would be wise to invest the extra money in a sync generator with genlock.

Tally lights and intercom systems, sometimes known as tally facilities, are an extra unit that can be added to some SEGs. They provide a signal to cameras having a tally light and intercom input/output. The tally system is interfaced with the SEG so that whenever a particular camera is put on the line, the tally light is activated so those on the set know which camera is the program camera. The intercom facilities allow the control room personnel to talk to the camera operators and others on the studio floor. Intercoms can be two-way or three-way. The two-way systems use a set of earphones with a boom mike attached to them to let the camera operators and director talk to each other. The three-way systems let the camera operators and director talk and hear each other as well as, in one of the two earphones, the on-line audio portion of the program.

Amplifiers

Processing amplifiers, known as **proc amps**, are one of the great fables of video. They are rumored to be the cure-alls for video instability and loss of quality. The legend goes that all you have to do is run your composite video signal through a proc amp when you're making copies from your master tape and, suddenly, your tapes will be of broadcast quality. That's not exactly true. I don't mean to belittle the place of a proc amp in studio operations since it *can* improve the level of your tape copies, but it's not going to radically alter what you've already got on tape.

Called by various trade names, including **signal processor**, **video processor**, **helical scan processor**, and the like, a proc amp is inserted on the line between any two points through which a composite video signal is traveling, such as two VTRs. All the control signals, including horizontal and vertical sync and blanking, are replaced by the proc amp, which regenerates the control pulses on the tape, usually with EIA sync pulses. A good proc amp will also **limit** or **clamp** the video signal and the video pulses to a fixed level. This may slightly improve the picture quality.

If you are transferring your half-inch video tapes to one- or two-inch VTRs for editing, or if you're planning to make third-generation copies from an edited composite master tape, a proc amp should be used to provide strong control pulses. It will *not*, however, remove the picture noise that has accumulated by the time you've reached the third generation.

At this time proc amps cost about $1500 but you should consider one if you're producing a large number of copies from a second-generation master tape.

A **time-base corrector** can also be inserted on the line between VTRs, as when editing, to improve signal quality by ensuring the stability of the picture. Time-base correctors sense the horizontal pulses of the incoming composite video signals and adjust the rate of their occurrence to close tolerances. Each new scan line is started at exactly the correct time, often by the use of digital computer techniques. A tape processed by a time-base corrector will produce an absolutely stable picture for broadcast: It will not, however, improve the actual resolution or quality of the picture; it simply makes it stable on the screen. At the moment, time-base correctors cost approximately $10,000. Time-base correction is the best way to stabilize recordings made on U-Matic or other formats for broadcast or duplication.

TIME-BASE CORRECTOR

Most TV studios, stations, and networks record news footage, public affairs shows, and other locally originated events on U-Matic cassettes, edit the program on cassette, then run the master cassette signal through a time-base corrector before broadcast. As long as good quality cameras are used, it is difficult to tell the difference between cassette record/cassette edit/time base broadcast and a 2-inch quad recording.

Another item that is inserted into the video system in some studios is the **pulse distribution amplifier**. This amplifier is used if you are running sync signals to a greater number of cameras than your sync generator has outputs, or if you are running the sync signals great distances along cables and find that the signals are too weak by the time they reach their cameras. In either case, one set of sync signals is run into the pulse distribution amp, where it is regenerated, reshaped, and boosted back to its proper level, before being fed to four or more output receptacles. With a four-output sync generator and a four-output pulse distribution amp it is possible to service up to seven cameras with sync.

The pulse distribution amplifier should not be confused with the **video distribution amplifier**, which does what its name implies: It amplifies the video signal from any particular source so that the signal can be distributed to a number of outlets. If you have one video camera and want to display its output on more than four or five monitors, feed the camera signal into the VDA and then feed the signal from the amp to any number of monitors. If you were making tape copies and wanted to record more than one copy at a time, the signal from the master VTR would be sent to the VDA, and the signal the VDA produces would then be sent to the VTRS used to make the copies.

One final amplifier you may have use for is an RF amplifier. This unit is fed the 75-ohm signal from the RF generator of the VTR which it then amplifies so that it can be fed to about forty television receivers. RF amplifiers are available from most large consumer electronics dealers such as Lafayette and Radio Shack. Their cost depends on the number of receivers they will service with a signal, the general range being from $20 to $100.

Film Chains

Other studio equipment that comes under the general category special effects are the **film chain**, **multiplexer**, or **telecine** units, which allow you to insert film or slides into your video programming through your SEG.

The phrase *film chain* comes from the days when a camera and its associated equipment was known as a *camera chain*, and so a movie projector and camera combined became the *film chain*. A multiplexer scans movies and con-

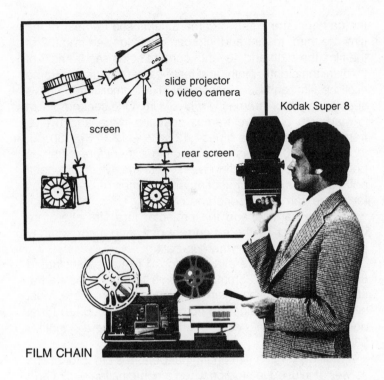

slide projector
to video camera

screen

rear screen

Kodak Super 8

FILM CHAIN

verts them into a video signal, which can be added to your program. The unit consists of a camera pointed into the lens of a movie projector in such a way that the projector throws its image right into the camera lens-vidicon assembly. The camera signal is then fed to the SEG for mixing or it can be fed directly into a VTR or monitor.

The process sounds simple, but the film chain is a specialized unit that cannot be adequately replaced by simply pointing a camera and movie projector at each other, lens to lens. This can be done in a pinch if the speed of the film is regulated until an acceptable, but not perfect, signal is obtained. The problem is that 16mm film, which is the standard used by television, runs at twenty-four frames per second whereas the video camera runs at sixty fields per second. A certain amount of flicker would be evident if the camera merely recorded the film being projected—the film frames would not match the camera fields one to one.

To overcome this, the motion picture projectors used in film chains have been modified to project the first frame twice, the second frame three times, the third frame twice, and fourth frame three times, and so on to the end of the film. The period of each projection is shorter than in normal screen projection and with twenty-four of the first of each two frames plus thirty-six of the second of each two frames being projected in one second the total equals the field rate of the video camera. To explain it from the point of view of

SPECIAL EFFECTS

the camera: The first two camera fields see frame 1 of the film, the third, fourth, and fifth camera fields see frame 2 of the film, the sixth and seventh camera fields see frame 3 of the film, the eighth, ninth, and tenth camera fields see frame 4 of the film, and so it goes until sixty camera fields have seen twenty-four "frames worth" of film. One second of film is made up of twelve frames scanned twice and twelve frames scanned three times—$12 \times 2 + 12 \times 3 = 60$. Although the projector is not projecting sixty *different* frames per second (it is projecting twenty-four fps, which is what it normally does), it is projecting sixty frames per second as far as the camera is concerned.*

Both the camera and the projector in a film chain are locked together by the ac current powering them, so they start and run together when operating.

Film chain units are available for all film sizes, including 8mm, Super 8, and 35mm, as well as 16mm. Prices start at about $1000 for a film chain and skyrocket from there. Most high-quality film chains run from $3000 to $6000, and that doesn't include the cost of the color camera—another $3000 minimum.

A multiplexer is a film chain that performs other functions as well. It is used in studios when a combination of film and slides is needed as part of the programming. Slide projectors can be positioned lens to lens with the video camera, but often both slides and film are needed. It is too expensive to get a video camera for every film projector or slide projector used, so a multiplexer is used instead. This system of mirrors and prisms allows the same camera to be used to receive projections from a number of slides and film projectors. It's a tremendously expensive unit requiring exact adjustments to operate properly. With it you can switch the camera input from film projector 1 to film projector 2 by moving the position of the mirrors. In the same way it's possible to switch between two slide projectors—which is necessary if you're showing a series of slides, since you must use two projectors to show two slides in a row or else you'd have a blank screen during the time a single projector changed slides.

You don't need a multiplexer if you want to use slides as part of your program material, however. Point the projector into the lens of the video camera, adjusting the f-stop of the camera and the wattage of the projector so that the light

*The European video standard uses 50Hz as the vertical field rate, or fifty fields per second. So the Europeans don't have to get involved in this special projection equipment. They simply speed up their movie projectors from 24 fps to 25 fps and each frame is scanned twice by the camera to get fifty frames of film in fifty fields of video. The only drawback is a slight pitch change in the film's audio soundtrack.

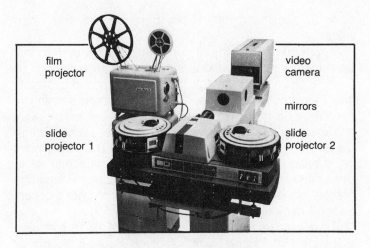

film
projector

video
camera

mirrors

slide
projector 1

slide
projector 2

MULTIPLEXER

levels are satisfactory. If you want to make sure that the strong light from the projector won't burn out the vidicon, project the slide on a small screen and point the camera at the screen, keeping the axis of the projected beam and camera lens as close to each other as possible to avoid distortion. Another method is to get a small rear-screen projection unit, or a piece of frosted glass, and put the projector on one side and the camera on the other. This system is good if you're showing only one slide rather than a series. Try using the slide as background and keying objects or people into it.

Super-8 Sound

Inexpensive portable video equipment has succeeded because it is an alternative method of creating audio/visual communications that is more easily manipulated and cheaper to obtain than broadcast TV gear. Recently, an alternative to the alternative has developed: Super-8 Sound movies.

Kodak's Ektasound Super 8 and other Super-8 Sound systems allow single-screen recording (sound and picture recorded simultaneously), and their introduction heralds an important development in portable visual communications. The time is upon us when the individual video maker must make a choice: to invest in portable color video equipment or to shoot wild footage on the much lighter and more portable Super-8 Sound equipment.

There are some significant considerations in the choice of Super-8 Sound versus portable video. On one hand, the Super-8 camera, sound projector, and sound editor setup are a quarter of the cost of even a black-and-white video camera and deck, and the Super-8 equipment is much smaller. On the other hand, processed film costs $3.00 a minute, whereas reusable video tape is $.50 a minute. And there are aesthetic differences between film and video visuals.

A compromise worthy of consideration is a system that combines Super 8 and video. An event can be filmed on Super-8 Sound, the footage transferred to video cassettes and then edited on video. The result is programming of excellent quality, with the warmth of TV film, but which, when hardware outlays are taken into account, costs considerably less than making color video programs. The pivotal point in the system—a color-capable camera—is what makes the price difference since the least expensive color video camera that works costs about $6000, but you can get a decent Super-8 camera for around $200.

The new-wave rock movement of the late 1970s spawned a feature film, "Punk Rock," shot on Super-8 Sound by Donovan Letts while he was working as a D.J. at rock clubs in London. The footage was transferred to video cassette, edited on the U-Matic editing system, then transferred to 35mm film for theatrical distribution. The technical result looked more like "Birth of a Nation" than like "Star Wars." Nonetheless, it worked.

There are some technical drawbacks to Super-8 Sound at this writing: The stability of the film as it is drawn through the film gate in the Super-8 camera can be less than rock steady. The result, when the film is transferred to video through a film chain is a slightly shakey picture. Also, if the film is projected or handled prior to video transfer, it comes out scratched and blurred. But if you shoot Super-8 Sound with a good Super-8 camera, transfer the film immediately after it is processed, then watch the playback on video and make editing decisions from the video transfer, much of the technical difficulties can be avoided, and the program you end up with is often hard to distinguish from 16mm TV news or movie footage.

You can have the film transferred professionally for about $10 an hour, plus tape, or you can do the transfer yourself. Kodak sells a Super-8 color film chain (film played on it comes out as a video signal) for about $1000. This, combined with a good Super-8 Sound camera ($300 and up), is worth thinking about as an excellent way to make "video."

Production Techniques

Special effects equipment really puts you into the world of mission control where you can sit behind the console and create a sense of electronic destiny that is peculiar to the medium. At your fingertips are buttons and sliders that literally let you put people and things in or out of the picture. In network television the complexity of the equipment used, plus the presence of unions on almost every level of the broadcast industry, requires that more than one person push each button. (Limitless ethnic jokes can be inserted here.) On the SEG board you'll find a director who says, "Switch to camera 2"; possibly a technical director who then repeats "Switch to camera 2"; and finally a union man who can be heard muttering something about switching to camera 2, by which time the man on camera 2 has gone out for coffee. With small format video you can have your own personal TV station and run it any way you like. This applies as much to the techniques you use as to the subjects you choose to tape.

The first thing to do is position your equipment properly in the area in which you will work. Special effects generators and their related equipment cannot be readily carried around. There are semiportable units, but even these require a source of ac and a good deal of muscle to transport.

The barest minimum of equipment required for mixing and switching are two video cameras, a switcher with built-in interlace sync, a video tape recorder, at least two monitors (one for preview, one for program), and a microphone (preferably more than one mike with an audio mixer). This equipment must be deployed so that it doesn't get in the way of the event being taped but, at the same time, is set up so that it can be operated conveniently.

If you are working with a three-person crew—two people on the two cameras and you working the controls—the VTR, monitors, SEG, and audio mixer all have to be placed within

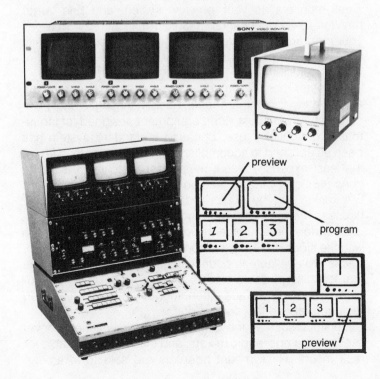

preview

program

preview

SPECIAL EFFECTS

285

your reach. A couple of small tables, one in front of you and one to the left or right, plus a chair are needed. Try using metal typewriter tables with fold-out leaves, and sit on a padded typing chair with casters. Set the switcher and audio mixer on the table in front of you with the monitors behind them so they can be viewed as you look in the direction of the SEG and camera-subject area. Place the VTR on the side table so you can turn to it when you want to start or stop the tape and are able to glance at it from time to time to see that it is running properly.

Equipment Setups

As you work with this minimum of equipment you'll almost immediately feel the need for two things: better monitoring facilities and some method of mounting the equipment so you're not totally surrounded by a mass of wires and components. Monitors are available in sets of two to four, the screens placed next to each other in a row. They are usually 3–5 inches (7.6–12.7cm) measured diagonally and are easier to work with than larger monitors. A set of four monitors in a row is ideal—one for each of two cameras, one switchable over to an effects preview so you can work out effects prior to inserting them into the program, and one for the master program signal going to the VTR. If you don't mind switching back and forth between the various camera and effects previews on one monitor, you can work with three or even two monitors, but four is best since it allows you to monitor most of the signals coming into and going out of the SEG at any moment. You may want to use a set of four small monitors as your preview system and one larger monitor—8–12 inches (20.3–30.5cm) in screen size—as your program-out monitor. A sheet of slightly colored plastic on the screen of the program-out monitor will make it instantly identifiable.

The other major improvement you can make is to mount all of the equipment in a rack. **Rack mounting** is easily accomplished with most video equipment designed for industrial or educational use, because a standard system has been established to accomplish just that. Except for a few of the least expensive SEGs, most video equipment is 19 inches (48.3cm) wide or less and conforms to the standard EIA racks sold by audio, video, and electronics supply houses. The racks are open-front cabinets just over 19 inches wide. The equipment slides into the rack so it's flush with the sides and only the front control panel can be seen. Racks are available in countless configurations from those that look like portable closets to space-age desks that include a working surface extending from the front and tilted rack area above the desk surface.

Monitors, such as the four-monitor set mentioned above, SEGs, proc amps, audio mixers, and the like are available for rack mount. Rack mount adaptors are sold to fit around

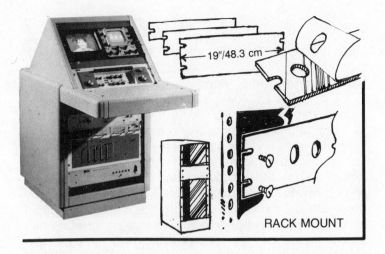

RACK MOUNT

smaller pieces of equipment so they can be racked. The only piece of equipment you can't rack at this point is your VTR, since it must be operated in a horizontal position and does not conform to the 19-inch width.

The backs of the racks open up so that you can run the camera cables out of them to the cameras, the SEG-to-VTR line to the VTR, and so forth. Although racks are expensive, ranging from $100 to $200 for a basic unit, they make things much neater and easier to use.

It is possible to purchase blank rack panels in various heights from 1–10 inches (2.5–25.4cm) to custom-mount equipment that is 19 inches wide or less and doesn't have a rack mount adaptor sold for it. The blank panels are made out of aluminum with a paper covering on which you draw the holes you need, then drill, and finally strip off the paper.

The semiportable video production unit I mentioned above is really just a combination of various pieces of video equipment—SEG, audio mixer, and a set of monitors—mounted on a portable rack unit with a metal front cover that closes and locks and a set of handles on the sides for carrying.

Having set up your production center, the next step is to start using it. Once more, I'd like to emphasize that the more convenient and neat your equipment layout, the easier it will be to work with. If it's a drag to sit down behind your production equipment you'll find yourself avoiding it.

It's possible to produce highly professional tapes with an SEG, audio mixer, two monitors, VTR, and two cameras. A third camera will be an additional help, even if you have no one to work it. Just set it up to cover the total scene you're taping while the other two cameras are worked by operators. A fourth camera might also be considered, not to cover the action, but set in another area to focus on title cards and logos.

SPECIAL EFFECTS

287

If you're starting with a two-camera system, it's handy to use one of those cameras as a portable unit with an ac camera adaptor—and the other on a tripod. Take advantage of the light weight of the portable camera and the techniques used when working with it.

You must learn to operate a full production system giving as little attention as possible to the mechanics involved and as much attention as possible to aesthetic decisions. Initially, you won't find this easy since even the least complex SEG requires a complete understanding of its controls before it can be used effectively—and when you get into setting up inserts or keying effects you'll find that they can't be done with just the flick of a switch. You'd do well to begin using the equipment prior to actually making tapes. Set up two or more cameras, turn everything on, and practice switching back and forth between cameras, using the faders (it's often hard to remember which is going to fade in and which is going to fade out), setting up inserts and then punching them into the program line; generally trying to discover all the variables available to you with your SEG. As you work the controls and try various combinations, keep a grease pencil or china marker handy and mark directly on the control panel any instructions, level settings, or directions of fader movement that might be unclear during the pressure of actual taping. Professional engineers do it. Also make sure that your monitors are properly labeled—either with a grease pencil right on the screen or with identification strips below the screen so that you can look up at the monitors and instantly see which is program, preview, camera, etc.

As you work with your equipment, try to establish a setup procedure that can be followed during the preparation for actual recording. Make a list of the things that must be done before the action starts:

1. Video heads cleaned
2. Tape threaded or cassette loaded
3. VTR on and in record mode
4. Cables connected
5. Cameras properly connected to the SEG
6. SEG controls set at the proper positions so you open with the correct camera
7. Monitors attached to the correct SEG outputs
8. Audio controls set
9. Audio connections made to the VTR
10. Mikes in place.

It may even be advisable to run a test tape, turning on all your equipment and running through the camera switching procedures on the SEG, then playing back the tape to see that everything is working properly before you begin the program.

Once you start to record the event, you are in the realm of live television. You can stop, rewind, and begin the segment over, but the advantage of using production equipment is that you can make a completely edited, titled, effect-laden show without recourse to post-production assembly, thus getting a first-generation master tape—the most desirable form of master tape when it comes to making stable, high-quality copies.

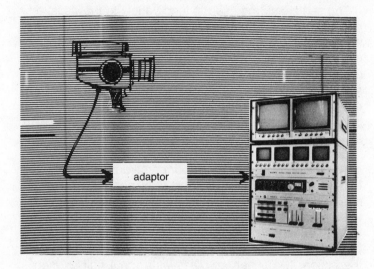

PORTABLE CAMERA
TO SEG

Working with the SEG during actual recording requires a well-developed sense of pacing. Do not work yourself into a corner by switching cameras every ten seconds. You must realize that you're producing a show for later viewing; make the recording of the message or contents your primary concern. Effects and tricky camera angles shouldn't get in the way. It's better to start off by switching only occasionally from camera to camera, by using keying, inserts, fades, and wipes as frugally as possible. There is no need to do anything if your subject is interesting and the camera view of that subject is good—just let the tape run in much the same way you would if you were taping with a single camera and deck. As you review your recorded program you'll begin to discover when the camera angles should have changed, what the sense of timing should have been, how your camera operators are responding to the subject they're taping. Once you acquire this knowledge, start taping again with a sense of what was right and wrong. You may find that you need to switch from camera to camera more frequently. You may feel that inserts or prerecorded materials run in through the genlock would help the flow of the program. That is what

SPECIAL EFFECTS

289

you are trying to establish: a flow that will make the electronics the easel for your video portrait rather than a distracting sideshow.

Most major television productions are done over the course of several days and include a run-through to check for camera angles and various potential staging problems. While you may not be able to hold your subjects for that long a period of time, there is no reason why you can't do a run-through prior to actually recording the tape, even if it's a condensation of the events as they will take place on the completed tape. This is called **blocking out** the action.

The SEG should be used to lend variety to the visual display of your program material. Switching between two long shots doesn't do this, but switching from a long shot to close-up might, under certain circumstances. Begin to understand camera angles as methods of examining your subject and communicate this understanding to your camera operators. Rotating assignments between personnel will help—let the camera operators run the SEG while you operate a camera.

Eventually you may want to invest in an intercom system so you can speak directly to your camera people, although behind-the-scenes conversation may necessitate the removal of the production system from the actual taping area to some other room where your talking won't be picked up by the mikes. This can be done by dividing a room with a partition that includes a glass window so you can watch the action (two sheets of glass with about half an inch—1.3cm—of space between them to buffer the sounds in either area) or by setting up the equipment in a separate room and using the monitors to direct the camera work rather than watching the action live. If you do this, set up the camera and monitor so you get a view of the entire scene including the other cameras.

Even when you become proficient with the SEG you'll find that running both the audio and video controls simultaneously is difficult. If possible, have someone else work the audio mixer while you work the SEG. Again, it all depends on the number of people you can count on to work with you.

Don't try to operate all this equipment in the dark. Invest in a couple of good lamps, such as the extension or small high-intensity lamps sold for use in workshops, and set them up over your equipment. You'll be amazed how much easier it is to clean the heads, thread the tape, and see the controls on your SEG if you have very bright lighting available when and where you need it.

Production with an SEG

What follows is a sample production run-through using three cameras, a special effects generator, an audio mixer,

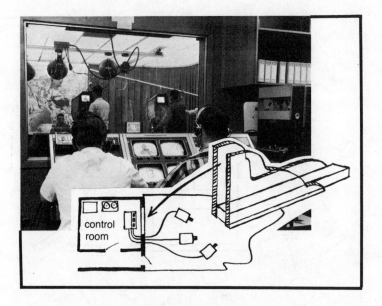

two microphones, and a lighting unit. The subject/action is an interview with two people sitting in a living room in chairs about three to four feet (.91–1.2m) apart. Not the most exciting set of audio/video events, but what you should concern yourself with here is not the subject/action but the behind-camera procedures.

1. The area in which the action will take place is defined. In this case, two chairs are placed facing each other but slightly turned toward the camera area in front of them.

2. The area in which your equipment will go is defined. In this case, it is in a space about ten feet (3.05m) away from and directly in front of the subject/action area.

3. The SEG, monitors, and VTR are connected to each other and set up so that the director can view them and have convenient access to their controls, but still see over them toward the live action.

4. Two of the three cameras are put in position. In this case, one camera is set back as far as possible to take in the full view of the subject/action area, one is placed closer to the subjects to get close-up shots. These two cameras are then connected to the SEG, one to bus A, one to bus B.

5. The lighting units are now set up. The cameras and SEG are turned on and the lighting is adjusted on the subject/action area until it's satisfactory.

6. The audio mixer is next set up. It is connected to the audio input of the VTR. Then the two mikes to be used (lavaliers or mikes on stands) are connected to the mixer and deployed to the subject/action area. Test the mikes.

7. At this point, both cameras and mikes should be numbered to avoid later confusion. The easiest way is to label left camera and mike as 1, and right camera and mike as 2.

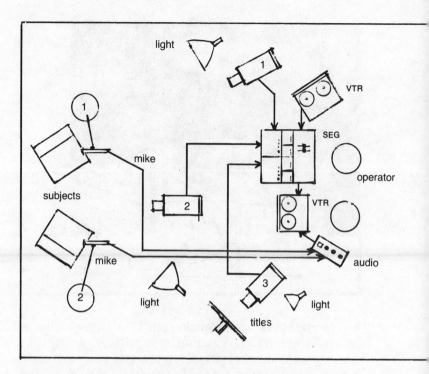

Mark on both the SEG panel and mixer panel with a china pencil which is camera 1, camera 2, mike 1, mike 2.

8. The nature of the program will determine how the third camera (if you have one) is used. In some instances you may want to run it through the external keyer, in others you may want to put it on bus A or bus B and use it for the titles and credits of the program. Set up the third camera, determine its use, and check to see that it's working.

9. If your SEG has a genlock, and if you're planning to insert prerecorded tapes into the program, set up the second VTR with the prerecorded tape on it, connect genlock, VTR, and SEG to each other and cue up the beginning of the tape segment to be inserted.

10. Turn on the entire system and do a run-through, checking both audio and video, camera switching, lighting, and recording.

Now you're ready to tape.

Naturally, this shooting script is only a sample of what might actually take place. The action would be followed by the cameras and director; if the guest got up and did an imitation of a mountain goat in heat you'd swing into action to catch every golden minute of it. I'm assuming also that the equipment available is limited (camera 1 has a wide angle lens, camera 2 has a zoom, camera 3 has a fixed lens of intermediate length). But if you run over this shooting script and punch holes in it ("Well, there *I'd* use a medium shot of the guest from behind the host and get an angle that

TAPING

Audio	Video
Dead	Titles of show, from camera 3. Cameras 1 and 2 should be off the line. Fade up camera 3 to bring title card onto line. VTR should be in record and up to speed for at least six seconds before action starts.
Host says hello and introduces guest.	Switch camera 1 (long view of subjects) onto the line.
Audio mixer should have both channels open.	x (cross-fade) to camera 2 (medium shot) to get host's face as he talks. Switch to camera 1, swing camera 2 to guest, switch to camera 2, hold.
Host begins to talk to guest about favorite subject.	x from camera 2 to camera 1 to give tight two-shot of host and guest talking.
Guest begins to talk about recent mountain climbing expedition.	Set up prerecorded tape on VTR B through wipe-effects bus. Start tape. Put camera 1 on wipe bus, punch in effects channel.
	Wipe left to right (host disappears first, then guest) to prerecorded mountain climbing tape.
Audio track of tape on VTR B up on mixer just under guest's rap.	Camera 2 in for close-up of guest.
Guest winds up rap.	Switch to camera 2.
Host asks guest details of mountain climbing technique.	x to camera 1 long shot.
Guest answers questions.	Hold for long shot on camera 1. Switch off VTR B. Set up camera 2 for close-up of host. Switch to camera 2.
Guest answers questions.	Switch to camera 1. Set up camera 2 tight on guest. Switch to camera 2.
Guest winds up rap.	x to camera 1 long shot.
Host thanks guest for stopping by and then turns to camera to thank folks for watching. (Audio down on guest mike, as host finishes fade master audio out.)	Hold camera 1 long shot. Set up end titles on off-line camera 3. As host finishes talking x to camera 3 by punching it up on camera 2 bus, then fading. Hold fade for superimposition of camera 1 shot under titles. As titles end fade out camera 1, then camera 3.
	Let VTR run for a few seconds on no-signal, then stop VTR.

included the host's left shoulder"), you'll begin to experience the video decisions that must be made during taping, and which, if executed properly, will make good tape.

Unexpected Effects

There are a number of visual effects that can be created with your equipment which are the result of the electronics you're using. The most common one is creating abstract patterns on the screen through the use of video feedback (sometimes called **howl**). In its simplest form this is accomplished by placing a camera in front of a monitor that is displaying the camera output. Zooming in on the monitor screen and moving the camera slightly from side to side will result in a swirl of circles and pulsating black-and-white waves. This feedback loop (camera-monitor-camera-monitor-camera, to infinity) can be further developed by pointing two or three cameras into the same monitor display screen and mixing the feedback through an SEG; by superimposing an object on camera 1 with the feedback loop of camera 2 and then keying the whole thing; by gen-locking a prerecorded feedback with live feedback into a new mix; and so on.

The nature of the vidicon can also produce effects. Any bright source of light viewed by the vidicon camera will produce a **halo**—a ghostly fog—around the light. This means that the light is too bright for the vidicon and could burn the target area. All the same, it can be used as an effect—a foggy haze surrounding a person's head can be accomplished by placing a spotlight behind the head pointing at the camera. Other camera tricks using the vidicon include starting to record before the vidicon is warmed up so that the AGC for the vidicon brings the picture slowly into view as if you were flying through clouds; adjusting the internal camera controls to produce a negative picture; rotating the camera on the tripod to get a sideways or upside-down picture.

As you use your equipment, pay attention to the unexpected effects as well as the designed ones. An unwanted effect (such as feedback) may be at another time just the thing to create an interesting, pertinent visual.

16

HOME VIDEO

VIDEO
RECORDS

The success of audio cassettes over open reels of tape, as well as the continued dominance of plastic records as the most popular method of listening to recorded music, have pretty much established that, in the consumer market, the fewer buttons there are to push, the happier people are. This fact has been taken into account by the manufacturers of video equipment, many of whom also produce audio equipment, and the result has been the decision that mass-market home video must supply the consumer with a self-contained "pop-in and play" tape or a video disc.

At the moment there are a number of video tape recorders that use self-contained tape configurations. Any of these cassette recorders can be interfaced with the cameras, monitors, SEGs, and other equipment used with half-inch open-reel machines. They are different only in their internal mechanisms, and the housings of the video tape—most being designed to make playing a tape a matter of dropping in a cassette rather than threading a reel of tape into a VTR.

The two major home video systems sold in the US are the **Betamax** format and the **VHS** format. These systems use a video cassette as the tape housing, and they allow the same ease of use (no manual threading) that their audio cassette player counterparts do. The video tape runs between two reels, supply and take-up, on a horizontal plane. The tape is on the left reel before play/record. It passes to the right reel during the recording/playback process and must be re-wound on the left reel before it can be used again. Along one of the lengthwise ends of the cassette case is an open-

VIDEO CASSETTE
RECORDERS

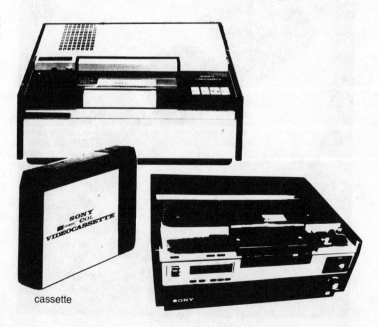

cassette

ing through which the tape can be seen as it travels from reel to reel. When the cassette is inserted into the video cassette player, a loop of the tape is pulled out of the cassette and wrapped around the video head drum, where it is scanned for play or record. Two motors, each driving a shaft running through the center holes in the left and right reels, run the tape through the machine (out, around the head drum, and back to the supply reel). All this is done automatically. In an alternate method, the tape is curved slightly into the cavity of the cassette and the head drum (on a disc) is inserted into the cassette for scanning. Either way, the tape is connected at both ends—to the supply and take-up reels—and the cassette must be rewound after use.

The differences between the Beta and VHS formats are minimal, but they *are* different, which means a cassette from one format cannot be played on a machine of the other format. (Copies of one format can be made on the other, however, by running the video signal out of one deck and recording it on the other deck.) Both Beta and VHS offer extended play modes, which allow you to record twice as much information on the same length of tape; you alter the tape speed, rather than the tape length. At this writing, the Beta can record up to three hours on a cassette, the VHS from four to six hours. Some models in each format have pause, memory search, timers, and other gimmicks. Sony has introduced a portable version of the Betamax deck, so a whole portable video system can be assembled using this small cassette format.

The recording quality of both Beta and VHS is not great. These units are made for home use, and it is assumed that duplicate copies, edited versions, and other second-generation situations are not of prime interest in that market. The signal-to-noise ratio is poor, and it is doubtful that picture stability would survive mastering on Betamax or VHS followed by editing a second-generation composite master and making third-generation copies.

Nonetheless, both Beta and VHS are ideal for home use or playback of productions made on other formats. Tape is relatively cheap, the cassette is small, and the picture is fine—even when blown up for viewing on a video projector.

To set up a home system, all you need is a Beta or VHS deck and your TV set. The signal from your TV antenna or cable box feed is sent through the deck, which has a VHF-UHF tuner built into it, and then to the TV. To record an off-air broadcast, you tune in the signal on the tuner in the deck, monitoring the signal through the RF-out-to-TV circuit on your television set. You watch while your recorder records what you're watching. If you want to watch one program and record another that is being broadcast at the same time, the set-up is different. You need an antenna signal splitter,

which produces two sets of TV signals from the antenna. One signal is fed into the deck, the other directly to the TV set. To record, use the deck's tuner to select the program you want to record for future viewing, then momentarily monitor the signal on the TV just to check that all is as it should be, using the RF-out-to-TV. Then switch off the RF-to-TV circuit and tune in the channel you wish to watch; the deck will record the other program independently of the TV set, taking the signal from the antenna.

Color and b&w cameras are available as accessories for the Beta and VHS decks, but they are of extremely low quality (just check the prices), and should not be used for taping material you may want to transfer to U-Matic to edit for broadcast.

Most of the cassette systems on the market or near the marketing stage will eventually be replaced by the video disc or some other form of audio/video display. Certainly, as was the case with audio, a few tape systems will become standardized. But the cost of raw tape is considerably more than discs, the raw materials for which cost only a few pennies, and the process by which discs can be made lends itself better to mass-production. It is doubtful, therefore, that tape copies of prerecorded shows can ever be produced as cheaply as discs, or some other format yet to be developed.

17

R,B, AND G

COLOR PIX

How soon will we have color TV?.... CBS says it's here now. DuMont says 20 years. RCA says it won't take 20 years and it's not here yet.... Best news so far... that RCA has come up with a color tube.

Bob Gardner's "Channel Chatter,"
TV Guide (April 8, 1950)

Color television is no longer in the experimental stage; it is here.

"As We See It,"
TV Guide (February 1, 1958)

... color television currently the runaway best seller with $2.35 billion in sales in 1971.

Consumer Electronics
Teledyne Report (Third Quarter, 1972)

If there is one common denominator between all credit plan Americans it is their color TV set. In fewer than twenty years this single piece of electronic equipment has become central to the vast majority of households across our great land; entertaining, informing, and selling us our must-haves from cars to hygiene sprays. Color television is the basis against which all other television is judged; color television represents the electronic marvels that have been developed and perfected by the major broadcast networks during the past twenty-five years to produce a picture that is of consummate technical quality.

The fact that color television has become the image of TV has lessened the impact of the advent of video equipment. Personal video is, at this point, a b&w medium, and color broadcasting is the standard by which we judge "television." The effect has been that the medium has not succeeded in taking its place as yet another form of television communication. Video people may be into the content and have accepted the present limitations of form, but those who are interested only in watching television, not in making it, have been kept from fully appreciating the implications of video. It doesn't match up to what they're used to watching on their TV sets. Of course there are more problems than just the lack of color: poor resolution, faulty stability, inadequate production facilities, and the like tend to make it difficult for video to even approximate broadcast standards. But color is the key. If color had been available when personal video was first introduced, the rate at which the equipment would have been disseminated would probably have been far greater.

As it stands now, color video will be available at prices comparable to b&w within the next few years. It is already available at reasonable cost, especially as compared to the price tags on broadcast standard equipment, but before it will replace b&w as the standard form of personal video, the equipment will have to be made smaller and the cost will have to be cut in half. Our faith in the Japanese being what it is, we have no doubt this will occur.

Once color video is generally available, the level of technical knowledge necessary to fully appreciate and utilize video equipment will become much greater. Not only will you have to understand the basic principles of television and their applications in b&w video, you'll also have to come to grips with the principles involved with the addition of color to the broadcast signal. Fortunately, that's just what it is: an addition. Because the color standard in this country was devised so that it would not make b&w receivers obsolete, the workings of color are firmly based on the precepts established for b&w electronic television.

Color Principles

Color is a visual phenomenon that has to do with the way we see light and the reflection of that light by objects around us. To explain color, let's start with black and white. Black is the absence of all color in an object or an area; white is the presence of all colors in an object or area. Gray is something we see that can be produced only by combining black and white. The shade of gray depends on the percentages of black and white present. Black, white, and shades of gray are considered to be the **luminance** of objects and of light.

Luminance is a more technical term for *brightness*. We can consider black and white as the two extremes of the brightness scale, white being very bright and black being not bright at all. This makes the shades of gray between black and white of various gradations of brightness. No matter what an object's color, it has a certain brightness value, which can be measured as a combination of black and white. If color were suddenly removed from all objects in the world, they would still have brightness values, or shades of gray varying from black to white. All light and all objects have a particular luminance, they have a particular brightness that can vary from very bright (represented as white light) to very dark (no white light, thus black).

While the luminance of a particular light or an object reflecting that light has to do with the brightness value we can assign to it, light and objects reflecting light have a visible

COLOR PIX

wavelength at which that light reaches us. That wavelength can be described as *color*. Color is an arbitrary value that has to do with our visual perception of the **saturation** and **hue** of light or of objects reflecting light.

You've seen the color wheels used to divide the light spectrum into a continuously changing progression of colors, from red to yellow to green to blue to violet and back to red. All color has a hue, which can be placed somewhere within this spectrum, or it has no hue at all, in which case it is either black, white, or gray. Hue is used as a method of determining the color of an object.

We can determine the saturation of any particular color using this same color wheel. Strong colors, such as red, have a high degree of saturation; less strong colors, such as light blue, have a lower degree of color saturation. Saturation is defined as the amount of difference between a color and gray. A vivid, brilliant color has a high saturation. A pale color has a low saturation. White, black, and gray have no saturation. So saturation can also be defined as the extent to which light is colored.

Chrominance or **chroma** is used in video as a description of just how "colored" a color is in relation to black, white, and gray. Chroma is a technical word for the presence of color in a light or an object reflecting light. There must be chroma if a light or object is to have a saturation level and a hue.

To run through all this again, a source of light or an object reflecting that light has a certain luminance or brightness value, which can be expressed in terms of how bright or dark that light or object is, and this is usually graded on a scale that goes through shades of gray from black to white. Chrominance is a value placed on a light or object to describe what we call color. The vividness of the chroma of a light or an object is defined as the saturation of that color and is the strength of the color as it strikes our eyes. Saturation is a combination of hue and luminance. The ratio of the primary colors present in a color and the saturation of those particular colors combine to produce a color of a certain value which is a hue. Every color has a certain degree of saturation and a certain hue—as in pale (saturation) pink (hue) or dark (saturation) blue (hue). Every color, when the chroma values are taken away, has a certain luminance level that corresponds to a shade of gray between black and white.

It has been found that three colors can be combined in specific ratios to produce the effect of all other colors. These three hues—red, blue, and green—are called **primary colors**. (In nature, the primary colors are red, blue, and yellow—not green. We are speaking here of the artificial production of color.) Various ratios of two or three of these colors will give us all possible hues. At the same time, each of these colors can be evaluated in terms of the amount of white light (luminance) it contains; in other words, it will have a saturation, which can be expressed as bright red, dark blue, or brilliant green.

There is one other quality of the three primary colors that is of great importance: *No two of these colors can be mixed together to produce the third.* You can mix red and green to get yellow, red and blue to get violet, and so on. But there is no way of getting blue from red and green, red from blue and green, green from red and blue. This is what makes possible the three-color television principles by which various combinations of the primary colors are added together to obtain all other colors. It also permits use of these three primary colors as electronic signals, since mixing any two of these signals together will not produce the third signal.

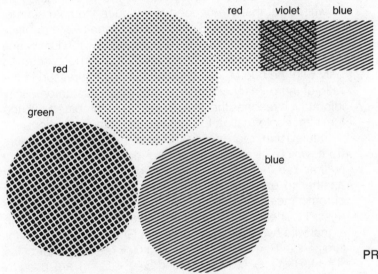

PRIMARY COLORS

The Color Signal

The color video signal is a combination of the b&w video signal and chroma. The b&w signal is an electronic reading of the brightness values of a scene ranging from black to white through the shades of gray. It is also one of the values used to determine the "coloredness" of a light or an object reflecting that light. The color video signal is this luminance signal together with a color signal. If the TV receiver is not capable of receiving and processing the color signal, it can still process the luminance signal and display the signal in terms of black, white, and gray. This is known as the **compatible color system** and is the standard color system for this country, called the **NTSC** (National Television Systems Committee) color standard.

In order to produce color and remain compatible with black and white, the color signal must include the luminance values, hue values, and saturation values of the scene being transformed from light to electronic pulses. An examination of the operation of the color camera will make clear how this is done, but before we get to that, a short explanation of the color signal is in order.

The color signal produced by the color camera is different from the b&w signal produced by a b&w camera; the main difference is the presence of the **chrominance signal**, the portion of the total color signal (luminance plus chrominance) that contains the color information in an electronically encoded form. The chrominance signal is an objective evaluation of the **chromaticity** of the scene in terms of the saturation and hue of the light reflected by that scene to the color camera. Chromaticity is expressed in terms of the ratios of the primary color combinations needed to reproduce the hue values of the scene plus the ratio of the colors to white light that are needed to produce the saturation of the colors in the scene. Since the luminance (white-to-black factor in terms of brightness) is conveyed by the luminance portion of the signal, the chromaticity has to be concerned only with the hue of the various colors as expressed in the ratio of the primary colors involved—once this chromaticity information reaches the color TV set it is recombined with the luminance information to produce the color saturations.

In television, red, green, and blue are the primary colors from which all other colors are determined, and since neither red nor blue nor green can be combined with another to produce the third, it is possible to reduce the chrominance signal from three separate signals (the redness, blueness, and greenness ratio of the scene) to one signal with two alternating portions. Given that Y stands for luminance, R = red, B = blue, and G = green, the formulas $R=Y+(R-Y)$; $G=Y+(G-Y)$; and $B=Y+(B-Y)$ are used to obtain the value of G without actually broadcasting the G signal, but just by using the R and B signals. These R and B signals are broadcast as the chrominance portion of the total video signal in the same wavelength at a reverse polarity to each other.

The total compatible color signal has two parts, the luminance information and the chrominance information. Since this total color signal must contain all the information necessary for b&w reception, including sync and blanking, the portion of the signal that deals with color must be within the band of frequencies used for transmission of the b&w signal but must not interfere with that signal. To accomplish this the color portion of the signal has been assigned a frequency of 3.58MHz, an odd multiple of the line scanning frequency that is high enough within the band of video frequencies of the total signal so that it will not create inter- ference with the video picture signal. The 3.58MHz is the **carrier frequency** of the color information and is known as the **color subcarrier**.

This 3.58MHz signal is produced by an oscillator located in the color camera or by a separate generator, and it is modulated by the color information. Then the modulated signal is combined with the rest of the composite video signal to produce a composite color video signal with the color information encoded into the signal but not interfering with the luminance. When the color signal reaches the color TV set it is decoded and the various parts of the color control signal are combined with the rest of the control signals to provide a color display on the screen of the color TV re- ceiver.

Sync pulses are needed to maintain the color signals in relation to each other so that the color values of the scene remain constant until they reach the receiver, just as other sync pulses are needed to maintain line and field relation- ships between camera scanning and receiver display.

Sync Pulses

Color values are much more sensitive to getting out of whack than any other part of the signal. They get **out of phase** if the relationships of the color components are not carefully controlled. When a color signal is out of phase, the combinations of red, blue, and green have been screwed up and the resulting hue is incorrect when reproduced on the receiver screen. To ensure that the color signal remains in phase, a reference mark for the color signal is placed at the start of each line. This reference is known as the **color burst**, and it occurs before the beginning of each line, just after the horizontal sync pulse has been fed to the receiver cathode ray tube controls. This color burst is sometimes called a **burst signal** or **burst flag**. It's about 9Hz of the 3.58MHz subcarrier without any color information and is used as the reference point to establish the phase of the subcarrier be- fore the subcarrier starts carrying the color values of that particular field.

The addition of color to video is a complex process. For- tunately, the operation of color equipment is not as compli- cated as the theory and principles behind that equipment. So, like the rest of video, it's possible to utilize color without understanding too much about what you're doing. When you get as far as special effects and color processing, how- ever, it is necessary to have some grasp of the basic elec- tronic events involved in getting the color signal from the scene to the color TV receiver.

Color Cameras

There are, in theory, four types of color video cameras, but only three of the four are in actual production and use at the moment. They're all based on the same principle: dividing the incoming light into its luminance values and primary color values.

Given any scene, the light striking that scene and reflected back at the color camera can be rated in terms of its brightness as it comes through the lens into the camera (as is done with monochrome video cameras) and in terms of its chrominance (as is done with color video cameras).

History

The first electro-optical color camera was introduced by RCA in 1940. It was a **three-tube color camera**, which produced the color video signal explained above. Three pickup tubes were placed behind the lens, and the incoming light image was split into the three primary colors of the scene, each including the luminance value of the color, and each of these colors sensed by one of three pickup tubes.*

Several devices were suggested to split up the light image coming through the lens so that it could be supplied to the three vidicon pickup tubes. These included semi-reflecting surfaces, prisms, and **dichroic mirrors**. Dichroic mirrors reflect the light of one color while they allow the light of all other colors to pass through. Today they are used

Color Separation

almost exclusively as the method of dividing the color spectrum of the light image into the three primary colors. The mirrors are placed behind the lens and in front of the target areas of the three vidicon tubes, which are mounted parallel to each other. The first dichroic mirror reflects the light energy in the blue region of the light spectrum onto the first vidicon (designated the B vidicon), allowing the longer wavelengths of red and green to pass through its surface. The remaining light image hits the second dichroic mirror, where the red light energy is reflected toward the second (R) vidicon, and the green passes through to strike the faceplate of the third (G) vidicon.

Three-Tube Color Camera

Each vidicon in the three-tube camera scans the light image of the picture that corresponds to a primary color for its luminance and chromaticity. Each vidicon then produces an electronic signal, and the three are combined to form the total color video signal.

It was found that three-tube color cameras needed very precise adjustments and tube alignments since the luminance (b&w) signal consisted of three fields of information

*The original pickup tubes in color cameras were Image Orthicons, first 3-inch and then 4½-inch, the latter being more sensitive. These tubes made the early color cameras large and bulky. With the invention of the vidicon tube, vidicons were used in color cameras to create the much smaller color camera in general use today. A further advancement in pickup tubes was the introduction of the lead oxide vidicon tube or Plumbicon (Philips' trademark name), which is the standard vidicon used in broadcast quality color cameras.

recombined after scanning; if one tube was out of kilter not only would one of the colors be slightly off, but the actual b&w picture signal would appear fuzzy and out of focus.

Four-Tube Color Camera

To correct this, RCA introduced the four-tube color camera in 1962. In addition to red, green, and blue pickup tubes this camera has a monochrome vidicon tube to pick up the luminance of the scene and supply the standard b&w signal. In this way, the luminance signal is independent of the color signal and can be easily adjusted and controlled. Within the camera the luminance signal is produced by the fourth tube; the three other tubes produce the color information of the scene, which is encoded onto the 3.58MHz carrier, and the two signals are combined to form the composite color signal. The addition of a separate luminance tube has been maintained in further camera developments.

Four-tube color cameras are pretty much the broadcast industry standard. They have drawbacks, the major one being the size of their housing—it must be large enough to

COLOR PIX

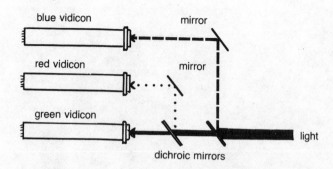

blue vidicon
mirror
red vidicon
mirror
green vidicon
light
dichroic mirrors

THREE-TUBE COLOR
SYSTEM

hold the four vidicons and their circuitry—but they produce excellent color.

Two-Tube Color Camera   In 1964, Toshiba created a stir by introducing the first two-tube color camera, which was used to broadcast the 1964 Olympic Games from Japan. Then, in 1968, Ampex, working in conjunction with ABC, also came up with a two-tube camera. Since then a number of these cameras have been put to use by the networks when they have required semiportable equipment, but as yet they haven't been accepted as having the full quality that the four-tube design provides. On the other hand, the two-tube camera has been adopted for nonbroadcast applications because of its smaller size and lower cost.

The two-tube camera has one vidicon to sense the luminance of the incoming light image and a second vidicon tube, known as a **color dissector tube**, **coloring tube**, or **chrominance tube**, depending on the camera's manufacturer, to provide the color difference signals used to produce the chroma portion of the total video signal. The target area of this tube has a series of red, green, and blue stripes across it known as a **stripe filter**. The stripes are incredibly narrow and they repeat across the target area. They're capable of reading the light image coming through the lens sequentially (one reading following the next) for red, green, and blue light values. This tube and the luminance tube are mounted behind an optical mirror system, which separates the luminance and chrominance components of the light image coming through the lens.

One-Tube Color Camera   A **one-tube color camera** is just off the drawing boards. This camera is as portable and lightweight as the b&w portapak cameras now in use; perhaps eventually they will be even smaller and lighter. The principle used to create this camera involves sequential reading of R, G, B, and Y values.

A one-tube color camera marketed by Magnavox for about $2,500 uses a color stripe tube manufactured by RCA. AKAI has introduced a two-tube camera, slightly larger and heavier than the Sony b&w portable camera, but

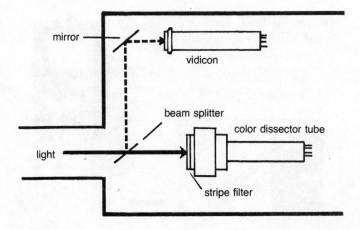

very obviously a sign of the future of color cameras for half-inch video. The $4,000 price is still steep compared to b&w cameras, but it is one half the cost of the more bulky three- or four-tube cameras also being manufactured for the small format video market. Inexpensive color cameras are now available for $2,000–$5,000.

Operation

The operation of a color camera is more complicated than that of a b&w camera. To begin with, there is just *more* of the color camera. In addition to the camera itself there is a separate camera control unit to which it is connected and from which the video signal is fed to the SEG, VTR, or monitor. The first control you'll notice on the camera or control unit, placement varying with the manufacturer, is the **white set** or **white level set**, which is used to establish the luminance level for the camera. The camera is pointed at a pure white card or other white surface and it senses what "white" is. Next, you'll find a gain set or gain control divided into two controls—one allowing you to set the video signal strength to either automatic or manual, the other to adjust the gain for the best signal-to-noise ratio. Another control is the **registration setting**, which lets you adjust the three picture signals coming from the vidicon tubes if they get out of line. On some color cameras you'll also find switches for red, green, and blue that let you eliminate one or all of these color signals from the total signal to test whether the vidicons are working properly. Behind-the-lens filter systems are featured on some color portable cameras. They allow adjustment to available light.

A color camera requires, at the very least, a thorough reading of the instruction manual that comes with it. It's probable that the low priced one- and two-tube cameras will eventually make many of these adjustments automatically, as is the case with b&w cameras.

COLOR PIX

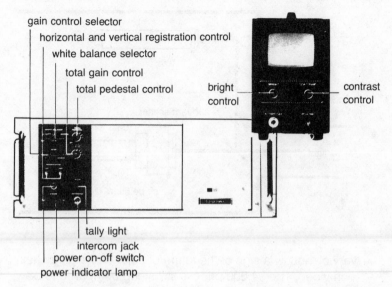

gain control selector
horizontal and vertical registration control
white balance selector
total gain control
total pedestal control

bright control

contrast control

tally light
intercom jack
power on-off switch
power indicator lamp

CAMERA CONTROL UNIT

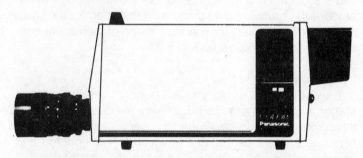

STUDIO COLOR CAMERA

Color Display

It was about ten years between the time of the introduction of the color camera and the perfection of a practical method of displaying the color signal. This may give you some idea of just how complex a process it is to reconstruct chrominance values on the face of a cathode ray tube. Of course, it's easier to understand now that it's been done. In 1949, RCA introduced the **shadow mask color tube** for the display of color pictures. The shadow mask has remained the standard color tube ever since, although the Japanese have recently come up with some alternative systems.

To understand the basics of color display, let's work backward from the face of the CRT to the antenna input signal. When you look at a color picture on a TV set, you're actually seeing a combination of red, green, and blue phosphor dots, about 1.5 million of them, coating the screen at the rate of 1000 of each color per scanning line. The dots are grouped in triangles, called triads, each point of the triangle being one of the colors.

Set about half an inch (1.3cm) behind the screen, parallel to the raster surface, is a sheet of metal with .5 million small

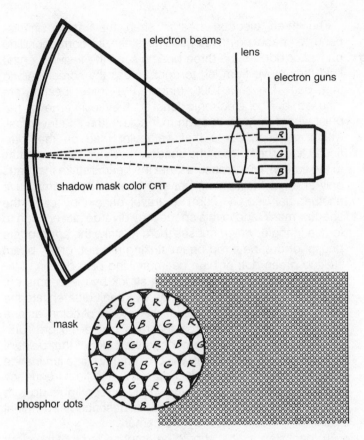

electron beams

lens

electron guns

shadow mask color CRT

R

G

B

mask

phosphor dots

RCA COLOR SYSTEM

holes punched in it, each hole located exactly behind one of the triads. This is the shadow mask.

At the far end of the tube is the electron gun assembly, similar to that in the b&w receiver, but consisting of three guns rather than one. Just in front of the three guns is an electronic lens, which focuses the beams sent out by each gun. Originally, guns were arranged in the same triangular configuration as the dots on the screen, but a significant variation, known as **in-line color** has been introduced in which the guns are set next to each other in a horizontal row.

The guns are designated red, green, and blue and are fed the corresponding chrominance values. These values have been decoded from the composite color video signal sent into the set and combined with the luminance values. With three electron guns, each producing an electron beam of one of the three primary colors, and each line of the raster sensitive to those primary colors and capable of producing them visably by glowing either red, blue, or green, it becomes a little clearer how color TV works. It's an additive process in which the three primary colors are produced in different ratios to create the actual color (including luminance) of the scene being transmitted.

COLOR PIX

Scanning

The three electron beams scan the raster simultaneously—each being focused to strike its corresponding phosphor dot. As the three beams scan the first line of the field they move from left to right across the screen. Along that path are the 1000 triads. The three beams are focused by the electronic lens so they converge at the point where they go through that hole in the shadow mask which corresponds to that particular triad on that particular line. The holes in the shadow mask are the only paths to the screen, which ensures that the beams strike the triads only and not the spaces between them. The beams have another half inch (1.3cm) to travel once they clear the shadow mask, and in this space they diverge just enough to form a triangle, which causes them to strike the appropriate phosphor dot, be it red beam striking red dot, green beam striking green dot, or blue beam striking blue dot.

The phosphor dots glow when struck by the beams but they're so close together that the eye can't differentiate the individual glows. Rather, it sees the mixture of colors as one glow, which, depending on the ratio of the mixture, produces a color that corresponds to the color of that point of the original scene before it was broken down into luminance and chrominance. As the standard line and field frequency of beam scan takes place, the whole screen lights up with the gradations of various colors to describe the original scene as it was sensed by the camera.

To backstep a bit, it should be emphasized that the electron beams are also striking the screen with a strength that corresponds to the brightness value of the scene—they're transmitting Y as well as R, G, and B. Because color TV is a combination of luminance and chroma and because the luminance is not affected by the chroma, black and white reception of the color signal is possible. In color TV, however, the luminance serves a purpose more important than simply providing brightness values; it gives us our sense of color saturation by supplying the percentage of white light that is combined with the hue to produce the total chroma effect of color, be it brilliant red, pale blue, or whatever. In b&w reception the receiver just ignores the 3.58MHz color carrier and reproduces the standard luminance portions of the video signal.

The electronics of the color set that control the three electron guns duplicate b&w set electronics with the addition of circuitry to sense and control the color portion of the signal. The signal is still received, stripped of its carrier frequencies, amplified, and divided into sync, control, and picture signals. In the case of color, the color signal must also be stripped of its subcarrier frequency and then used to control the strength of the electronic beams and their phase as they scan the raster.

THE VIDEO PRIMER

I have no intention of even beginning to discuss the electronics and associated problems involved in getting the color signals to the tube so that people aren't green and trees aren't purple. If you have a desire to get inside a set and see how it all works, I seriously suggest a course in TV repair.

During the last few years, variations of this basic color CRT have been introduced, including one by Sony that differs radically from the RCA system.*

The Sony color system, called Trinitron color, was first shown in this country in 1968. Since then it has come to be considered the finest color television available, especially for displaying the less-than-fine NTSC signal. It has also allowed Sony to produce color TVs without having to pay the several million dollars a year in patent licensing rights to RCA that all those using the RCA three-gun shadow mask system must do. But it isn't so much that RCA has met its match on a technical level. It's that Trinitron produces a much brighter, sharper, flatter, squarer color picture than the RCA system does. Watching a Trinitron set is like watching a small movie screen rather than a fish bowl.

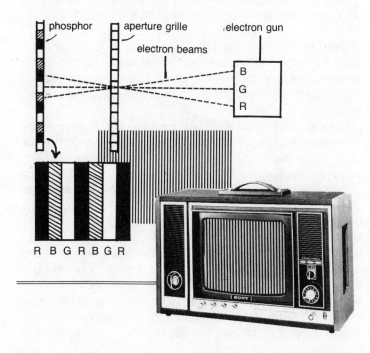

TRINITRON COLOR SYSTEM

*The Sony system is based on the invention of an American Nobel prizewinner named Ernest Lawrence, who created a one-gun color TV over twenty years ago. Sony leased the patents from Paramount Pictures, who have been holding them all these years, and added their own improvements to make the invention viable.

The Sony Trinitron doesn't have a shadow mask or three-gun assembly. Instead, it is coated with red, green, and blue phosphors in vertical stripes, one next to the other repeating in threes across the screen, rather than triads of phosphor dots. Behind the screen is an **aperture grille**, which, like the shadow mask, provides the spaces through which the electron beams travel. But instead of being made of tiny holes, the aperture grille is just that, a grille with vertical slits cut in it to correspond to the vertical stripes of phosphor. The beams are sent toward the face of the tube to converge at the slits in the grille and then pass through to diverge and hit three points on their corresponding phosphor stripes. The beams travel in a horizontal direction so they make glow only those points that they strike on the vertical stripe. The slits in the aperture grille are wider than the diameter of the holes in the shadow mask. It is still possible to make sure that the beams hit only the proper phosphors, but more of each beam gets through the grille, producing a brighter picture.

Trinitron Color Display

At the far end of the Sony tube is one electron gun containing three cathodes to produce three beams—red, green, and blue. The convergence problems associated with the three-gun system are virtually eliminated because all three beams are generated by one electron gun. It's unlikely that the color will ever get out of register, and if it does it can be easily corrected. In the traditional three-gun system, however, getting the colors (the beams of each of the guns) back into register is a matter of adjusting upward of a dozen controls.

Color Monitors

Color monitors have the same internal circuitry as color TV receivers, except they allow the insertion of a composite color video signal into their circuitry in the same way that b&w monitors work. Color monitor/receivers also have a composite color video signal output, which allows the recording of color signals on a color capable VTR. The same connecting configurations are used in color as in b&w. Color monitors are equipped with either eight-pin connectors or UHF line connectors or both. Externally, using a color monitor/receiver or color monitor is the same as using a b&w monitor. To display a color signal, connect the monitor to a color VTR or color camera. To record a signal off the air onto a VTR, tune in the signal to your satisfaction and then proceed as you would if recording off a b&w monitor/receiver.

The cost of a color monitor/receiver is appreciably more than a b&w monitor/receiver. A b&w monitor costs about $100 more than a regular b&w set. The cost of a color monitor can be figured at about twice the retail price of the same screen size in a regular color set. The best low priced color monitor/receiver on the market today is Sony's 17-inch

(43.2cm) Trinitron color monitor/receiver, a superb bit of workmanship and the only piece of equipment I'm going to recommend in this book.

Color VTR

Standards

The color standard established by the EIAJ pertains to video tape recorders. EIAJ color describes two basic parameters for all VTRs conforming to it. First, all recordings made on EIAJ color machines must be compatible with EIAJ b&w machines so they can be played back in b&w and so recordings made in b&w can be played in b&w on EIAJ color machines. In other words, the EIAJ color standard incorporates the EIAJ b&w standard in the same way that color and b&w television signals are compatible. Second, the EIAJ color signals are compatible with the NTSC color standard just as EIAJ b&w is compatible with NTSC b&w. Recordings made on any EIAJ color VTR can be displayed on any color TV (via a color RF generator) or on any b&w TV (via a color or b&w RF generator) made in this country; similarly, off-air recordings can be made of the NTSC color signal on EIAJ color VTRs.

EIAJ-standard color VTRs are readily available; most of the major Japanese video manufacturers feature one to half a dozen models in their line. They are identical in appearance and operation to EIAJ b&w machines, with the exception of the color controls and the additional internal circuitry to sense the incoming color portion of the signal and to encode it on the video tape along with the rest of the picture and sync information.

Sony, for one, has made their color VTR as simple to operate as their b&w deck. Color recording and playback are totally automatic functions of the deck. Panasonic's color decks are a bit more complex in their user control provisions and necessitate setting of audio and video levels and choosing a function selector switch for either b&w or color record. Other firms have followed the same design principles with the result that color VTRS are just as easy to use as b&w VTRS. In fact, you couldn't make any adjustments if you wanted to. Both types of VTRS deal with the video signal as an electronic waveform with a range of signal components grouped together within certain frequencies. That the composite color video signal contains additional electronic information (the 3.58MHz subcarrier with its chroma information) does not mean it is any more difficult to record the signal except in terms of the internal circuitry needed to sense the incoming color subcarrier and include it with the other pulses sent through the control track and video heads to the tape.

Color Recording   The only practical concern you must face when working with color VTRS and color video recording is not to record b&w signals with the color circuitry engaged or else a certain amount of stray color noise will appear in the picture during playback. Also, when you play a color tape on a b&w machine, that machine must have a **color killer circuit** to remove the color portion of the signal so it doesn't create noise on the playback picture.

PORTABLE COLOR
CAMERA

Recording color off-air signals requires more attention than b&w. You must be sure that the gain controls are set properly on the color VTR and that the monitor/receiver controls are adjusted for optimum reception and color. Too much gain on the VTR will result in distortions of the color portions of the signal, too little gain will produce a noisy picture.

Recording a color signal from a color camera is no more difficult than recording a b&w camera signal. But the validity of the color does depend on the proper handling of the camera.

The real problems in color recording arise when it's necessary to use more than one color camera plus a color SEG to produce a composite color signal.

Color Effects—Processing

Color switchers and color special effects generators differ from their b&w counterparts because they can handle noncomposite video signals that include the color subcarrier chroma information. This means you can't use a b&w SEG for color production work. You must get a color-capable SEG. This type of equipment operates in the same fashion as b&w equipment; there are the usual sets of buses, buttons, preview and program facilities, fader controls, and insert patterns. The cost of a color SEG, however, is almost three times that of a good b&w SEG and none of them includes a sync generator.

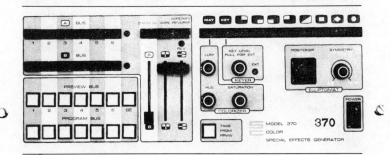

Color SEGs do include extra features not found on b&w equipment, mostly controls centering around the key and matte functions. The keying function on color SEGs is called chroma keying. A chroma keyer permits you to work with the various portions of the chroma values of the video signal without disturbing the luminance. Hue and saturation values are used to create special effects.

To key a person into a scene, use two cameras, one on the person, the other on the scene. Place the person in front of a blue background—chosen because it contains no red,

Chroma Keying

COLOR PIX

a color present in human skin, and because human skin tones have little blue in them. Then use the chroma keyer to sense the blue portion of the picture signal, using the blue vidicon, and develop a keying signal to allow the blue portions of the scene to be dropped out of the scanned signal. Because of the non-red quality of the blue background and the non-blue quality of human skin, the person is left in the scene while the background around him disappears. Of course, blue clothing cannot be worn unless you want to have that part of the person missing as well. Fill in the missing blue portions with the signal from the second camera—the person will be surrounded by the scene. Some SEGs also allow electronic generation of backgrounds with circuitry known as **color background generators** to produce any solid color background into which any object can be keyed.

Color Sync Generator

The color information of the video signal must be maintained in its proper relationship during its journey from color camera to SEG to VTR to display. And, when more than one color camera is used, the problems of maintaining the color phase and sync are increased. A separate color sync generator is used. This sync generator must be highly dependable since the rest of the color equipment and processing depends on it. A color sync generator provides the standard vertical drive, horizontal drive, horizontal blanking, vertical blanking, and sync pulses (usually as four outputs: line drive, field drive, mixed sync, and mixed blanking), plus two extra pulses (the color sync controls), which are crucial to the color relationships of the chroma portion of the signal. They consist of a burst flag pulse and the 3.58MHz subcarrier, which are fed to all the color cameras and to the SEG.

Color Cables

The same type of cable is used in color as in b&w video, but there are many more problems associated with cable connections. This is especially true when using the standard coaxial cable to carry the color signal. If you have two lengths of cable coming from two color cameras to a color SEG, and if those cables are of different lengths, it is likely that the color signals from the cameras will be out of phase (their timing will be off in relation to each other) by the time they reach the SEG. This can be corrected in two ways. You can either use equal lengths of cable or add lengths of cable until the phase is proper again; or you can get a **subcarrier phase shifter** control, which allows you to adjust the phase of the subcarrier as it is being sent to the color cameras, thus providing a total adjustment of the phase of the subcarrier throughout the system. You may have to do both. Phasing is a major concern in color and may require

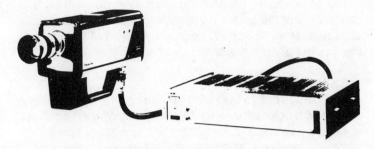

the use of additional control equipment plus a total overhaul and rewiring of the system's connections if the resulting phase problems show up on the final display of the composite signal.

Color Lighting

More lighting is needed when making color video than in b&w. It is always necessary to have artificial lighting when you're working indoors, and this lighting must conform to certain specifications.

Every source of light has a color quality, a certain color given off by that light. This quality is expressed as the temperature of the color, **degrees Kelvin** (°K). To prevent the lighting itself from affecting the color values of the scene, lights must be of 3200°K for color video. An ordinary incandescent household bulb does not achieve this temperature, and if you were to use such bulbs to illuminate a scene everything and everyone in the scene would have a reddish tint.

The best lighting for color is a quartz unit with quartz lamps rated at 3200°K. These are available in a number of wattages and can be used with any quartz lighting fixture. There are some incandescent bulbs available that are 3200°K, such as those made by GE with a power of 500 watts. These bulbs can be used in inexpensive aluminum scoops.

Lighting a scene for color can be a lot of fun, even though you must take care that your lighting has the proper color temperature. Recording in color lets you create lighting and color effects not possible in b&w. But you'll need a great deal more light for color recording, meaning either more lighting units or higher lamp wattages. The average color camera requires about 200 to 300 footcandles of light to give the best pictures.

Taping in color outdoors presents less of a lighting problem, but if all your color seems to be off in bright sunlight, you'll need a #85 filter over the camera lens to lower the temperature of the sunlight to 3200°K.

COLOR PIX

Color Lenses

The same lenses are used in color as in b&w video. It is important that the lenses be of as high a quality as possible with at least an f2 speed, because you'll want to minimize the amount of light needed on a scene.

B&W to Color

You can "treat" b&w video recordings or b&w camera signals with color information by using a **colorizer**. The effect is slightly psychedelic, but it can be used to tint scenes; to introduce captions, subtitles, or titles in color over a b&w picture; and to create other interesting effects. The colorizer is inserted between any two pieces of equipment (camera/ VTR; VTR/monitor) and it changes the various shades of gray present in the scene to colors whose hue and saturation are continuously variable, thus allowing the operator to produce any part of the color spectrum. The number of colors that can be produced at one time by a colorizer is referred to in terms of **steps**. A three-step colorizer allows you to introduce three colors into the scene, each color or combination of colors corresponding to a certain level of brightness in the gray value. A six-step colorizer might permit six colors. The most complex colorizers presently available have twelve steps. Each step has red, green, and blue controls from which you can produce any color. A good colorizer will cost about $1000.

Video will undergo an important change as a media device when color equipment is as inexpensive and easy to use as b&w equipment. Working with color video is fabulous, even if you're just getting a color capable VTR to record color broadcasts off the air (possible now since color VTRs aren't much more expensive than b&w VTRs). When a cheap color camera can be linked to a cheap color portable deck, video will take a huge stride forward. As personal media, color video is a much closer approximation of the present standards of network broadcasting. In addition, advanced color technology will make video a more desirable item for millions of people in this country who are hung up on their color home movies.

18

UNITED VIDEO

STANDARDS

The number of standards that exist in the world of television attest to man's ability to build a better mousetrap coupled with his belief that no matter how the mousetrap works, it must be the best because he thought of it. It's really unfortunate that television is the only medium in which there can be no agreement on the standards by which it functions in the various parts of the world. Records, audio tape recorders, film, and other media have been standardized, or if they vary at all it's minimal. At the moment there are three major television system standards in existence, none of which is immediately compatible with the others. Because of this variance in standards, the video equipment we've been discussing in this book is limited in its ultimate universal compatibility and therefore, in its potential as an easy exchange of information between the peoples of various nations.

At this time, a tape recording made on small format equipment in Great Britain or Europe can't be played on equipment here in the US, and vice versa. A tape recording made on equipment in South America can't be played on either US or European equipment. In fact, the EIAJ Type #1 standard discussed earlier (pp. 260–61) applies only to the United States, Canada, and Japan; a variation of the standard applies in Europe, that variation having been necessitated by the broadcast standards of the countries involved.

The initial point of departure between various TV systems has nothing to do with TV at all, but with the electrical current used in various parts of the world. In America and Japan, electrical current is described as 110v–120v 60Hz. This means that the strength of the current alternates at 60 cycles per second. In Europe, parts of Asia, and Russia, the electrical current is 210v–240v 50Hz. (Occasionally you'll find 135v 50Hz.) In other words, the voltage of their current is double the strength of ours, while their current alternates at fifty cycles per second. Europe is vibrating at a slower pace than we, although the quality of that pace is more impressive. The implications of those lost ten cycles reach beyond electrical theory; they affect the total life of those involved and also reflect a certain attitude toward the quality of life.

We know that the vertical and horizontal frequency of the video scan rate is based on the cycles per second of the alternating current. In countries where the ac is 50Hz, the frequency rate of the television system is also 50Hz, fifty fields per second. Therein lies the basic problem in achieving compatibility between various TV and video systems.

A tape made on a US machine will not play back on a 50Hz machine. In fact, a 60Hz machine won't operate on a 50Hz current even when the voltage is reduced to the proper level, and the same is true going in the opposite direction.

The development of television took place in a number of countries at the same time, principally the US, England, and Germany, with France kicking in later on. Although all of those working on television eventually adopted the same basic electronic principles on which the system would function, such as scanning and sync, the application of those principles in building equipment took on the electrical standards of the locale.

John Logie Baird and his apparatus, 1925

日光や照明器具の光線が直接映面に入ら

The basic principles of television were initially applied to producing black-and-white cameras and TV sets. In America the system eventually adopted was a camera and receiver capable of scanning 525 lines per field and changing fields at the rate of 60 per second. This meant that American TV sets, under the basic principles of electronic television, have a line frequency of 15.7KHz, a field frequency of 60Hz, a frame frequency of 30Hz, and each scanning line takes about 53 milliseconds to complete. It wasn't until 1940, by the way, that this 525 line/60Hz standard was established. Prior to that the standard had been set temporarily at 343 lines in 1935, although it was raised to 441 lines shortly thereafter. But it was decided in America that raising the line standard to 525 was necessary to provide enough detail in picture resolution to make TV a viable

NTSC Standard   consumer item, and so the 525-line standard was set and made official in 1941 by an FCC committee known as the National Television System Committee (NTSC).

BBC Standard   The British Broadcasting Company in London began transmissions in 1929, but their first service was more or less experimental even though the public was "invited" to buy receivers. It was an electro-mechanical system with a picture resolution of 30 lines and a field rate of 25Hz. Then, in 1936, an all-electronic system was adopted and the standard was set at 405 lines/50Hz, which has remained in effect since then as the standard for VHF b&w television in England.

The suspension of all television broadcasting during the second world war effectively halted the establishment of TV systems in other countries until the late forties and early

CCIR Standard   fifties. It was at that time that two other standards were introduced, the first being a 625-line/50Hz system, which has become, with country-to-country variations, the standard for Europe. This system is known as the **CCIR** standard (*Comité Consultatif International des Radiocommunications*) and it is the basis for European color television. In addition, the French introduced what remains the most highly defined system in the world, an 819-line/50Hz standard.

Up to this point, the only concern was the transmission and reception of monochrome picture signals. The concept of color television was alive in the laboratories of RCA, CBS, and other films involved in the development of TV, but it wasn't until the early 1950s that any steps were taken to introduce color broadcasting on a commercial level.

Color Standards
There were two questions that had to be dealt with before color could be introduced. The first was what kind of color system was going to be used and the second was whether it

would be compatible with the existing b&w systems. Again, different countries took different stances on these issues. Although John Baird had demonstrated color TV in the mid-thirties, it was left to the American TV companies to come up with the first workable systems. CBS, under the laboratory leadership of Dr. Peter Goldmark (the man who invented the long-playing record), developed an electro-mechanical method of producing color in which whirling discs, divided into segments of transparent filters in series of three primary colors (red, green, and blue), were placed in front of both the camera and the home screen. The rotation of the discs was synchronized between the camera and receiver, and color resulted.

While CBS was pushing their system, RCA was developing a totally electronic method of producing color using a three-color tube camera (1940) and a special color TV tube (1949). It's interesting that industry thinking was divided on the issue of whether TV should be the RCA electro-optical event (electric camera using special optics to divide up the colors before scanning) or the CBS electro-mechanical event. Since the concept of television evolved from the mechanical age to the electro-mechanical age of the late 1800s and early 1900s, the development of an electro-mechanical color system is not surprising, even in the forties. There is no doubt that there was a good deal of political lobbying involved in the presentation of the various color systems; the adoption of any one system as the standard meant that the company that developed it would be in a patent-holding, rights-licensing situation.

Eventually, the major concern with color television dealt with whether it would be compatible with already existing b&w broadcasts; whether a b&w receiver would be capable of receiving color broadcasts in b&w or whether it would be necessary to buy a new color TV set just to get any broadcasts at all; or whether the channels would be split up, some with color and some without, which would necessitate a TV set that could receive both via different systems.

The CBS-Goldmark system would have necessitated the latter; the RCA system would not. In 1950, the FCC through the NTSC adopted the CBS color standard. Fortunately for the future of television, the TV industry told the FCC to jump in the lake and the standard was withdrawn, to be replaced in 1953 with the present color standard, referred to now as **NTSC color**. This RCA system, although more complex than the CBS system in terms of electronics, could be received on b&w TV sets as b&w or on color TV sets in color.

NTSC Color Standard

During this same period, a variation on the RCA system was developed in America by the Hazeltine Corporation. With the acceptance of the RCA system as the US standard, Hazeltine abandoned work on their system, but it was

country	tv standard	scan lines
Algeria	French	819
Argentina	West European	625
Australia	West European	625
Austria	West European	625
Belgium	French,	819
	West European	625
Bermuda	American	525
Brazil	West European	625
Bulgaria	East European	625
Canada	American	525
Canary Islands	West European	625
Chile	American	525
Colombia	American	525
Costa Rica	American	525
Cuba	American	525
Cyprus	West European	625
Czechoslovakia	East European	625
Denmark	West European	625
Dominican Rep.	American	525
Ecuador	American	525
Egypt	West European	625
El Salvador	American	525
France	French	819
Finland	West European	625
Germany	West European	625
Ghana	West European	625
Gibraltar	West European	625
Greece	West European	625
Guadeloupe	East European	625
Guatemala	American	525
Haiti	American	525
Hawaii	American	525
Honduras	American	525
Hungary	East European	625
Iceland	West European	625
India	West European	625
Indonesia	West European	625
Iran	American	525
Iraq	West European	625
Israel	West European	625
Italy	West European	625
Jamaica	West European	625
Japan	American	525
Kenya	West European	625
Korea	American	525
Kuwait	West European	625
Lebanon	West European	625
Liberia	West European	625
Libya	American	525
Luxembourg	French	819
Malta	West European	625
Mauritius	West European	625
Mexico	American	525
Monaco	French	819
Morocco	French	819
Netherlands	West European	625

Nicaragua	American	525
Nigeria	West European	625
Norway	West European	625
Okinawa	American	525
Panama	American	525
Peru	American	525
Phillippines	American	525
Poland	East European	625
Portugal	American	525
Rhodesia	West European	625
Roumania	East European	625
Ryukyu Islands	American	525
Samoa	American	525
Saudi Arabia	American	525
Sierra Leone	West European	625
Singapore	West European	625
Spain	West European	625
Sweden	West European	625
Switzerland	West European	625
Syria	West European	625
Thailand	American	525
Trinidad & Tobago	American	525
Tunisia	French	819
Turkey	West European	625
United Kingdom	British	405/625
Uruguay	American	525
U.S.A.	American	525
U.S.S.R.	East European	625
Venezuela	West European	625
Virgin Islands	American	525

American: Based on NTSC with local variations in technical application of system.

French: Based on SECAM with local variations in technical application of system.

West and East European: Based on PAL with local variations in technical application of system.

British: Based on PAL with local variations in technical application of system.

picked up by Telefunken in Germany, who continued to refine it and in 1963 announced that it had been perfected. Then, in 1967, it was accepted as the standard for color in West Germany and very soon after as the standard in England. Known as the **PAL** color system, it was compatible with b&w and color receivers and based on the 625-line/50Hz CCIR standard. This created some problems in England, where the standard was 405 lines/50Hz. To get around it, the BBC continued to broadcast programs in b&w on their earlier system, which was located on the VHF band, and at the same time set up the UHF band for PAL 625-line broadcasting, which could be received in both b&w or color, depending on the receiver.

The differences between the NTSC and PAL systems were

STANDARDS

not tremendous since PAL was essentially a modification of NTSC. NTSC is jokingly referred to as standing for "never twice the same color," and in a way this is true; NTSC requires a good deal of adjustment at the color receiver to set color values, hue, and saturation. PAL, which stands for "phase alternation line," was designed to provide corrections at the point of transmission, which ensured that errors in color and hue did not take place. So instead of having color, hue, and saturation controls as on the NTSC receivers, PAL receivers rarely have more than a color saturation control.

After seeing both of these systems, the French came up with their own color system known as SECAM, which they put into operation in 1967. Just as the British had had to bypass their 405-line standard with PAL, the French allowed their 819-line standard to remain only for b&w transmissions and also went from VHF to UHF bands to adopt the 625-CCIR standard as the line and frequency base for their color broadcasting.

SECAM Color Standard   SECAM (*Séquential Couleur à Mémoire*) was a complete innovation in color broadcasting, varying significantly from the NTSC and PAL systems and ensuring that the color was totally controlled at the point of broadcast, with no color controls on the TV receiver.

This is where color broadcasting stands at the present time, with the US, Canada, and Japan using NTSC; Great Britain and West Germany using PAL; France, the USSR, and parts of the Middle East using SECAM.*

Color Signal Transfer

Because of the number of systems in use and the desirability that programs be exchangeable by producers using the various systems, any number of techniques have been invented and tried as methods of changing a 625/50Hz signal to 525/60Hz, or a 525/60Hz signal to 625/50Hz. There are three problems associated with the transfer of signal from one standard to another: the change in the number of lines, the change in the number of fields, and the change in

*There are many technical variations not only between color systems but also between the same systems as they are used by different countries. Channel band widths, color subcarrier frequencies, modulation techniques for the audio portion of the signal, and other factors vary from country to country, especially those using the same general system of CCIR/PAL or CCIR/SECAM, with NTSC being the only stable system in the three major countries where it is used. As an added kick, portions of South America have taken on 626-line/60Hz systems while the Italians, at the present time, are waiting for the development of a dual PAL/SECAM receiver capable of automatically switching between the two before they allow color TV to be introduced into their country. They may have a long wait.

the color encoding system. The earliest attempts to transfer signals dealt only with b&w and were fairly successful, if somewhat crude. One favorite was to simply make a kinescope of the original signal in which the movie camera was set slightly out of focus to more or less eliminate the original scanning lines. The film would then be projected on a film chain using the second standard. Since then, other methods, especially those for color signal transfer, have been suggested, including the use of a special cathode ray tube that holds each line and field longer than normal to create an overlap, which tends to eliminate the difference in field frequencies. The problem of conversion is essentially mathematical—fitting 625 into 525 or 50 into 60, for example—and the eventual answer to standards conversion appears to be through the use of computers, which will process one standard and reevaluate the signal in terms of the second standard. At the moment, prototypes of these electronic systems are in use, although the hope is that they will become even more precise. A number of BBC video tape productions have been shown on US TV using electronic conversion, the only drawback being an occasional and apparently unavoidable sluggishness in the field rate so that the 50 fields can be repeated to fill the 60 field rate of the US system.

Converting Equipment

Small format video equipment has been seriously affected by the variations in standards among different countries. Video makers in the US are, therefore, thankful that the Japanese use NTSC standards. There are very few units available that will work on more than one standard, although it is possible to convert certain components from one standard to another with the replacement of circuits and other modifications.

Because the number of lines in a TV picture is determined by the video signal and not by the mechanics of the recorder, it is possible to use the same machine to record b&w tapes by simply converting the mechanics from 60 Hz to 50 Hz. All you have to do is alter slightly the rate at which the heads rotate within the VTR. Sony makes a 117/50 Hz machine, and a number of video experimenters have introduced simple modifications that allow portable decks to record either 50 or 60Hz at the flip of a switch. But tapes made on either field frequency cannot be played at the other field frequency. The same is true with monochrome cameras and monitors—both can be easily converted from 50 to 60Hz or from 60 to 50Hz. The modification can include a switch to give a dual standard device. As for color, each of the color systems uses a different method of encoding the color signal and it is impossible to interchange tapes—

although PAL and SECAM color tapes can be shown in b&w on any 50Hz VTR.

Image Buffers   There are **image buffers** available for the transfer between b&w standards. Sony made one in the past, although it is not listed in their more current catalogs. The price tag of $6000 makes it rather inaccessible to most video people.

You can use your US standard equipment in Europe as long as you record with a camera and not off-the-air broad-
Step-Down Transformers   casts. With battery-operated portable units a **step-down** transformer (220v–110v) is needed for US operation in Europe. This allows you to power the ac adaptor to recharge the batteries; it doesn't alter the ac frequency rate from 50 to 60Hz. This isn't crucial to the portables since they don't operate on ac anyway, and even when they're being run on their ac adaptors all the adaptor does is change the ac to dc before feeding the power to the deck and camera. Since dc does not alternate, the concern with 50Hz/60Hz is elimi-nated. With ac-powered VTRs the problem can also be solved, but the equipment needed is more complex. To run a 110v/60Hz machine on 220v/50Hz, you need two pieces
AC-DC Converters   of equipment: an **ac-to-dc converter** (usually capable of
and Inverters   converting 220v/50Hz to 12v dc) and a **dc-to-ac inverter** capable of turning the dc signal into 110v/60Hz ac for use to power the VTR. Both units are expensive and must be of the highest quality possible. The quality of the inverter is espe-cially important because the 60Hz alternations of the current must be exact and stable. With an ac-dc-ac converter/inverter combination you can run any piece of 110v/60Hz equipment off 220v/50Hz mains. A similar converter/inverter setup can be used to run European equipment in the US. The total cost of such a system is over $500.

There are a few pieces of specially constructed equip-ment that are also available for small format use, such as the 110v/50Hz VTR already mentioned, which can solve some of the problems of using 60Hz equipment in Europe. The most
Receivers and Monitors   readily available are b&w and color monitors made by Sony, which can be adjusted to either 50Hz or 60Hz operation and a TV receiver made by Crown, which can be switched to either 50 or 60Hz.

If you're planning to use your equipment in Europe, it's advisable to stick to the portable system. First of all, the only item you'll need other than your standard equipment is a step-down transformer (50-watt transformer, which costs about $10), and even that can be eliminated if you have your video dealer convert your ac adaptor to 220v/50Hz by rewiring the internal transformer and replacing the fuse. Some units come with a built-in voltage selector, which re-quires only a fuse change to switch ac. If you really want to get into it, you could buy an ac adaptor in Europe from Sony
THE VIDEO PRIMER or Panasonic, which would supply the same dc voltage to

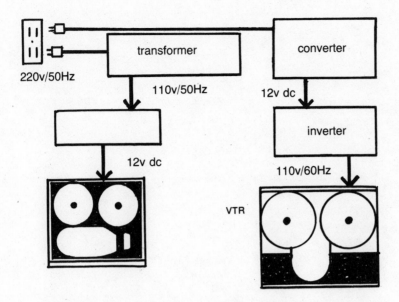

your portable unit but would accept the local ac voltage and frequency values.

If you do get your portable equipment to Europe, try watching local TV there through your camera monitor. You'll see a flicker rate of the local TV broadcast that corresponds to a 50-field frequency being scanned by a 60-field camera.

The possibility of eventual standardization among TV systems around the world is bleak. Too much money and time have been invested by the various TV services for them to consider changing standards. Eventually, however, it will be possible to transfer images from one standard to another more easily and effectively than it is today. With the availability of these standards and converters at reasonable prices, the need for a world standard will become superfluous.

APPENDICES

AC—Alternating current; also called line voltage. Common household electricity in the United States. Can be used by electronic circuitry as a point of reference. The voltage, or strength, of an alternating current rises to its maximum and then falls (through zero) to a negative level that is equal in amplitude to the maximum. This alternation takes place sixty times per second with common ac.

ACHROMATIC—Monochromatic; literally, without color; black-and-white.

ACTIVE MIXER—Any audio mixer that contains electronic circuitry to amplify the signals inserted into it before they are fed out of the mixer to the tape recorder. Because signals sent into a mixer from a microphone suffer a certain amount of **insertion loss,** their strength is diminished by the electronics of the mixer itself. An active mixer, with its complement of preamplifiers, can amplify the signal back to its original strength.

AC-TO-DC CONVERTER—An electronic unit that rectifies the flow of an alternating current to a direct current.

AGC—*See* **automatic gain control.**

ALC—*See* **automatic level control** and **automatic light control.**

ALTERNATING CURRENT—*See* **AC.**

AMP—*See* **ampere, amplifier.**

AMPERAGE RATING—The number of amps a particular electronic unit needs to function.

AMPERE—Amp. Unit of measurement of the electrical current used by a particular circuit; is equal to watts divided by volts.

AMPHENOL CONNECTOR—A microphone cable connector noted for its sturdy connections, which are secured by means of a screw-down collar; more commonly found on p.a. systems and mike-to-cable connections than on mixers, video monitors, and VTRS.

AMPLIFIER—Amp. An electronic circuit that strengthens any electronic signal sent through it.

AMPLITUDE—The strength of an electronic signal as measured by the height of its waveform when displayed on an oscilloscope.

ANGLE OF VIEW—That portion of a scene visible through a particular lens; usually expressed as the horizontal width of a scene as viewed by a lens; determined by **focal length** of the lens.

APERTURE—The opening at the camera-end of the lens through which the light image collected by the lens is allowed to pass to hit the vidicon **target area.** The size of the aperture determines the amount of light that will get to the vidicon; measured in **f-stops.**

APERTURE GRILLE—A metal screen located just behind the inside of a TV display tube's screen surface; used to limit the points at which the electrons hit the phosphor coating of the screen; a Sony invention, the function of which is similar to that of a **shadow mask,** the purpose of both being to ensure the reproduction of a true color TV picture.

ASPECT RATIO—The size of the TV picture area; the aspect ratio of television is four units of width to every three units of height; this is expressed as a 3 × 4 or 3:4 aspect ratio.

ASSEMBLY EDITING—A method of electronic editing in which various taped segments are retaped in a determined sequence to produce a coherent whole.

ATTENUATE—To turn down or reduce the level of a signal. *See* **fade out.**

AUDIO AND CONTROL TRACK HEAD—A two-function, stationary head on the VTR; the top portion is the audio head, which records and plays back the audio signal, while the bottom portion is the control track head, which pulses the control information onto the tape during record and induces it off during playback. *See also* **audio head.**

AUDIO CUE—The identification of an event by the use of sound; a word, noise, or other sound that alerts those producing either an audio or video tape to the fact that something is about to happen. In video productions, certain words in the script are used as "cues" to denote shifts in action, camera position, microphones, or other technical events; in electronic editing, audio cues are often used to signal edit points.

AUDIO DUB—To rerecord the audio portion of a video tape without disturbing the video portion of the signal; *also,* to make a copy of an audio tape.

AUDIO HEAD—The recording and/or playback unit on an audio or video tape recorder; it receives the audio signal, pulses that signal onto the magnetic tape moving past it, and/or induces that signal back off the tape for reproduction.

AUDIO-IN—Input jack that delivers an audio signal to a particular piece of equipment; an input receptacle that receives an audio signal.

AUDIO MIXER—An electronic circuit capable of accepting a number of audio signals from various sources (microphones, tape decks, turntables) and combining them at relative signal levels to form one composite signal; a unit that "mixes" together various sounds into one total sound and enables the operator to control the level of each sound and the overall level of the total sound; used when two or more sound sources are combined to be recorded on a single **audio track.**

AUDIO-OUT—The output jack that carries an audio signal from a particular piece of equipment; the output receptacle that delivers an audio signal.

AUDIO TRACK—That portion of the video tape on which audio information, sound, is recorded.

AUTOMATIC GAIN CONTROL—AGC. An electronic circuit that adjusts the incoming signal to a predetermined level; an automatic volume control; usually denotes an audio function. *Compare with* **automatic light control.**

AUTOMATIC LEVEL CONTROL—ALC. When used to describe an audio signal control, means the same as **AGC.**

AUTOMATIC LIGHT CONTROL—ALC. The vidicon camera control that automatically adjusts the target voltage to compensate for variations in light levels.

A/V—Industrial term for audio/visual; *also* audio/video.

AVAILABLE LIGHT—Sources of lighting (both natural and artificial) present in the scene to be recorded, before you show up with your portable equipment.

BACKING—The plastic ribbon, usually mylar onto which is coated the oxide formulation of both audio and video tape.

BACK-LIGHT—Light placed behind objects in a scene and pointing toward the camera to provide a rim of light, which outlines the object and creates a sense of depth by setting off that object from the rest of the scene.

B&W—Black-and-white; **achromatic** monochromatic.

BARN DOORS—Metal flaps that attach to the sides of a lamp housing and control the shape of the light being cast onto an area.

BARREL DISTORTION—The characteristic distortion of a scene by a **wide angle lens:** a rounded and out-of-proportion look around the edges of the scene; caused by objects being too close to the lens.

BASE-LIGHT—The basic "sunlight" that illuminates the scene so that the camera vidicon **signal-to-noise ratio** and the **aperture** setting of lens are acceptable for recording.

BEAM—A coherent flow of electrons.

BEAM ADJUSTMENT—A control on vidicon cameras that regulates the amount of current flowing in the **beam.**

BETAMAX—A video tape recorder standard developed by Sony and marketed by several Japanese and American electronics manufacturers as mass-market, inexpensive, consumer video recorders. The Beta format uses half-inch tape in a cassette housing and is not compatible with other consumer video cassette formats, notably that of its main competitor, **VHS.**

BI-CONCAVE—A lens configuration in which the lens element has an inward curve on both sides.

BI-CONVEX—A lens configuration in which the lens element has an outward curve on both sides. A magnifying glass is the most common example of a bi-convex lens.

BI-DIRECTIONAL—Describes a microphone that accepts sound waves from two different directions, while **attenuating** sound waves from any other direction. A microphone with a specific selective pickup pattern, as compared to an **omni-directional** microphone. *See also* **uni-directional.**

BINDER—Chemical adhesive used to hold the iron oxide particles on the video tape **backing** material.

BLACK CLIPPING—A video control circuit—found in cameras and VTRS—that regulates and contains the **black level** of the video signal so that it does not disturb or appear in the sync portion of the signal.

BLACK LEVEL—The bottom level of the picture signal, below which are the **sync, blanking,** and other control signals that do not appear as picture information.

BLANKING—Suppression. The process—and the period of time that process takes—during the **scanning** of the **raster** area when the beam is shut off. Line blanking takes place when the beam is returning from the end of one line to begin another; similarly, the process and the period of time in which the beam finishes scanning one field and retraces its path to the top of the raster area to begin scanning the next field, called field blanking.

BLANKING CIRCUIT—*See* **line blanking** and **field blanking.**

BLANKING SIGNAL—The pulses added to the video signal to indicate that the signal from the beam to the target area should be cut off; signal takes place as the beam goes into **flyback.**

BLOCK OUT—To draw a sketch, write down, or run through the action a scene or series of scenes will contain prior to taping that scene.

BODY BRACE—A metal frame worn over the upper torso to which a camera is attached and which supports that camera.

BOOM—Can be a mike boom, light boom, or camera boom; a mike or light boom is a long piece of metal piping at the end of which a light or mike is attached to allow either mike or light to be positioned over the heads of subjects in a scene while remaining outside the camera **angle of view;** a camera boom is a complex piece of heavy-duty equipment, which allows the camera and operator to be raised to selected heights.

BOOST—To turn up; to increase in volume; to make a signal stronger.

BREAK-UP—Disruption of the video signal by extraneous electronic signals producing a noisy, distorted, or otherwise imperfect video picture. Used to describe various problems that produce an incoherent video picture.

BRIGHTNESS RATIO—An indication, expressed as a ratio, of the difference between the whitest and the blackest object in a scene; the range from brightest white to darkest black as it occurs in the scene being recorded. Too wide a range between brightest and darkest can lead to an unacceptable **contrast ration** when the scene is displayed on a TV screen.

BRIGHTNESS VALUE—**Luminance.** The relative brightness of a particular object in a scene; the point on the **gray scale** at which the object is between absolute black and absolute white, either of which can be used as a point of reference to determine the brightness value of the object; essentially a relative determination made by the observer.

BRING UP—*See* **fade in.**

BURN—To destroy the light conversion function of a certain portion of the vidicon target area by exposing it to a light source that is too intense; *also* the result, an image permanently impressed on the target area, which appears as black spots during display.

BURST FLAG—A pulse produced by a color sync generator; when present, it causes the signaling color camera to produce a **burst signal,** or **color burst.**

BURST SIGNAL—*See* **color burst.**

BUS—One complete **channel** of an audio or video mixing system, including inputs, **gain** controls, and an output; two or more buses are required for video signal switching.

CABLE—A grouping of wires in a protective sheath used for the transmission of electrical power and/or signals; *also* **CATV.**

CAMERA—The eye of the video system; an instrument capable of absorbing the light values of a scene and converting them to a corresponding series of electrical pulses through the use of a cathode ray pickup tube such as the **vidicon;** a light-sensitive **cathode ray tube** (and its associated electronic circuitry and lens optics), which translates the light values of any scene it views into a set of voltage variations that can be used to re-create those light values on another cathode ray tube such as that used for TV display.

CAMERA CHAIN—The camera and its associated electronics; used to describe the camera, camera control unit, and power supply of large studio camera systems.

CANNON CONNECTOR—A particular brand of audio **jack** that features three leads—two for the signal and one for the overall system grounding; a very secure type of connecting jack often found on high-quality microphones, video monitors, and VTRS.

CAPACITOR—A component used in audio and video circuits to store and release voltages within the circuit.

CAPSTAN—A rotating shaft on the VTR which is turned by a motor and which, in turn, governs the speed of the tape as it proceeds from the supply reel to the take-up reel.

CAPSTAN SERVO—The control of the tape to assure accurate and uniform speed during record and playback by passing the tape between a spinning post in the tape path, which is an extension of the VTR motor and is called a **capstan,** and a rubber pinch roller. Internal electronic circuitry in the VTR senses any fluctuations in tape speed and advances or retards the tape to a predetermined rate as it passes between the capstan post and the pinch roller.

CAPSTAN SERVO EDITING—Head override editing. A method of **electronic editing** in which a new video signal replaces an already existing signal without disrupting the picture (except for a switch from signal 1 to signal 2 at the point of addition). The motor speed of the capstan—which controls the speed of the video tape—on VTR B is controlled by the **vertical sync** pulses on the video tape on VTR A during the "editing" of the signal on tape A onto tape B so that no disruption of the flow of signal information occurs during the switch of signals.

CARDIOID—One of many possible **pickup patterns** of a directional microphone. As the name suggests—cardioid means heart-shaped—sound waves coming to the microphone's rear and sides are rejected, those directly in front of it are received.

CARRIER FREQUENCY—The particular wavelength of a **frequency** on which a signal is impressed for transmission in a coherent fashion to a receiver, where it is stripped of its carrier frequency, amplified, and reproduced; can apply to either audio or video.

CATHODE RAY TUBE—CRT. A vacuum tube with a cathode and heater element at one end capable of producing electron beams. The beams flow down the length of the tube, where they either hit a phosphor coating on the face of the tube and make it glow or hit an oxide coating and produce a voltage—depending on the purpose of the cathode ray tube. Both a vidicon and a TV picture tube are cathode ray tubes.

CATV—Community antenna television; a system whereby TV signals are received from off-the-air broadcasts or are otherwise generated and are then sent along a **coaxial cable** to TV receivers; originally developed in the 1940s as a method of providing TV reception in rural, mountainous areas; currently being developed for commercial uses. Using available technology, more than eighty channels can be sent by cable to any TV set.

CCIR—*Comité Consultatif International des Radiocommunications.* French-based European and British standardization committee responsible for many European audio and video standards, including the 625-line, 50Hz TV standard.

CCTV—Closed circuit television; any form of television that is originated and displayed locally; non-network, non-cable TV.

CERAMIC MIKE—*See* **piezoelectric.**

CHANNEL—The space on the **frequency** waveband assigned to a particular television broadcast; width varies from country to country; in the US, spectrum space is about 6MHz wide for each channel.

CHARACTER GENERATOR—A device that electronically displays letters or numerals on a TV screen.

CHROMA—*See* **chrominance.**

CHROMA KEYING—The electronic introduction of a color background into a scene; process differs from b&w keying because the presence of color makes it possible for the operator of the keying unit to introduce the background by adjusting the color values.

CHROMATICITY—A subjective evaluation of the **hue** and **saturation,** the "coloredness," of an object.

CHROMINANCE—Chroma. The **hue** and **saturation** of an object as differentiated from its **brightness value (luminance).**

CHROMINANCE SIGNAL—That portion of the total video signal which contains the color information. Without the chrominance signal, the received TV picture would be in b&w.

CHROMINANCE TUBE—*See* **color dissector tube.**

CIRCUIT BREAKER—An electromechanical safety control that monitors the flow of ac from its source to wall outlets. Excessive power demands will trip open (break) the circuit, thereby stopping the flow of power to the outlet. A circuit breaker is reset by hand once the condition that tripped it has been corrected. *Compare with* **fuse.**

CLAMPER—An electronic circuit that sets the video level of a picture signal before the **scanning** of each line begins; ensures that no spurious electronic **noise** is introduced into the picture signal from the electronics of the video equipment.

CLAMPING—The action of the electronic **clamper** circuit.

CLEAN EDIT—An **electronic edit** of a video picture free of **noise,** distortion, or other disruption in the signal when it changes from picture 1 to picture 2. In a clean edit, the picture is instantly replaced by a subsequent picture.

CLIPPING—A circuit that removes the positive and negative overmodulations of a **composite video signal** so that they do not interfere with the picture or sync information.

CLOSE-UP—A relative determination of a camera **angle of view;** usually, a shot that shows the subject of a picture in great detail.

C-MOUNT—A mounting plate for vidicon video cameras and 16mm movie cameras that accepts a certain type of lens. Both the lens and mounting plate have matching nut-and-bolt threads, the mounting plate hole is of a standard diameter and the lens has a threaded collar on it that can be screwed into the hole. A type of lens mount for personal video cameras.

COAX—*See* **coaxial cable.**

COAXIAL CABLE—A one-ground, one-**conductor** cable that can carry a wide range of frequencies as far as 1000 feet with little or no signal loss.

COAXIAL CONNECTOR—F-connector. A specially designed cable connector used in cable TV and other 75-ohm cable applications.

COLOR BACKGROUND GENERATOR—An electronic circuit used in **chroma keying** to produce a solid color background of any desired **hue** and **saturation.**

COLOR BURST—**Burst signal.** A set of high frequency (3.5 MHz) pulses at the beginning of each scanning line which determines the **phase** of the color signal.

COLOR CAMERA—A video camera capable of changing both the **brightness values (luminance)** and the **chrominance** values (**hue** and **saturation**) of a scene into a series of electronic pulses.

COLOR DISSECTOR TUBE—**Coloring tube; chrominance tube.** A **cathode ray tube** designed to separate a scene's **hue** and **saturation** values into their **R,B,G** components for electronic encoding as part of the color video signal.

COLORING TUBE—*See* **color dissector tube.**

COLORIZER—Electronic circuitry used to generate a chrominance signal in relation to the gray values of a b&w video signal. Each gradation of gray from black to white is assigned a color value. The result is an artificially, and often inaccurately, colored picture.

COLOR KILLER CIRCUIT—An electronic circuit used in a VTR to suppress the 3.58MHz color **carrier frequency** when a b&w tape is being shown; the same circuit in a b&w VTR used to suppress the color carrier frequency when a color tape is being played back in b&w. Without a color killer, the color signal would appear in the displayed b&w picture as random **noise.**

COLOR PHASE—The proper timing relationship within a color signal. Color is considered to be in phase when the hue is reproduced correctly on the screen.

COLOR PICTURE TUBE—A **cathode ray tube.** The screen-end of which is capable of glowing with the three primary television colors—red, blue, and green. Its cathodes produce three electronic beams (each corresponding to one of the three colors) and its **raster** area is coated with three different types of phosphor (each one corresponding to one of the three colors).

COLORPLEXER—Encoder. Electronic circuitry which processes three separate color signals—red, blue, and green—coming from the pickup tubes into one composite encoded color video signal.

COLOR SUBCARRIER—The carrier wave on which the color signal information is impressed; contains the **color burst** and alternating phase color information. In the US the color subcarrier is 3.57954MHz, usually rounded off to 3.58MHz.

COLOR SYNC—A control signal necessary for the operation of color cameras, SEGs, and monitors; consists of a 3.58MHz burst (which sets the **color phase** and placement before each line is scanned) and a 3.58MHz color subcarrier.

COMPATIBLE COLOR—A TV broadcast system that produces a color signal that can be received by either a b&w or color set. The **luminance** values (the basis of b&w reception) and the **chrominance** values (the basis of color reception) are broadcast as different portions of the total signal so that the luminance values are not dependent on the chrominance values for reproduction.

COMPONENT—Any portion of a total electronic system; a component can be the transistor on a circuit board or the circuit board in a TV camera or the TV camera in a studio system, and so on.

COMPOSITE MASTER—An original program produced by editing various portions of other recordings onto a new reel of tape; in **electronic editing** the resulting tape is one **generation** away from the master materials from which it was recorded.

COMPOSITE SYNC—The total **sync** signal, containing both horizontal and vertical scan controls.

COMPOSITE VIDEO SIGNAL—Video signal containing both picture and **sync** information.

COMPRESSION—An audio term similar to video **clipping;** the automatic adjustment of volume variations to produce a punchy, meaty sound. Elimination of audio overmodulations produces a sound lacking in dynamics—it is never soft or loud but always at the same level.

CONCAVE—A lens configuration in which the lens element has an inward curve.

CONDENSER—Describes a type of microphone element that uses two condenser plates to convert sound waves to voltage variations.

CONDUCTOR—A strand of wire(s) capable of carrying an electronic signal along its length; a length of cable that conducts a signal from one point to another.

CONLOCK—*See* **genlock.**

CONTRAST RANGE—*See* **contrast ratio.**

CONTRAST RATIO—Contrast range. The range of gray between the darkest and the lightest (brightest) value in a scene; expressed as a ratio of dark to light, such as 20:1, and used to evaluate the picture on a TV screen. *See also* **brightness ratio.**

CONTROL TRACK—The lower portion along the length of a video tape on which sync control information is pulsed and used to control the recording or playing back of the video signal on a VTR.

CONTROL TRACK HEAD—*See* **audio and control track head.**

CONVERGING MENISCUS—A lens configuration in which the lens element has an outward curve on one side and an inward curve on the other.

CONVEX—A lens configuration in which the lens element has an outward curve.

CORNER INSERT—A second video picture signal added to an area of the first video picture signal. Corner inserts are achieved by halting the horizontal and vertical **scanning** of the first picture in a predetermined area and inserting the second picture's scanning portions into that area.

CPS—Cycles per second; *see* **Hertz (Hz).**

CRIMPING—A mechanical method of attaching a **jack** to the end of a cable—the sleeve of the jack is squeezed around the cable so that it will stay on; most often used in **coaxial cable** installations.

CROSSFADE—To fade out one video signal and fade in another as a simultaneous movement; can be written as X on taping scripts.

CROSS-PULSE GENERATOR—An electronic circuit, available as a separate component or built in to high-priced TV studio monitors, which shifts the video picture on a TV screen so that the horizontal and vertical control pulses are visible on the screen, making it possible to adjust tracking and screw and thereby achieve as stable a picture as possible.

CRT—See **cathode ray tube.**

CRYSTAL MIKE—See **piezoelectric.**

CUT—To instantly replace one picture with a second picture.

CUTTING ON THE ACTION—A production or editing technique in which two events are set in contrast to each other; as event A is taking place on the screen, the camera switches to event B before event A has ended; often used to develop plot, create tension, or produce contrast.

CUTTING ON THE REACTION—A production or editing technique in which one event is followed by a scene that displays the results of that event; after event A has taken place, the camera cuts to event B to show the impact of event A on the plot or characters.

CUTTING TO TIGHTEN—An editing procedure used to shorten a series of shots. Used to eliminate excess footage and to produce a coherent whole.

dB—See **decibel.**

dBm—**dB** rating that indicates the number of decibels a signal is above or below one milliwatt.

DC—Direct current. Electrical current which, unlike ac, maintains a steady flow, does not reverse directions, and cannot, therefore, be measured in cycles per second (**Hz**).

DC-TO-AC INVERTER—An electronic unit that converts direct current to alternating current; used with an ac-to-dc converter to change one ac standard to another. For instance, 220-volt 50Hz is changed to 12-volt dc through the converter and the 12-volt dc is then changed to 120-volt 60Hz through the inverter.

DECIBEL—dB. A unit for the subjective evaluation of the volume of any particular sound measured in relation to other sounds; an evaluation of the strength of a signal in relation to a predetermined reference level. Decibels increase logarithmically.

DEFEAT—Turn off.

DEFINITION—The sharpness of a picture evaluated subjectively in terms of its **resolution,** or the amount and clarity of its detail.

DEFLECTION CIRCUIT—A complete set of **deflection coils** consisting of one coil to control vertical scanning and another to control horizontal scanning; *also,* the electronics used to power a deflection system.

DEFLECTION COIL—An electromagnetic coil wound around the cathode end of the **cathode ray tube** to produce a magnetic field, which controls the movement of the electron beam. *See also* **focus coil.**

DEGAUSS—To demagnetize.

DEGREE KELVIN—*See* **Kelvin.**

DEMODULATE—To strip a signal of the carrier frequency onto which it was modulated; a signal is demodulated after it has been broadcast, prior to its being displayed.

DEPTH OF FIELD—The distance between the closest object in focus and the farthest object in focus within a scene as viewed by a particular lens; can vary with the quality and mm of the lens or with its **f-stop** setting.

DEPTH OF FOCUS—The allowable latitude of lens **image plane** to vidicon target area which ensures that a given picture remains in focus. Depth of focus is adjusted by moving the vidicon closer to or farther away from the end of the lens. Not normally used in video as a focusing adjustment.

DEPTH PERCEPTION—A subjective evaluation of the distance between objects viewed with regard to their size and the planes they describe.

DIAPHRAGM—The element in a microphone vibrated by sound waves entering the mike. The vibrations of the diaphragm are converted into voltage variations to produce the audio signal. *See also* **iris diaphragm.**

DICHROIC DAYLIGHT CONVERSION FILTER—A lens **filter** that balances the color values of objects in direct sunlight so that they will match the values of scenes taped in artificial light.

DICHROIC MIRRORS—Special mirrors that reflect some wavelengths of the light spectrum, while allowing others to pass through their semitransparent surfaces; used in color TV cameras to divide light into the three primary colors for pickup tubes.

DIELECTRIC—Describes an insulator placed between conductors to prevent them from touching and thus shorting out the signal being carried.

DIN—*Deutsche Industrie-Norm.* German standard for electronic connections. DIN plugs can be three-, four-, five-, or six-pin plugs, depending on their use, although they all have the same outer diameters and appearance.

DIRECT CURRENT—*See* **dc.**

DISSOLVE—A slow **crossfade;** one picture gradually fades out, the next picture gradually fades in; can be written using the symbol $\overset{x}{\underset{x}{}}$.

DIVERGING MENISCUS—A thicker version of the **converging meniscus** lens configuration. One side curves inward and the other side curves outward, but the edges of the curves do not meet at the rim as they do in a converging meniscus lens.

DOLLY—Wheels on the feet of a tripod; *also,* the action of moving a camera toward or away from a scene, as to dolly in or dolly out.

DOLLY IN—*See* **dolly.**

DOLLY OUT—*See* **dolly.**

DRIVE PULSES—*See* **sync.**

DROP-OUT—Loss of a portion of the video picture signal caused by lack of iron oxide on that portion of the video tape or by dirt or grease covering that portion of the tape.

DROP-OUT COMPENSATOR—Circuitry which senses signal loss produced by **drop-out** and substitutes for missing information the signal from the preceding line—if one line drops out of a picture, it is filled in with the preceding line, resulting in no visible drop-out on the screen. Drop-out compensators are built into a number of VTRs currently on the market.

DUBBING—Duplicating an audio and/or video signal, such as a **composite master** tape, to make additional tape copies. Dubbing puts the resulting copy or dub one generation away from the tape from which it was recorded. Can also refer to erasing an audio track and recording a new track in its place. *See* **audio dub.**

DYNAMIC MIKE—A type of very sound-sensitive **uni-** or **omni-directional** mike, which can stand rough handling.

ECHO—The single repetition of a sound after that sound has taken place: hello-hello; an audio effect in which a tape loop, echo chamber, or other device causes a sound to be repeated a short time after it has been made; *compare with* **reverberation.**

EDIT IN—The point at which, or the process whereby, one video signal replaces another during the editing process.

EDITING—More accurately, in video, electronic editing. The assembly of various segments into a deliberately composed order achieved by rerecording the desired segments in the chosen order on a new video tape to produce a composite program, a new whole.

EDIT OUT—The point at which, or the process whereby, an inserted video signal stops and the signal it replaced for a predetermined length of time resumes. The other end of an **edit-in** when **insert editing** is being done. With currently available editing decks, edit-ins are more electronically stable than edit-outs (except in the case of computer-controlled quad machines).

EDITING DECK—A specially constructed video tape recorder which has, in addition to play and record circuitry, circuitry and controls for assembly and/or insert editing; an editing deck is used in conjunction with a second video tape recorder to record a master program tape (on the editing deck) from various tape recorded segments being played back on the second VTR.

EFFECTS BUTTONS—The push-button controls on a special effects generator that indicate the special effects (**inserts, wipes, keying,** etc.) available on that SEG and which are engaged when an effect is desired.

EFFECTS CHANNEL—That bus in a three-bus **switcher** set aside to produce special effects.

EIA—Electronic Industries Association. The people who determine audio and video standards in the US.

EIA SYNC—Also called EIA RS-170 sync. The standard waveform for broadcast equipment in the United States as established by the **EIA.**

EIAJ TYPE #1 STANDARD—That standard established by the Electronic Industries Association of Japan for half-inch helical scan video tape recorders.

EIAJ TYPE #1 RECOMMENDED COLOR STANDARD—The color standard established by the Electronic Industries Association of Japan to be compatible with the **EIAJ Type #1 Standard;** color tapes can be played back in b&w on EIAJ Type #1 b&w VTRS, and b&w tapes can be played back in b&w on EIAJ Type #1 color VTR.

EIDOPHOR SYSTEM—Trade name of a large video projection system.

EIGHT-PIN CONNECTOR—A type of jack commonly used for the VTR-to-monitor connection; provides a full set of audio and video connections—one ground and one lead each for audio-in, audio-out, video-in, and video-out.

ELECTRET CONDENSER—A type of very sensitive microphone requiring dc power (usually supplied by a battery built into the mike).

ELECTRON GUN—The assembly at the end of the **cathode ray tube,** which produces the electron beam for **scanning;** consists of cathode, heater, and grids.

ELECTRONIC EDITING—Repositioning video signal segments on a reel of video tape without physically cutting the tape; a rerecording of the video signal segments in different order. Electronic editing implies that the edited version of the program will be one **generation** removed from the recordings from which it was assembled.

ELECTRONIC VIEWFINDER—Viewfinder; viewfinder monitor. A small TV screen attached to the video camera, which allows the operator to see a given scene exactly as it is viewed by the camera.

ELECTROSTATIC FOCUS—Method of focusing the electron beam in a **cathode ray tube** without using **deflection coils;** its advantage over deflection coils is that it improves the quality of the picture while requiring less power to operate.

ENCODER—*See* **colorplexer.**

EQ—*See* **equalization.**

EQUALIZATION—eq. The normalization of an electronic signal, either audio or video; adding eq in audio means reshaping the frequency response to emphasize certain **frequency ranges** and eliminate others.

EQUALIZER—An audio or video circuit that provides **equalization** either automatically (usually in the case of video) or manually (often available in higher-priced audio mixers).

EQUALIZING AMP—A video circuit preset to provide a selected **equalization** to the video signal.

ERASE HEAD—Either an audio or video **head** that erases the signal on a video tape prior to the recording of a new signal on that tape; simply, an electromagnet which disturbs the signal previously recorded on the tape.

ESTABLISHMENT SHOT—A long shot used to set the scene by showing the environment in which the action to follow takes place; for example, a long shot of a burning house preceding a medium shot of firemen with hoses.

E-TO-E—Electronics-to-electronics. Monitoring the output signal of a VTR while it is recording is an E-to-E process. The signal monitored has not yet been recorded on tape, rather, a sample of the signal is being fed from the VTR directly to the monitor. With E-to-E it is not possible to be certain that the signal is being recorded on the tape.

EXTENDER—A lens accessory that lengthens the barrel and in so doing reduces the minimum focusing distance of the lens and increases the effective f-stop.

EXTERNAL KEYING—A **keying** effect accomplished when a particular camera is assigned to supply the key signal through an SEG; *compare with* **internal keying,** in which any of the cameras can supply the key signal.

EYE-LIGHT—A small pencil-beam spotlight which, when directed at subject's eyes, produces a glitter that appears natural on the screen.

FADE—To vary the strength of a signal, as in **face in** or **fade out;** can be used with reference to both audio and video.

FADE IN—Bring up. To increase the strength of a signal, be it audio or video.

FADE OUT—**Attenuate.** To reduce the strength of a signal, be it audio or video.

FADER—A sliding pot control with which an audio or video signal is **faded.**

F-CONNECTOR—*See* **coaxial connector.**

FIELD—The electronic signal corresponding to one passage over the **raster** area by the **scanning spot;** consists of 262.5 lines in the American TV system; two fields interlace to make one **frame,** so a field is half a frame.

FIELD BLANKING—Field retrace period; field suppression. That period of time during which the field **scanning spot** returns from the bottom to the top of the **raster** area; used to add control pulses to the video signal; occupies about 15–20 lines of the 525-line system.

FIELD FREQUENCY—The number of fields scanned per second; 60 fields are scanned per second in the American TV system.

FIELD RETRACE PERIOD—*See* **field blanking.**

FIELD SUPPRESSION—*See* **field blanking.**

FIELD TIME BASE—The pattern of and points at which a **field** changes; 60Hz is the field time base of the American TV system; the field time base must be kept as steady and regular as possible to ensure the best possible picture.

FILL-LIGHT—The illumination used to light shadowy areas in a scene to establish the proper **brightness ratio** or **contrast ratio** within the scene.

FILL SIGNAL—That signal, or portion of a signal, used in keying, wiping, or inserting to replace portions of the primary signal; any video signal used to replace another video signal in a special effects application.

FILM CHAIN—A special motion picture projector combined with a video camera to transfer movies to video tape.

FILTER—A glass element whose ability to transmit light varies with its design; used to exclude certain wavelengths or types of light; sometimes needed for color and b&w recording.

FIRST GENERATION—The original recording of a tape segment. The first time the signal is recorded on tape, that tape is called first generation. Every subsequent recording of the already recorded segment will be a **generation** removed.

FLUID HEAD TRIPOD—A tripod whose camera-mount consists of two metal plates. The upper, rotating plate rests on a bed of fluid, and the movement provided is very smooth.

FLYBACK—Retrace. The movement of a **scanning spot** from the end of a line or a field to the beginning of the next line or field.

FM—Frequency modulated; modulated RF; sideband FM. Descriptive of a signal that has been impressed onto a radio carrier wave in such a manner that the carrier frequency changes as the original signal does.

FOCAL LENGTH—Distance between the optical center of a lens and the **image plane,** which, in the case of the video camera, is the pickup tube's target area. The distance is measured in millimeters (mm) and determines the angle of view of the lens.

FOUR-TUBE COLOR CAMERA—An early color camera system using one tube to sense each of the primary colors (red, green, blue) and a fourth to sense black and white values (**luminance**). Required more exact adjustment because four signals had to be coordinated to produce one picture, and was a larger unit than the later **three-tube, two-tube,** and **one-tube color cameras.**

FOCUS—The greatest possible resolution of an object, when the object seems to be sharp and well defined; the act of moving a lens to make the image sharp and defined; to bring an electron beam or a light ray to its minimum size.

FOCUS COIL—An electromagnetic coil surrounding the **cathode ray tube** and producing a magnetic field, which controls the flow of the electron beam from cathode to target area so that it strikes the target area as the smallest possible spot; works on the same principle as the **deflection coil.**

FOCUS CONTROL—Focus ring. A calibrated lens control, which focuses light rays going through that lens by moving the internal elements of the lens; the control that governs the **focus coil** in a video camera; the setting of the distance between vidicon and end of lens. *See also* **depth of focus.**

FOCUS RING—*See* **focus control.**

FOG FILTER—A lens filter that lends the effect of fog to the scene.

FOLLOW FOCUS—The continual adjustment of the lens to

keep an object in focus while either object, camera, or both are moving.

FOOTCANDLE—Ft-c; **lumens** per square foot. The measurement of the intensity of light in an area; the amount of light cast on a particular area, varying with the distance of that area from light source.

FRAME—A complete TV picture, composed of two **fields;** a total **scanning** of all 525 lines of the **raster** area; occurs every 1/30 of a second.

FRAME FREQUENCY—The number of **frames** occurring in a given period of time; usually 30 fps (frames per second) or one half the **field frequency** of 60Hz.

FREQUENCY—The number of times a signal vibrates each second; expressed as cycles per second (cps) or, more commonly, as **Hertz (Hz)**.

FREQUENCY MODULATION—*See* FM.

FREQUENCY MODULATOR—An electronic circuit that produces a carrier wave signal on which the audio or video signal is impressed.

FREQUENCY RANGE—*See* **frequency response.**

FREQUENCY RESPONSE—Frequency range. The span of frequencies, from the highest to the lowest, of which a piece of equipment is capable. Most circuits have limits to their frequency ranges measured in cycles per second. For instance, the frequency range of audible signals that the human being can hear is described as being between 20 and 15,000 Hz.

FRESNEL LENS—Fresnel spot. A specially constructed lens, which produces a soft-edged concentration of light; used as a lens in a spotlight lamp housing.

FRICTION HEAD TRIPOD—A tripod whose camera mount consists of two metal plates, the lower stationary, the upper rotating; generally does not provide smooth camera movement, though more expensive models utilize ball bearings to offset this problem.

F-STOP—A calibrated control (f 1, f 2, f 3.5, f 4, etc.), which indicates the amount of light passing through a lens to the target area; a control that can be adjusted to vary the size of the lens **iris diaphragm.**

FT-C—*See* **footcandle.**

FUSE—Similar in purpose to **circuit breaker** in that it insures against danger of power demands in excess of what circuit is equipped to deliver. When a fuse "blows," it stops the flow of ac current to the ac wall outlet. Unlike circuit breakers, fuses must be replaced when they blow. Fuses are available in 15, 20, and 30 amp ratings.

GAIN—Amount of signal amplification; turning up the gain means increasing the strength of the signal, turning down the gain decreases the signal strength; used in reference to both audio and video to denote the relative strength of the signal in question.

GAIN CONTROL—A circuit that changes the strength of the signal passing through it. In audio, a gain control is often called a volume control, since increasing the gain of an audio signal makes it louder and decreasing makes it less loud. In video, gain control increases or decreases the strength of the video picture.

GAP—The small space in an audio or video head across which the magnetic field is produced when recording and induced on playback; the audio and video heads are small, horseshoe-shaped electromagnets and the gap is the space between the arms of the horseshoe shape, which this tape must contact for recording/playback.

GENERAL LIGHTING—The overall lighting of a scene produced by available light and **base-lighting** units.

GENERATING ELEMENT—The component that enables the sound waves entering the head of a microphone to be used to produce an electronic signal composed of voltage variations corresponding to the sound wave; some of the most common generating elements are **ribbon, condenser, crystal,** and **dynamic** elements.

GENERATION—Each time a signal is recorded from a camera or other source (such as off-the-air broadcast); the first (original) recording is said to be first generation, the first rerecording (that is, the second time the information is recorded) is said to be second generation, and so on. The more generations, the more **time-base** error and the more **noise** there will be on the video and audio tracks.

GENLOCK—Conlock; circuitry that locks the **sync generator** used to control cameras and the SEG by way of the sync signal from a prerecorded tape on a VTR so that the signal from that tape can be mixed through the SEG with live camera signals.

GRAY SCALE—The range, divided into steps from black to white within which a camera is capable of resolution; how faithfully the light values of a scene (changes of brightness from black to white) can be rendered by any piece of electronic equipment, such as camera or monitor; that set of gray tones—from black to white—on a test pattern, consisting sometimes of five steps, sometimes of ten, depending on the pattern. *See also* **contrast ratio,** with which it is frequently interchangeable.

GROUNDED OUTLET TESTER—An inexpensive device used to test an ac outlet to ensure that its polarities have not been reversed and that it is properly grounded.

HALO—The black area around a very intense source of light, as seen by the camera and monitor. When a match is struck and held in front of the camera, a halo is visible around the flame; a halo visible through the viewfinder of a vidicon camera may mean that the light source is bright enough to **burn** the target area.

HEAD—Audio or video; a small electromagnet that pulses magnetic signals onto a video tape moving past it or induces those signals off a recorded tape; audio heads are usually stationary, video heads move in reverse of the tape's direction in most VTRs.

HEAD ALIGNMENT—The positioning of the audio or video **heads** so that they describe the correct path at the correct angle across the video tape; heads that are out of alignment won't record or play back properly.

HEAD CLOGGING—Occurs when the **gap** of an audio or video **head** gets filled with dirt, grease, or oxide so that it can no longer record or play back a signal; cured by cleaning the heads.

HEAD DRUM ASSEMBLY—That portion of the VTR in which the video **heads** and their related mechanical and electronic controls are located. In **helical scan,** the head drum assembly is the large circular unit around which the tape wraps as it passes the video heads.

HEAD DRUM SERVO—One method, now largely obsolete, of controlling the video tape during playback so that the video **heads** contact the tape with the proper timing (sync) to retrieve the information on the tape. The control track pulses are used to control the rotation of the video heads. *See* **capstan servo,** which is the far more common tape control setup.

HEAD DRUM SERVO EDITING—An editing method that is an inferior variation on **capstan servo editing.** Instead of the tape being slowed down or speeded up to maintain sync, the speed of the video **heads** is varied by the use of the **head drum servo** controls.

HEAD OVERRIDE EDITING—*See* **capstan servo editing.**

HEAD SHOT—A close-up view of a subject, encompassing the subject from top of head to top of shoulders.

HEADS OUT—A reel of tape wound so that the beginning of the program is at the beginning of the tape; a rewound tape. *Compare with* **tails out.**

HELICAL SCAN PROCESSOR—*See* **processing amplifier.**

HELICAL SCAN VIDEO TAPE RECORDING—A type of

video recording in which the video **heads** and the tape meet at such an angle that the resulting pattern on the tape is a long, diagonal series of tracks from the video heads, each diagonal stripe containing the full information for one **field** of video picture; named after the helical path the tape describes between supply reel and take-up reel.

HELICAL WIND—The screwlike configuration of the tape across the video **heads,** from the plane of the supply reel, through the plane on which the heads are rotating, to the plane of the take-up reel.

HERTZ—The international electronic term for cycles per second (**cps**), most commonly written **Hz.** The number of times per second an electrical event is repeated.

HIGH END—The highest frequency portion of a video or audio signal; in audible audio signals, the high end is the treble portion of the signal.

HIGH END NOISE—Any spurious noise occurring in the high frequencies of a signal.

HIGHLIGHT—An area of great brightness on a TV display; a very bright portion of a picture.

HIGH RESOLUTION—Descriptive of a camera or monitor capable of displaying a great number of scanning lines (1000–2000), which produce a picture that is very detailed, defined, and sharp.

HI-Z—*See* **impedance.**

HORIZONTAL BLANKING—Line blanking. *See* **blanking.**

HORIZONTAL FREQUENCY—*See* **line frequency.**

HORIZONTAL RESOLUTION—A subjective evaluation of the number of vertical lines (usually from a test pattern) that can be seen in a horizontal direction. The better the horizontal resolution, the sharper and less blurry the picture will be. *See also* **resolution.**

HORIZONTAL SYNC—The sync pulses that control the horizontal line-by-line scanning of the target area by the electron beam. *See also* **line frequency, line scanning.**

HOT—Live wire; a conductor carrying a signal is said to be a hot conductor; the wire carrying the signal, as opposed to the ground wire.

HOT CONDUCTOR—*See* **hot.**

HOT SPOT—An area of intense heat or light, which produces a washed-out area on the camera pickup tube. Some lights (such as common incandescent bulbs) produce uneven floods of light with some areas brighter (hotter) than others; this results in hot spots appearing in the TV picture.

HOWL—Positive feedback; in video, the wild, swirling effect that results when a camera is pointed into a monitor displaying the picture that camera is producing; in audio, any high-frequency feedback caused by a microphone being too near the speaker reproducing the sound picked up by the mike.

HUE—A term used to describe the position of a color in a range that runs from red to yellow to green to blue to violet, and back to red; all colors have a hue.

HZ—*See* **Hertz.**

IC—Integrated circuit. A type of very small electronic component.

ICONOSCOPE—An early (1923) type of camera pickup tube, no longer used.

IMAGE—The picture on the TV screen; whether color or b&w, the image is usually measured by its **luminance** values.

IMAGE BUFFER—Electronic circuitry capable of modifying a 625-line 50Hz video signal to the 525-line 60Hz standard (or the reverse).

IMAGE DISSECTOR TUBE—Early camera pickup tube, no longer used, invented by Philo Farnsworth in 1927.

IMAGE ORTHICON—A camera pickup tube developed prior to the **vidicon;** the first really reliable and sensitive camera pickup tube; larger and bulkier than a vidicon.

IMAGE PLANE—The point behind the lens on which the image collected by the lens is cast; in video, the image plane is at the surface of the vidicon **target area.**

IMAGE RETENTION—Lag. The tendency of the vidicon pickup tube to retain an image on its **target area** after it has stopped **scanning** that image. Extreme image retention results in the image being **burned** into the target area.

IMPEDANCE—The resistance of a component to the flow of a signal toward it; expressed as high or low impedance, hi-Z or low-Z.

INDUSTRIAL SYNC—*See* **random interlace.**

INLAY—Keyed insert; static matte insert. An insertion effect in which the **fill signal** is static and of a predetermined shape. *Compare with* **overlay.**

IN-LINE COLOR—A color TV tube system in which the three electron guns producing the primary colors of the color signal are set next to each other in a straight line rather than in a triangle, as has traditionally been the case in color TV manufacture.

INSERT—A general effects term meaning the introduction of a secondary signal into an already existing picture; accomplished by **keying, wiping,** or **crossfading.**

INSERT EDIT—The insertion of a segment into an already recorded series of segments on a video tape; the inserted segment replaces a pre-existing segment, which must be of the exact length. Insert edits demand that the segment be **edited in** (as in **assembly editing**) and then **edited out** at the end of the segment, since video information already recorded follows the edited-in segment on the original tape.

INSERTION LOSS—The loss of signal strength that occurs when a piece of video or audio equipment is added to the path of the signal flow from origin to display; can be corrected by using an amplifier to build up the signal strength again.

INTERCUTTING—A production technique in which a cut is made from a scene (long shot) to a detail of that scene (close-up) to clarify or emphasize a point.

INTERFACE—To connect two or more components to each other so that the signal from one is supplied to the other(s). Feeding a signal between units that run on different standards is the most frequent form of interfacing, as in connecting a half-inch **helical scan** VTR to a two-inch **quadraplex** machine.

INTERFERENCE—Unwanted signals entering the signal path.

INTERLACE—A scanning method in which the lines of two **fields** are combined into a **frame** in such a way that all the lines of each field are visible as part of the frame; the positioning of 262.5 lines from one field with 262.5 lines from the next field to form a full 525-line frame. *Compare with* **random interlace.**

INTERNAL KEYING—A method of **keying** in which the key signal can be sent through the SEG from any one of the cameras already in use. *See also* **external keying.**

INVERTER—Circuitry to convert dc to ac; *see* **dc-to-ac inverter.**

IPS—Inches per second; the customary unit of measurement of tape speed on an audio or video tape recorder.

IRIS DIAPHRAGM—Iris. An adjustable set of metal leaves over the **aperture** of a lens, used to control the amount of light passing through the lens. Iris openings are measured in **f-stops.**

JACK—Male cable connector, or plug, as in **phono, mini, phone** plug.

JACKET—The protective and insulating housing of a cable.

JEEP—To convert a TV receiver into a TV monitor or monitor/receiver by rewiring the internal circuitry and adding input and output jacks for video and audio.

KELVIN—Also expressed as degrees Kelvin or °K; the unit of measurement of the temperature of light. In color recording, light temperature affects the color values of the lights and of the scene they illuminate.

KEYED INSERT—*See* **keying.**

KEYING—Keyed insert; **inlay.** The control of one video signal by the waveform of a second video signal when the two are combined to form a composite picture. The signal from source 1 fills in the scanning lines of the total picture of source 2 at the points where the picture goes above a preset gray level.

KEY-LIGHT—The spotlight or main light on a scene, which emphasizes the important objects in that scene.

KINESCOPE—An early and imperfect video storage and reproduction technique in which a film of a television program is made by placing a movie camera in front of a TV screen displaying that program.

LAG—*See* **image retention.**

LAVALIER MIKE—A small and unobtrusive microphone worn around the neck and resting on the chest cavity.

LENS—A series of optical elements, contained within a barrel or tube, which collect and focus light.

LENS CLEANING BRUSH—A very fine brush specially made for cleaning a lens.

LENS MOUNT—The assembly on the front of the camera to which the lens is attached. *See also* **C-mount.**

LENS PAPER—A paper specially made for cleaning lenses.

LENS SPEED—Measurement of the ability of a particular lens to collect light; the ability of that lens to work at different light levels; usually expressed by its lowest **f-stop** number.

LIGHTING RATIO—The brightness level of the **fill-light** compared to the brightness level of the **key-light,** or the shadowy areas compared to the brightly lit areas; measured as a ratio determined by the **f-stop** of the lens; a 1:2 ratio means that key is one f-stop brighter than fill; 1:3, a stop and a half; and 1:4, two stops.

LIMITER—A circuit that shapes a signal sent through it to conform to certain preset tolerances; used in both audio and video to regulate signal flow and prevent overamplification, distortion, and the introduction of spurious noise.

LIMITING—Controlling the strength of a signal to a predetermined level.

LINE BLANKING—The suppression of the video signal during the period of time when the **scanning spot** retraces one field and begins the next; the suppression ensures that no spurious electronic **noise** produced during this interval will disturb the visual integrity of the video picture.

LINE COUNT—The number of **scanning** lines actually used to carry the video picture signal; always less than 525 lines in the US TV system.

LINE DRIVE PULSE—The signal generated to control the **horizontal blanking** circuits.

LINE FREQUENCY—The number of lines **scanned** in one second; in the US system this is 525 × 60, or a line frequency of 15.7KHz.

LINE-IN—*See* **video-in.**

LINE LEVEL IMPEDANCE—An impedance of 600 ohms, considered a low impedance signal.

LINE MATCHING TRANSFORMER—An audio device used to match the **impedance** of a microphone to the input impedance of a mixer, tape recorder, or amplifier; a device that changes the output impedance of a mike from low-Z to hi-Z or vice versa.

LINE-OUT—*See* **video-out.**

LINE PERIOD—The length of time it takes for a line to be **scanned** and then **retraced** to the point where scanning of the next line will begin.

LINE SCANNING—The path over the target area of the electron beam, as it moves from the left edge across the area.

LINE SUPPRESSION—*See* **blanking.**

LINE TIME BASE—The control of the horizontal deflection of the **scanning spot** so that it starts to scan each new line at exactly the right moment.

LINE VOLTAGE—*See* **ac.**

LOAD—To place a termination across a video or audio line.

LONG LENS—A high focal length lens with a long barrel; performs similar function to the **telephoto lens** without the advantage of that lens' shorter barrel.

LONG SHOT—A camera angle of view taken at a distance and including a great deal of the scene area.

LOWELL LIGHT—A small, lightweight, portable lighting unit made by the Lowell Company.

LOW-LEVEL LIGHTING—A scene illuminated with under 50 ft-c of light; often results in a poor **signal-to-noise ratio** and/or poor **contrast ratio** in the recorded picture.

LOW-LIGHT LAG—A blurring, image-retention effect, which occurs when a vidicon tube is operating in insufficient light.

LOW-PASS FILTER—A filter, often used on two-way cable systems, which inhibits the flow of high-frequency information along a cable while it allows low-frequency information to pass.

LOW-Z—*See* **impedance.**

LUMEN—A measurement of light quantity, taken at the source of light against a predetermined constant. Lumens per square foot equal **footcandles.**

LUMINANCE—The brightness of an object and/or the brightness of the signal produced by that object as a video signal; thus, the degree of brightness—the position it occupies between dead black and pure white on the **gray scale**—of a signal. *See also* **brightness value.**

LUMINANCE SIGNAL—The black-to-white brightness values of a scene, which produce a b&w display picture.

LUX—An old-fashioned measurement of light quantity; taken at the surface that light source is illuminating. One footcandle equals 10.76 lux.

MACRO LENS—A magnifying lens capable of focusing down to a few inches.

MAGNETIC TAPE DEVELOPER—A special chemical solution applied to the control track edge of video tape to make **control pulses** visible to the eye and thereby allow precise cutting of the tape between pulses; necessary for physical tape editing.

MASTER MONITOR—High quality monitor equipped with such facilities as picture focus, internal and external **sync,** and **horizontal** and **vertical scanning** controls.

MASTER VOLUME CONTROL—An audio term, most often used with mixers and amplifiers to denote the final overall volume control of signal level.

MASTER VTR—When duplicating tapes, the deck that plays the original tape is called the master VTR, the deck that records the original signal on blank tape to produce the copy is called the **slave VTR.**

MATCHING TRANSFORMER—A circuit that changes the **impedance** of a TV signal, often from 75 ohms to 300 ohms; for audio, *see* **line matching transformer.**

MATTE—A film term sometimes used in video production work to denote a **keyed** effect, an **insert** of video signal information keyed from one source into a second video signal.

MEDIUM SHOT—Camera **angle of view** between **close-up** and **long shot;** a view of the head and shoulders of a subject, as opposed to head only (close-up) or full body (long shot).

METER—Unit consisting of a calibrated dial and swinging

needle, which gives a visual indication of the operation of the particular circuitry it is connected to.

MINI PLUG—Similar to a **phone plug** in design but much smaller; a plug introduced by Japanese electronic firms for use on miniaturized pieces of equipment.

MISTRACKING—Improper **tape path** and tape-to-head contact, resulting in bursts of **noise** appearing in the picture during display.

MODULATED RF—See **FM.**

MODULATION—The process of adding audio or video signals to a predetermined carrier signal; see also **demodulate.**

MONITOR—TV set without receiving circuitry used to display directly the composite video signal from a camera, video tape recorder, or special effects generator.

MONITOR/RECEIVER—A combination **monitor** and **TV receiver** capable of accepting composite video signals directly from source or those video signals broadcast as **RF;** also capable of producing a **composite video signal** output from a broadcast input signal, allowing user to record "off the air."

MONOCHROMATIC—See **achromatic.**

MONOCHROME SIGNAL—A black-to-white video signal containing **luminance** information only and capable of being received either by a b&w or color TV receiver and displayed as a b&w picture.

MOVING MATTE INSERT—See **overlay.**

MULTIPLEXER—An optical system allowing a number of film and slide projectors to feed video information into the same video camera.

NEUTRAL DENSITY FILTER—A filter placed over a lens to reduce the brightness of a scene.

NOISE—Any unwanted signal present in the total signal; both an audio and video term to describe one signal interfering with another; usually created by some malfunction of either a component or circuitry that is part of the signal path; a signal inherent in certain audio or video components.

NON-COMPOSITE VIDEO SIGNAL—A video signal containing picture and **blanking** information but no **sync** signals.

NORMAL LENS—A subjective evaluation of the **angle of view** of a lens; a normal lens is one that is neither **wide angle** nor **telephoto.**

NTSC—National Television System Committee of the Federal Communications Commission. Determines the principles upon which the US television system functions.

NTSC COLOR—The color standard used in the US and set by the National Television System Committee; a compatible color signal, which can be received in black and white on b&w TV sets.

OCTOPUS CABLE—Any cable with a number of different **jacks** at one or both ends; a cable that allows two pieces of video equipment with dissimilar jacks to be **interfaced.**

OMNI-DIRECTIONAL—A microphone whose pickup pattern is such that the mike is sensitive to sound waves coming at it from every direction.

ONE-INCH VIDICON—A vidicon tube with a target area one inch in diameter.

ONE-TUBE COLOR CAMERA—A color-capable video camera, which produces a color signal through the use of only one pickup tube.

ON THE LINE—A term used in video production with a special effects generator to identify the signal that is leaving the SEG for broadcast or recording; the camera signal being fed out of the SEG to the VTR is said to be on the line.

OPEN UP—*See* **stop down.**

OPTICAL CENTER—The point in a lens at which the image is collected to be focused on the **image plane.** The distance between the optical center of a lens and the image plane is described as the **focal length.**

ORTHICON—An early version of the **image orthicon** TV tube.

OUT OF PHASE—*See* **color phase** and **phase.**

OVERLAY—Self-keyed insert; moving matte insert. An insertion effect in which the **fill signal** is a moving object, which determines its own parameters as it moves.

PAL—Phase alternation line. British-German color TV standard.

PAN—To follow action by swinging camera left or right.

PASSIVE MIXER—An audio mixer containing no active electronic components or circuitry; usually an inexpensive audio mixer capable of combining and regulating the level of various signals but permitting a loss in the strength of those signals since they are not amplified within the mixer.

PATCH CORD—Any cable with a **jack** at each end used to connect audio or video components to each other; in audio, from which the term originates, the traditional patch cord is a cable with a **phone plug** at each end.

PATCHING—The act of connecting two components to each other with a **patch cord** and/or **patch panel.**

PATCH PANEL—A plate on which a number of female receptacles is mounted, each receptacle the termination of a different audio or video signal; used with **patch cords** to make secure but temporary connections between components.

PEAK-TO-PEAK VOLTAGE—Ppv; Vpp. The total voltage produced by a signal, determined by adding together the positive and negative extremes (the peaks) to which the voltage modulates.

PEAK WHITE—The brightest, whitest portion of the picture signal corresponding to the highest frequency the signal attains.

PEDESTAL—**Black level.** The minimum level the blackest portions of the displayed signal are allowed to reach.

PEDESTAL TRIPOD—A professional tripod, often with a hydraulic shaft.

PERSISTENCE OF VISION—A phenomenon of the brain working in conjunction with the eyes which makes the present system of television possible; the retention by the retina of any image it sees for a short period of time. If the image is replaced by a slightly different image before the first image fades, the distinction between the two separate images is lost and motion is perceived.

PERSPECTIVE—The angle at which we see things; our viewpoint. A lens produces its own perspective of the scene it is viewing, which may or may not agree with our sense of perspective.

PHASE—The timing of a signal in relation to another signal; if both signals occur at the same instant, they are in phase; if they occur at different instants, they are **out of phase.**

PHONE PLUG—Variety of **jack** often used as a microphone connector.

PHONO PLUG—Variety of **jack** most often used with audio amplifiers. Also known as RCA plug.

PHOSPHOR—A chemical coating used on the inside face of a cathode ray display tube. When hit by electrons, the phosphor glows according to the strength of the electron beam.

PHOTOCONDUCTOR—Any unit that permits the flow of an electrical current corresponding to varying light input.

PHOTOFLOOD—A self-contained light bulb, which gives a very bright, intense light without the use of external lenses or lamp housings.

PICKUP PATTERN—A determination of the directions from which a microphone is sensitive to sound waves; varies with the mike element and mike design. The two most common pickup patterns are **omni-** and **uni-directional.**

PICKUP RESPONSE—The tendency of a microphone to receive or to reject sound coming from different directions; the sensitivity of the mike to various frequencies.

PICKUP TUBE—*See* **cathode ray tube.**

PICTURE AREA—The area of a TV screen containing the video picture.

PICTURE LOCKING—Synchronizing the picture signal; the sync controls on a picture.

PICTURE SIGNAL—The picture information part of the composite video signal; the portion of the video signal above the **pedestal.**

PICTURE TUBE—A **cathode ray tube** designed to display the video picture signal.

PIEZOELECTRIC—A microphone element usually used in inexpensive, limited frequency response mikes; often called ceramic or crystal mikes.

PINCH ROLLER—A rubber roller that "pinches" or presses the video tape to the **capstan.** Together with the capstan, the pinch roller pulls the tape through the tape path on the video tape recorder.

PLANO-CONCAVE—A lens configuration in which the element has an inward curve on one side and a flat surface on the other.

PLANO-CONVEX—A lens configuration in which the lens element has an outward curve on one side and a flat surface on the other.

PLAYBACK—Function which induces the magnetic patterns on a video tape from that tape into the circuitry of a video tape recorder in order to reconstruct the composite video signal for display.

PLAYBACK AMPLIFIER—In audio, a circuit that amplifies the audio signal prior to its being reproduced through a speaker; in video, a circuit that amplifies the video signal in the VTR prior to its being supplied to a monitor.

PLAYBACK HEAD—Audio or video **head** used to obtain the signal from the video tape. Some heads are capable of playback and record, others of playback only; the video heads of most helical scan VTRS are both record and playback heads.

PLUMBICON—The trade name of Philips' special **vidicon** tube; more sensitive than a normal vidicon; used in some color cameras.

PLUS DIOPTER—A special lens accessory that fits over a camera lens to make that lens capable of extreme **close-ups.**

POLARITY—The positive or negative orientation of a signal;

in video, the polarity of the picture is black/negative, white/positive; reversed polarity would result in a negative picture.

POLARIZING FILTER—A special **filter** with polarizing properties; a filter which, when placed over the lens, can be rotated so that it cuts down the amount of reflected light coming into the lens.

POLAR RESPONSE PATTERN—*See* **pickup pattern.**

POP FILTER—A sponge rubber or plastic foam cap placed over the end of a microphone to reduce sibilance, breathy sounds, popping *p*'s and *b*'s, and other unwanted vocal effects.

PORTS—Air ducts built into microphones to control the **pickup pattern** and frequency response characteristics.

POT—Audio term for a volume control knob.

PREAMPLIFIER—An electronic circuit which maintains or establishes an audio or video signal at a predetermined signal strength prior to that signal's being amplified for reproduction through a monitor or speaker.

PREVIEW—The monitoring of a video signal prior to its being processed through an SEG.

PRIMARY COLORS—The three colors used in color TV, no two of which can be combined to produce the third; red, green, and blue. Not to be confused with the three primary colors of nature: red, blue, and yellow.

PROC AMP—*See* **processing amplifier.**

PROCESSING AMPLIFIER—Proc amp, signal processor, video processor, helical scan processor. A unit inserted on the line between any two components through which a composite video signal travels; serves to stabilize the composite signal, regenerate the control pulses, and, in certain models, change the gain and pedestal to improve contrast.

PROJECTION—In video, the production of a video picture signal as a strong light image that can be cast on a screen to attain a display area larger than a **cathode ray tube** is capable of giving.

PROXIMITY EFFECT—Associated with certain types of microphones; the closer the sound is to the mike, the more the bass frequencies are exaggerated.

PULSE—The variation of a constant signal for a specific period of time.

PULSE DISTRIBUTION AMPLIFIER—An amplifier designed to boost the strength of the **sync** as well as other control signals to the proper level for distribution to a number of cameras, special effects generators, and the like.

PUNCH UP—To engage a function button, as in punching up an effect on a special effects generator.

QUADRAPLEX—A system of video tape recording using two-inch tape and four rotating video heads. The heads pass the tape at an angle perpendicular to the tape's path.

QUARTZ BULB—A small lighting element which produces a great deal of light for a long period of time.

RACK MOUNTING—A method of mounting equipment, standardized by the EIA so that the sides of the rack are just over 19 inches (48.3 cm) apart, and all equipment 19 inches wide or less can be mounted in the rack. Equipment designed to be rack mounted has screw mounting holes at the edges.

RADIO FREQUENCY—RF. The range of frequencies used to transmit electric waves; a broadcast of that frequency range assigned to a certain bandwidth of that frequency.

RANDOM INTERLACE—Industrial sync. A method of **scanning** in which the horizontal and vertical scan controls run independently of each other in a random relationship; there is no fixed **phase** between the two, and the result is an unreliable **time base,** especially if two or more cameras are being used.

RASTER—The pattern described by the **scanning spot** of the electron beam as it scans the target area of a **cathode ray tube;** the pattern of scanning in both the pickup and display tubes.

RBG SIGNAL—The **chrominance** information; red, blue, and green.

RECEIVER—An electronic component capable of collecting radio frequency broadcasts and reproducing them in their original audio and/or video form.

RECORDING AMPLIFIER—Amplifier used in a video tape recorder to set the level of the video signal prior to its being supplied to the video heads.

REGISTRATION—An adjustment associated with color TV to ensure that the electron beams pulsing the three primary colors toward the phosphor of the screen's **raster** are hitting the proper points on that raster; also, a similar adjustment of the tubes in color cameras.

RESOLUTION—A subjective evaluation of the amount of detail in a picture; *see also* **horizontal resolution, vertical resolution.**

RETRACE—*See* **flyback.**

REVERB—*See* **reverberation.**

REVERBERATION—The persistent repetition of a sound after the original sound has ceased: Hello-hello-ello-ello; caused by sound waves bouncing off objects and surfaces and thus reaching the ear or microphone later than the original sound. Reverberation is characterized by a gradual cessation of sound; not to be confused with **echo.**

RF—*See* **radio frequency.**

RF ADAPTOR—RF amplifier; RF modulator. A unit which accepts the composite video signal and modulates a **carrier frequency** to produce a predetermined TV bandwidth, thus producing a broadcast signal.

RF AMPLIFIER—*See* **RF adaptor.**

RF MODULATOR—*See* **RF adaptor.**

RIBBON MIKE—A microphone with a metal coil element, old-fashioned and not very durable since the "ribbon" tends to fray at the edges.

ROLL—Loss of **vertical sync** causing the picture to move up or down the screen.

ROLL OFF—The preset attenuation of a predetermined range of bass frequencies—used by some microphone manufacturers on their mikes to reduce the **proximity effect.**

ROTARY IDLER—Stationary guide along the tape path.

ROTARY ERASE HEAD—A set of **heads** on the rotating video-head assembly which erases the video signal during recording and editing; usually positioned one scan line in front of the video heads; produces cleaner edits than a stationary erase head.

ROUGH EDIT—A rapid assembly of various segments in the order they will appear in the final program; not a finished **master** tape; not a **clean edit.**

RPM—Revolutions per minute.

SAFE AREA—Ninety percent of the TV screen, measured from the center of the screen; that area of the display screen (and therefore of the camera **scanning** area) which will reproduce on every TV screen, no matter how the set is adjusted.

SAFE TITLE AREA—Eighty percent of the TV screen, measured from the center of the screen; that area of the display screen (and therefore of the camera **scanning** area) which will reproduce legible title credits no matter how the set is adjusted.

SAFETY—Extra copy of a video tape kept in case something happens to the original copy; usually a second **generation** copy, although on special effects generators with two program outputs it's possible to record two master tapes, one to be set aside as the safety tape.

SAND BAGS—Canvas sacks filled with sand used to provide stability and ballast to light stands, tripods, and other portable equipment.

SATURATION—One of the determinations of the color of an object or light; how vivid a color is; related to the strength of the **chrominance** signal.

SCANNING—The action of the electron beam as it traces a pattern over the **target area** of the camera pickup tube.

SCANNING SPOT—The point at which the electron beam strikes the **target area.**

SCHMIDT OPTICAL SYSTEM—An arrangement of lenses and mirrors in combination with a very bright **cathode ray tube;** used in video projection systems.

SCOOP—A large bowl-shaped unit—often made of aluminum—into which a light-source is placed so that it will reflect light over a wide area.

SECAM—*Séquential Couleur à Mémoire.* A color TV system developed by the French which differs radically from both the **NTSC** and **PAL** color systems.

SECOND GENERATION—*See* **generation.**

SEG—*See* **special effects generator.**

SELECTIVE FOCUS—The adjustment of the lens so that a particular object in a scene is in perfect focus. When a **telephoto** lens is used, all but that object will be out of focus, creating the familiar effect of object surrounded by blur.

SELF-KEYED INSERT— *See* **overlay.**

SEPARATE MESH—A mesh screen located in **vidicon** cameras, which helps control the path of the electron beam from cathode to **target area,** thus improving the **scanning** process and the resulting picture.

SHADOW MASK COLOR TUBE—RCA developed standard color tube; a color tube equipped with a metal sheet with .5 million small holes punched in it. The metal sheet (which is the shadow mask) is placed between the electron guns and the phosphor-coated screen.

SHOOTING RATIO—The amount of tape recorded as compared to the amount of tape actually used in the final, edited program. Expressed as a ratio—3:1 means three times as much tape was recorded as eventually comprised the finished program.

SIDEBAND FM—*See* **FM.**

SIGNAL PATH—The movement of the signal from point of origin to point of display; the course a signal takes through a component or series of components.

SIGNAL PROCESSOR—*See* **processing amplifier.**

SIGNAL-TO-NOISE RATIO—The amount of electrical signal inherently produced by a unit in operation compared to the level of the signal that unit is processing. The higher the signal-to-noise ratio (the more signal, the less noise), the better the quality of the resulting sound or picture.

SIMPLE EDITING—A term used by some manufacturers to indicate that the VTRS do not have **capstan servo** or **head drum servo editing** facilities; an imprecise method of electronic editing, which does not guarantee clean edits.

SINGLE-D—A type of **ported** microphone with reduced **proximity effect.**

SKEW—The tape tension between supply reel and first **rotary idler** of tape path around head assembly of a VTR; skew must be maintained properly or picture instability will result.

SLANT TRACK—The original, now rarely heard, term for **helical scan.**

SLAVE VTR—A video tape recorder used to record a copy of a video tape from another (**master**) video tape recorder.

SMEAR—A video picture in which objects are blurred at the edges and seem to be running or bleeding beyond the edges; often caused by a combination of poor lighting and the nature of the vidicon tube.

SNOW—Random **noise** on the display screen, often resulting from dirty heads.

SOLENOID—An electromagnetic circuit control.

SOLID STATE—Circuitry that does not contain vacuum tubes. Integrated circuits and transistorized equipment are often referred to as being solid state.

SOUND WAVE—Air set in motion by any physical force or entity. This motion is a vibration of a certain rate and strength that makes it unique unto itself. The rate of vibration is measured in **Hertz (Hz)** or cycles per second; the force of the vibration is measured in **decibels;** a final evaluation of a particular sound wave is the sound's pitch, but this factor doesn't enter into the evaluation of the sound as an electronic signal.

SPECIAL EFFECTS GENERATOR—SEG. A unit used in video production to mix, switch, and otherwise process various video signals to create a final signal known as the program signal.

SPECIFIC LIGHTING—Lighting used to illuminate an object in order to create a desired effect on the display screen; any lighting units that are set up especially for recording an event.

SPLICING—The physical cutting and rejoining of recording tape; the joint of the two ends of the video or audio tape is secured by tape coated with an adhesive substance on one side.

SPOTLIGHT—A lighting unit whose light can be focused into a beam and directed at a particular object or part of the scene.

STANDARD FOCUSING—Optical nomenclature used to describe a lens that can be focused by moving a section of the other barrel of the lens backward or forward until the image passing through the lens is shown in sharp detail on the image plane.

STANDARD (MINIMUM) SIGNAL—The **peak-to-peak voltage** of a signal whose amplitude is sufficient for its use within a system; for **non-composite video signals,** it is 0.7 Vpp; for **composite video signals,** 1.0 Vpp.

STATIC MATTE INSERT—*See* **inlay.**

STEP-DOWN TRANSFORMER—An electronic circuit which can change electric current from one voltage to another; the most common transformer of this type is 220v-to-120v. A step-down transformer does not change the **Hertz** of the current; thus, 220v 50Hz can only be changed to 120v 50Hz (to get 60Hz at 120v you'd need an **ac-to-dc converter** and a **dc-to-ac inverter**).

STEPS—Term used to describe the number of controls on a **colorizer;** the control for each color is called a step.

STOP DOWN—Open up. To close down a lens; to adjust the iris/aperture of the lens so that less light passes through the lens. A lens set at f2 which is then adjusted to f4.5 is said to be stopped down two steps; to stop a lens down all the way is to set it at its highest f-stop number.

STREAKING—Similar to **smearing;** occurs when objects in a scene bleed beyond their edge.

STRIPE FILTER—A **chrominance** tube system in which the target area of the tube is divided into sequential stripes for R, B, G, and Y (**luminance**), and can therefore derive a color signal by using only one pickup tube.

SUBCARRIER FREQUENCY—The frequency on which color information is **modulated** in the color TV system; in the US it's 3.58MHz.

SUBCARRIER PHASE SHIFTER—Special circuitry designed to control the phase relationships of the two portions of the encoded color signal so that they maintain their correct relationship during recording, transmission, and reproduction. A phase shifter allows the user to change the timing of the signals involved so that they are occurring at the correct time and are thus said to be in phase.

SUN SHADE—A metal cylinder attached to the end of a lens to keep light from entering the lens from the periphery of the **angle of view.**

SUPER—The superimposition of one video signal on

another using the **fader** controls of the special effects generator.

SUPER-CARDIOID—Variable-D. A microphone with a very directional pickup pattern, allowing sound to affect the element only if it's coming toward the front of the mike; *see also* **pickup pattern.**

SUPPLY REEL—The left reel on an open-reel VTR, which contains blank tape or a recorded program prior to its being run through the VTR.

SUPPRESSION—*See* **blanking.**

SURROUND SHOT—An **angle of view** peculiar to highly portable recording equipment; the camera can enter into the action and wander through it, giving the viewer the impression that the camera is part of the event.

SWITCHER—Term often used to describe a **special effects generator;** a unit which allows the operator to switch between video camera signals. Switchers are often used in industrial applications to switch between video cameras monitoring certain areas for display on one monitor; these kinds of switchers do not have **sync generators.**

SYNC—Synchronize; various drive pulses, both horizontal and vertical, which maintain the horizontal and vertical **scanning** procedures of the video picture signal from camera to display; *see also* **horizontal sync, vertical sync.**

SYNC GENERATOR—A pulse generator which produces the sync signals necessary to integrate the functioning of various pieces of video equipment in relation to each other and the video signal.

TAILS OUT—A tape that has been played but not rewound; a tape whose end is nearest the outside of the reel; the opposite of **heads out.**

TAKE-UP REEL—The right reel on an open-reel VTR which accepts the video tape after it has been run off the **supply reel** through the VTR's tape path.

TALLY—A system of audio intercommunication among various members of the video production crew. A **tally light** is set on top of each camera and glows when that camera is the one **on the line** as the program signal camera.

TALLY LIGHT—Part of a **tally** system; standard equipment on some studio cameras, which signals which of a multi-camera system is in use at any given moment.

TAPE—A medium capable of storing an electronic signal and consisting of **backing, binder,** and iron oxide coating. The orientation of the iron oxide determines whether the tape can be used for **helical scan video tape recording.**

TAPE GUIDES—*See* **rotary idlers** and **pinch roller.**

TAPE PATH—The circuit the tape runs from **supply reel** to **take-up reel** past the erase head, video heads, audio/control track head, and between capstan and pinch roller; standardized on half-inch machines by the **EIAJ.**

TAPE TENSION GUIDE—The first guide off the **supply reel,** adjusted to maintain proper **skew.**

TAPE TRANSPORT—Those mechanical components of the video tape recorder which move the tape from **supply reel** to **take-up reel** and back.

TARGET AREA—The face of the **vidicon** tube or other camera cathode ray pickup tube; also known as the **raster.** This area (opposite the cathode heater) is where the image formed by the lens is transformed into an electronic signal. On the outside face of the tube the image is read by the electron scanning beam. A circuit is completed at each point where the beam strikes the target, and because a voltage is being applied to the target area a certain resistance results, which gives a voltage variation or video signal.

TEARING—Occurs when **horizontal sync** is lost or distorted in a picture, resulting in some of the horizontal lines getting out of place, or phase, with the rest of the picture.

TELECINE—*See* **film chain.**

TELEPHOTO LENS—A lens with a large focal length (large number of millimeters) and a very narrow angle of view.

TELEVISION RECEIVER—TV set; capable of sensing and receiving broadcast video signals, stripping them from their carrier frequencies, and producing them as a light image picture on the face of a cathode ray display tube; *see also* **receiver.**

TENSION—The pull of the **capstan** assembly on the video tape to keep it against the video **head drum assembly;** used in conjunction with the **skew** control to keep tape properly in path.

TERMINATION—The insertion of a **load** at the end of a line carrying a signal; a video terminator is a 75-ohm resistor placed at the end of a line to keep the signal from bouncing back along the line; 600 ohms is commonly used to terminate an audio line.

THIRD GENERATION—Two copies away from the master tape; *see* **generation.**

THREE-TUBE COLOR CAMERA—A color-capable camera which produces a color signal through the use of three pickup tubes, each assigned to one of the primary colors. An early stage in the development of the color video camera, introduced by RCA in 1940.

TIGHT SHOT—*See* **close-up.**

TILT—To move the camera up toward the ceiling (tilt up) or down toward the floor (tilt down).

TIME-BASE—The relative accuracy of any portion of the video scanning process in record and/or reproduction as measured against the theoretical "time" at which a given scan element is supposed to occur. The time base of a **field,** for example, is sixty times per second, each of which should occupy no more or less than one-sixtieth of a second.

TIME-BASE CORRECTOR—A computer that evaluates the video signal to determine if each scan line, field, and frame is in the correct time position. If any of these elements is occurring too early, the time-base corrector will fill in the space left open by repeating a previous line or field of information.

TIME-BASE STABILITY—What **helical scan** hasn't got; the maintenance of the **scanning** process to very close tolerances.

TIN—To coat the end of a cable wire with solder before making a solder connection, thus ensuring a more secure connection.

TRACKING—The angle and speed at which the tape passes the video heads.

TRANSDUCER—The element of the microphone that changes the sound vibrations into electronic pulses.

TRANSMISSION ABILITY—An evaluation, usually expressed as a percentage, of the amount of light a filter will admit. A ninety-percent transmission filter will allow ninety percent of available light to pass through it, eliminating ten percent.

TRIPOD—A three-legged stand on top of which a camera is mounted.

TRIPOD DOLLY—*See* **dolly.**

TRIPOD HEAD—The top portion of a tripod where its legs meet and the camera is mounted; friction- or fluid-head tripod designs are available.

TRUCKING—Moving the camera left (truck left) or right (truck right) on a tripod with a **dolly,** or moving your body while holding a portable camera.

T-STOP—A rating of a lens in terms of its ability to transmit light; the **transmission ability,** or transmission stop, of a lens; more exact in terms of a lens's transmission ability than an **f-stop** rating, but no more helpful than the f-stop when working with a video camera; used on some lenses in motion picture and still photography in place of f-stop.

TV STORY BOARD—Sheets of paper with blank TV screens on them; used for roughing out the action of a program.

2:1 INTERLACE—A **scanning** system in which the horizontal and vertical control pulses are locked together so that they occur at the correct time in relation to each other.

TWO-THIRDS-INCH VIDICON—A **vidicon** with a target area two-thirds of an inch in diameter; the most commonly used vidicon in portable video cameras.

TWO-TUBE COLOR CAMERA—A video camera with one vidicon tube dedicated to **luminance** (black and white brightness values) and the other to color (**hue** and **saturation** of color).

UHF CONNECTOR—Standard type of video-in/video-out jack; commonly found on professional monitors; used to carry either composite or non-composite video signal.

U-MATIC—The Sony-originated video recorder system using three-quarter-inch-wide video tape in a cassette housing. Capable of producing broadcast-quality video recordings when used with a high-quality color camera and a **time-base corrector.**

UNI-DIRECTIONAL—Microphone **pickup pattern** which accepts only sound coming in front of it.

VARIABLE-D—*See* **super-cardioid.**

VARIABLE FOCAL LENGTH LENS—*See* **zoom lens.**

VARIABLE MIKE—A microphone with a number of ports in its casing; designed to correct the **proximity effect** and to produce a **super-cardioid** pickup pattern.

VDA—*See* **video distribution amplifier.**

VERTICAL BLANKING—Field blanking. The **blanking** of a signal during **scanning,** when the **scanning spot** is **flying back** from scanning one field to begin scanning the next field, and at which time blanking and sync pulses are introduced to the signal.

VERTICAL FREQUENCY—*See* **field frequency.**

VERTICAL INTERVAL EDITING—A method of **electronic editing** in which the edit takes place on the vertical interval between picture fields so the picture switches from one signal to the next without visible distortion.

VERTICAL INTERVAL SWITCHING—A method of switching video signals in a special effects generator; this replacement of one signal with another takes place during the **vertical retrace** period.

VERTICAL RESOLUTION—The number of horizontal lines on a test pattern that can be **scanned** by a camera or monitor so that they are distinctly visible; the number of horizontal lines a piece of video equipment is capable of processing per field as picture information. *See also* **resolution.**

VERTICAL RETRACE—The return of a **scanning spot** to the top of the target area to begin scanning a new field after completing its scan of the previous field; the time it takes for this to occur. *See also* **flyback.**

VERTICAL SCANNING—The field-by-field scanning of a picture, at the rate of sixty fields per second. *See also* **scanning.**

VERTICAL SYNC—The sync pulses which control the vertical field-by-field **scanning** of the target area by the electron beam. *See also* **sync, field, field blanking, field frequency.**

VHS—A video tape recorder standard championed by RCA and Panasonic as a consumer video recorder. The VHS format uses half-inch tape in a cassette housing, and it is not compatible with other consumer video systems, notably its main competitor, **Betamax.**

VIDEO—Television and the technical equipment and events involved in creating television; the visual portion of a signal containing both sight and sound information; an alternative to broadcast television.

VIDEO AMPLIFIER—A circuit that can increase the strength of a video signal sent through it.

VIDEO CASSETTE—A self-contained video module played on a specially designed video tape recorder; similar in design to an audio cassette; houses two reels—supply and take-up—with the tape running between them but connected to both.

VIDEO DISTRIBUTION AMPLIFIER—VDA. A special amplifier for strengthening the video signal so that it can be supplied to a number of video monitors at the same time.

VIDEO FREQUENCY—VF; a **composite video signal** unmodulated by a radio **carrier frequency.**

VIDEO GAIN—The **amplitude** of the video signal; the control on a VTR that determines the "volume" level of the video signal.

VIDEO HEAD—*See* **head.**

VIDEO-IN—Line-in. A jack through which a video signal is fed into a given component.

VIDEO-OUT—Line-out. A jack from which a video signal is fed out of a given component.

VIDEO PROCESSOR—*See* **processing amplifier.**

VIDEO SWITCHER—*See* **special effects generator** and **switcher.**

VIDEO TAPE RECORDER—VTR; an electro-mechanical

device capable of recording, storing, and reproducing an electronic signal which contains audio, video, and control information.

VIDEO WAVEFORM—The pictorial display on a special oscilloscope of the various components of the video signal; used to check the integrity of the signal and signal components.

VIDICON—A vacuum tube capable of changing light images into electrical voltage variations corresponding to the brightness of those images; a particular type of **cathode ray tube** used in some video cameras, containing a cathode assembly and a target area coated with antimony trisulfide; the least exhensive and most generally reliable pickup tube presently available.

VIEWFINDER—*See* **electronic viewfinder.**

VOICE COIL—The element in a dynamic microphone which vibrates when sound waves strike it; the coil of wire in a loudspeaker through which audio frequency current is sent to produce vibration of the cone and reproduction of sound.

VOICE-OVER—A voice speaking over other sounds; the addition of a narration to the original sound track during post-production.

VOLTAGE PEAK-TO-PEAK—Vpp; Ppv. *See* **peak-to-peak voltage.**

VOLUME UNIT METER—VU meter. A meter generally associated with the monitoring of the amplitude of a video or audio signal. *See also* **meter.**

VPP—*See* **peak-to-peak voltage.**

VTR—*See* **video tape recorder.**

VU METER—*See* **volume unit meter.**

WATT—A unit of electrical power; that amount of power required to maintain a current of one **amp** under the pressure of one volt.

WAVEFORM MONITOR—Special oscilloscope used to display the video waveform.

WHITE CLIPPING—*See* **clipping.**

WHITE LEVEL SET—White set. A camera control that establishes the **luminance** level for a color camera.

WHITE SET—*See* **white level set.**

WIDE ANGLE LENS—A lens with a very short **focal length;** a lens that has a very wide **angle of view.**

WIDE OPEN—Descriptive of a lens set at its lowest **f-stop** rating so that the **iris** is opened as wide as possible.

WILD FOOTAGE—Audio tape recorded out of sync with any

particular video picture for use in post-production as an audio track; video tape recorded without audio for use as visual material in post-production, to which narration will be added.

WIND SCREEN—Similar to a **pop filter;** a heavy foam rubber microphone cover, used outdoors to cut down on audible noise created by wind blowing across the top of the mike.

WIPE—Term used to describe the SEG effect of replacing a portion of video signal A with video signal B; also to erase a tape.

Y—The symbol for the **luminance** portion of a color video signal; the color video signal consists of R, G, B, and Y.

ZOOM LENS—A lens with a variable **focal length;** a lens whose **angle of view** can be changed without moving the camera.

ZOOM RATIO—A mathematical expression of the two extremes of **focal length** available on a particular **zoom lens.**

STANDARD ARTIST'S RELEASE FORM

In consideration of the sum of Ten Dollars ($10.00) and other good and valuable considerations, the receipt of which is hereby acknowledged, I hereby authorize Bonzai Video Productions to use my name, likeness, voice, and features for inclusion in Bonzai Video Production's video tape tentatively entitled *Frances, The Talking Portapak* (the title of which is subject to change at the discretion of the producer). I also authorize Bonzai Video Productions to use my name, likeness, portrait, picture, voice, and features for publication and related advertising of the above mentioned production and expressly waive any rights or claims I may have against your firm and/or any of its affiliates, subsidiaries, or assignees except as outlined in this contract.

I hereby agree and consent to the foregoing authorization and waiver.

____

Signature of artist

____

For Bonzai Video Productions

____

Witness

____

Date

SAMPLE PRODUCTION NOTES

Title: __________

Producer: __________

Production #: __________

Director: __________

Dates for Shooting: __________

Ass't. Dir.: __________

Character	Actor	1st day	2nd day	3rd day	4th day	5th day	6th day
1.							
2.							
3.							
4.							
5.							
6.							
7.							
8.							
9.							
10.							
11.							
12.							

DATE	SETS • SCENES • DESCRIPTION	CAST	LOCATION
1st Day	Set: Scs: Props: End of 1st day		
2nd Day	Set: Scs: Props: End of 2nd day		
3rd Day	Set: Scs: Props: End of 3rd day		
4th Day	Set: Scs: Props: End of 4th day		
5th Day	Set: Scs: Props: End of 5th day		
6th Day	Set: Scs: Props: End of 6th day		

SAMPLE
SHOOTING
SCHEDULE

Terms Used for Production Notations

Int.—Interior   Trans.—Transportation vehicles
Scs.—Scenes   Bits—Bit players
Props   Atmos.—Non-speaking actors
Ext.—Exterior   in scenes

APPENDICES

FADE IN:

1 EXT. SEASHORE—DAY—
   ESTABLISHMENT SHOT   1
   The author (a nice-looking chap) is walking
   along the seashore. He is carrying a portable camera and deck, both slung over his
   shoulder. He appears to be reading a copy
   of the Sony manual. Wind is blowing.

2 EXT.—CLOSE SHOT   2
   Author shakes his head and turns page of
   manual. Suddenly, he looks off camera to
   right, surprise on his face.

3 ANOTHER ANGLE   3
   Author backs away from camera.

                    AUTHOR'S VOICE
                 Oh no! I don't believe it.

4 LONG SHOT   4
   Author looks out to sea where a submarine
   has surfaced; it has Japanese letters on it
   and a Japanese flag. Figure appears on
   deck and, holding up bullhorn, begins to speak.

                   VOICE OF FIGURE
          What'sa matta, you can't understand
                     instructions?

                       AUTHOR
                  I beg your pardon?

5 CLOSE SHOT   5
   Author is fumbling with his gear, trying to
   get it ready to record.

                       AUTHOR
             Hold it just a second, will you?

6 EXT. OF SUB   6
   Figure on deck disappears back into sub,
   after shaking his head in dismay.

                  AUTHOR, OFF CAMERA
                 Hold it, hey! Hang on!

7 PAN SHOT   7
   From sub, back to beach, author still fumbling with equipment. Author suddenly takes
   deck and camera and tosses both into the sea.

8 CLOSE SHOT   8
   Author reaches inside jacket and pulls out
   Kodak Instamatic, takes aim, and a flash
   fills screen.
   Freeze flash, FADE.

THE VIDEO PRIMER
———————————   END OF ACT ONE

SAMPLE
VIDEO
TAKE SHEET

VIDEO PRODUCTION COMPANY NAME

Tape Catalog #____________________

Production Title ____________________

Date of Recording ____________________

- ☐ Original Master
- ☐ Standard
- ☐ Dupe
- ☐ b&w
- ☐ Composite Master
- ☐ Color
- ☐ Safety

Produced by ____________________

Camera by ____________________

Sound by ____________________

Location ____________________

Special Notations:

Take #	Description	Length of Sequence

- ☐ Copyright Free for Broadcast
- ☐ Artist Releases Signed

Advent Corporation 195 Albany Street Cambridge, MA 02139	Video projector
Adwar Video Corporation 100 Fifth Avenue New York, NY 10011	Video sales, service
AKAI 2139 East Del Amo Blvd. Compton, CA 90220	Video hardware
AKG North American Philips 100 East 42nd Street New York, NY 10017	Microphones
Ampex Corporation 401 Broadway Redwood City, CA 94063	One-inch and two-inch VTRS, processing equipment
Angenieux Corporation 1500 Ocean Ave. Bohemia, NY 11716	Lenses
Atlas Sound 10 Pomeroy Road Parsippany, NJ 07054	Mike stands, booms, mike accessories
Ball Brothers 1633 Terrace Drive St. Paul, MN 55418	Monitors
BASF Crosby Drive Bedford, MA 01730	Tape
Berkey-Colortran 1015 Chestnut Street Burbank, CA 91502	Lighting equipment, portable lighting kits
BSC, Inc. 8600 West Sunnyside Ave. Chicago, IL 60656	One of the first mail-order video equipment houses
Camera Mart 456 West 55 Street New York, NY 10019	Video sales, rentals
Canon Optics Division 10 Nevada Drive Lake Success, NY 11040	Lenses
Comprehensive Video Supply Corporation 148 Veterans Drive Northvale, NJ 07647	Excellent catalog, source for most video hardware and many hard-to-find items.
Concord Benjamin Electronics 40 Smith Street Farmingdale, NY 11735	Video equipment line

Conrac 600 North Rimsdale Ave. Covina, CA 91722	Monitors
C.T.L. Electronics 86 West Broadway New York, NY 10007	Video supplies, *Video Tools* catalog
DAK Enterprises 10845 Vanowen North Hollywood, CA 91605	Tape
Dynair 6360 Federal Blvd. San Diego, CA 92114	SEGs and other processing equipment
Dynasciences Corporation Township Line Road Blue Bell, PA 19422	Processing equipment
Electro-Voice 600 Cecil Street Buchanan, MI 49107	Microphones, speakers
Electronic Industries Association 2001 I Street N.W. Washington, DC 20006	EIA test patterns, standards manuals
Electronic Industries Association of Japan Video Technical Committee 4-2-2- Marunouchi Chiyoda-Ku, Tokyo, Japan	EIAJ standards manuals
F&B/Ceco 315 West 43 Street New York, NY 10036 and 7051 Santa Monica Blvd. Hollywood, CA 90038	Video, movie equipment sales
Fuji Photo Film Video Division 350 Fifth Avenue New York, NY 10001	Tape
GBC Closed Circuit TV 74 Fifth Avenue New York, NY 10011	Video sales
Video Display Equipment General Electric Electronics Park Syracuse, NY 13201	Video projector
Grass Valley Group PO Box 1114 Grass Valley, CA 95945	SEGs and other processing equipment

Inter Video 342, rue des Pyrenées Paris XXe, France	Video Sales
International Video Corporation 990 Almanor Avenue Sunnyvale, CA 94086	IVC one-inch VTRs and other equipment
Jerrold Electronics 401 Walnut Street Philadelphia, PA 19105	CATV amplifiers
JVC Video Division 58-75 Queens-Midtown Expressway Maspeth, NY 11378	Video equipment line
Lafayette Radio Electronics 111 Jericho Turnpike Syosset, NY 11791	Mail-order house, TV accessories, mikes, audio processing equipment
Magnavox CATV Division 133 West Seneca Street Manlius, NY 13104	Color cameras
3M Company Magnetic Products Division 3M Center	Tape
Motion Picture Camera Supply 424 West 49 Street New York, NY 10019	Video, movie equipment sales
Norelco Philips Broadcasting Equipment 91 McKee Drive Mahwah, NJ 07430	Color cameras
Nortronics Company 8101 Tenth Avenue North Minneapolis, MN 55427	Video recorder care kit
Panasonic VTR—CCTV Department 200 Park Avenue New York, NY 10017	Video equipment line
Gately Electronics Pro-kit Division 57 West Hillcrest Avenue Havertown, PA 19083	Portable mixers, kits, assembled
RCA Consumer Electronics 600 North Sherman Drive Indianapolis, IN 46201	SelectaVision

Riker Information Systems 101 Industrial East Clifton, NJ 07012	8mm film chain, other equipment
Howard W. Sams & Company 4300 West 62 Street Indianapolis, IN 46268	Technical video
Sanyo Electric 1200 West Artesia Blvd. Compton, CA 90220	Video equipment line
Sennheiser Electronics 10 West 37 Street New York, NY 10018	Microphones, audio accessories
Sharp Electronics Corporation 10 Keystone Place Paramus, NJ 07652	Video hardware
Shure Brothers 222 Hartrey Avenue Evanston, IL 60204	Microphones, audio mixers, audio processing equip- ment and accessories
Smith-Victor Corporation Griffith, IN 46319	Lighting equipment, portable lighting kits
Sony Corporation 47-47 Van Dam Street Long Island City, NY 11101	Video equipment line
Super Negozio Telestore Via del Trintone, 39-40 Rome, Italy	Video sales
Technisphere 215 East 27 Street New York, NY 10016	Video supplies
Teletape Video 76 Brewer Street London W. 1, England	Video sales
Thomas J. Valentino, Inc. 151 West 46 Street New York, NY 10036	Copyright-free production music

SELECTED
BIBLIOGRAPHY

Bensinger, Charles. *Petersen's Guide to Video Tape Recording.* Los Angeles: Petersen Publishing Company, 1973.

Bretz, Rudy. *Techniques of Television Production.* New York: McGraw-Hill, 1962.

Clarke, Charles G. and Strenge, Walter, eds. *American Cinematographer Manual.* Hollywood: American Society of Cinematographers, 1973.

Clifford, Martin. *Microphones: How They Work and How to Use Them.* Blue Ridge Summit, PA: Tab Books, 1977.

Crowhurst, Norman H. *Audio Systems Handbook.* Blue Ridge Summit, PA: Tab Books, 1969.

Eddy, William C. *Television, The Eyes of Tomorrow.* Englewood Cliffs, NJ: Prentice-Hall, 1945.

Faenza, Roberto. *Senza Chiedere Permesso, come rivoluzionare l'informazione.* Milan: Feltrinelli Editore, 1973.

Fredericksen, H. Allen. *Community Access Video.* Santa Cruz, CA: Johnny Videotape Publications, 1972.

Hutchinson, Thomas H. *Here Is Television, Your Window to the World.* New York: Hastings House, 1948.

Jones, Gary, and Squyers, Phil. *1973 Electronic Film/Tape Post-Production Handbook.* Dallas: Fratellitre Communications, 1973.

Lipton, Lenny. *Independent Filmmaking.* San Francisco: Straight Arrow Books, 1972.

Mattingly, Grayson, and Smith, Welby. *Introducing the Single Camera* VTR *System.* Washington, DC: Smith-Mattingly Productions, 1972.

Minus, Johnny and Hale, William Storm. *Your Introduction to Film: TV Copyright, Contracts, and Other Law.* Hollywood: Seven Arts Press, 1974.

Oliver, W. *Introduction to Video Recording.* London: W. Foulsham, & Co., 1971.

Pincus, Edward. *Guide to Filmmaking.* New York: New American Library, 1969.

Rosien, Arthur H., ed. *The Video Handbook.* New York: Media Horizons, 1972.

Shamberg, Michael. *Guerrilla Television.* New York: Holt, Rinehart & Winston, 1971.

Showalter, Leonard C. *Closed-Circuit TV for Engineers & Technicians.* Indianapolis: Howard W. Sams & Co., 1969.

Spottiswoode, Raymond, ed. *The Focal Encyclopedia of Film & Television Techniques.* New York: Hastings House, 1969.

Video Freex. *The Spaghetti City Video Manual.* New York: Praeger Publishers, 1973.

Weiner, Peter. *Making the Media Revolution: A Handbook for Video-Tape Production.* New York: Macmillan, 1973.

White, Gordon. *Video Recording, Record and Replay Systems*. London: Newnes-Butterworths, 1972.

Wortman, Leon A. *Closed-Circuit Television*. Indianapolis: Howard W. Sams & Co., 1972.

Zwick, George. *Beginner's Guide to TV Repair*. Blue Ridge Summit, PA: Tab Books, 1972.

Electronic Industries Association:

"EIA-RS-330, Electrical Performance Standards for Closed Circuit Television Camera 525/60 Interlaced 2:1" (1966).

"EIA-RS-170, Electrical Performance Standards—Monochrome Televison Studio Facilities" (1957).

Industrial Electronics Bulletin No. 1; "Closed Circuit Television Definitions" (1962).

Industrial Electronics Bulletin No. 8; "A Guide to Helical Scan Tape Recording" (1971).

Electronic Industries Association of Japan:

"EIAJ Standard CP-501, VTR connections" (1969).

"EIAJ Standard CP-502, tape reel standards" (1969).

"EIAJ Standard CP-503, VTR-to-monitor connections" (1969).

"EIAJ Standard CP-504, tape compatibility between VTRS" (1969).

"EIAJ Standard CP-505, symbol marking standards" (1970).

"EIAJ Standard CP-506, cable connections" (1970).

"EIAJ Standard CP-507, color tape compatibility between VTRS" (1971).

"EIAJ Standard CP-508, cartridge standards" (1972).

The above are available from the EIAJ in English translation.

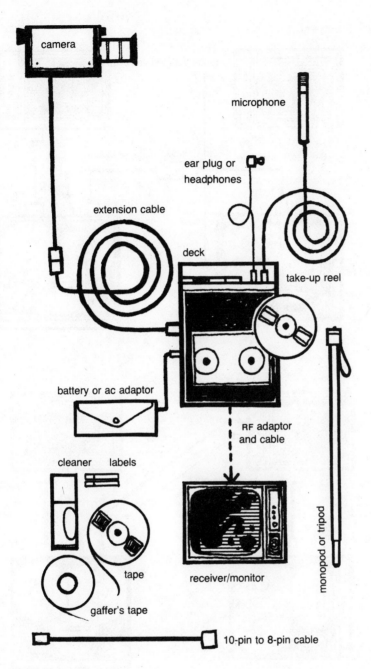

camera

microphone

ear plug or
headphones

extension cable

deck

take-up reel

battery or ac adaptor

RF adaptor
and cable

cleaner   labels

tape

gaffer's tape

receiver/monitor

monopod or tripod

10-pin to 8-pin cable

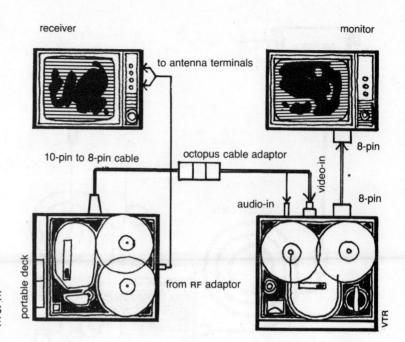

receiver

monitor

to antenna terminals

10-pin to 8-pin cable

octopus cable adaptor

8-pin

video-in

*

8-pin

audio-in

portable deck

from RF adaptor

VTR

EDITING FROM PORTABLE DECK WITH OCTOPUS CABLE

EDITING FROM PORTABLE DECK THROUGH MONITOR

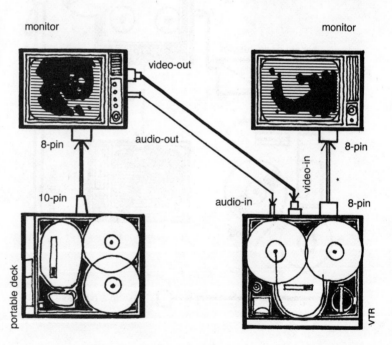

monitor

monitor

video-out

8-pin

audio-out

8-pin

video-in

*

10-pin

audio-in

8-pin

portable deck

VTR

*can substitute RF system

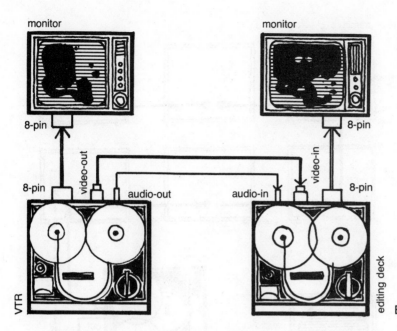

monitor

monitor

8-pin

8-pin

8-pin

video-out

audio-out

audio-in

video-in

8-pin

VTR

editing deck

EDITING FROM VTR

RF DISTRIBUTION SYSTEM

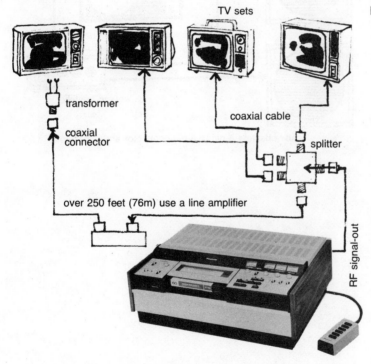

TV sets

transformer

coaxial cable

coaxial connector

splitter

over 250 feet (76m) use a line amplifier

RF signal-out

MULTIPLE MONITOR DISPLAY SYSTEM

 monitors

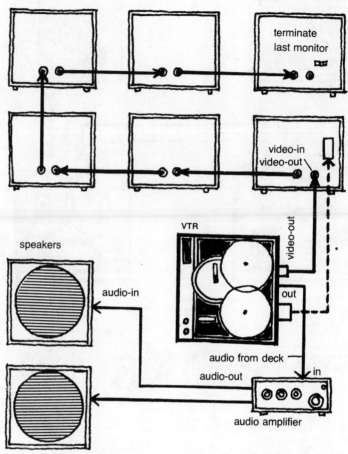

terminate last monitor

video-in
video-out

video-out

*

VTR

out

speakers

audio-in

audio from deck

audio-out

in

audio amplifier

*can substitute 8-pin to 8-pin connector and use monitor's audio

THE VIDEO PRIMER

THE VIDEO PRIMER